Green Energy and Technology

More information about this series at http://www.springer.com/series/8059

Walter Leal Filho · Vakur Sümer
Editors

Sustainable Water Use and Management

Examples of New Approaches
and Perspectives

 Springer

Editors
Walter Leal Filho
Faculty of Life Sciences
Hamburg University of Applied Sciences
Hamburg
Germany

Vakur Sümer
Department of International Relations
Selcuk University
Konya
Turkey

ISSN 1865-3529   ISSN 1865-3537   (electronic)
Green Energy and Technology
ISBN 978-3-319-12393-6   ISBN 978-3-319-12394-3   (eBook)
DOI 10.1007/978-3-319-12394-3

Library of Congress Control Number: 2014957149

Springer Cham Heidelberg New York Dordrecht London

Springer International Publishing AG Switzerland is part of Springer Science+Business Media
(www.springer.com)

Preface

In the second half of the twentieth century and particularly in its last quarter, the need for sound use and protection of freshwater resources, coupled with due consideration to the needs of different interest groups, have become more and more obvious. It is understood that less than 0.5 % of the world's water resources is freshwater available for human use, and that around one-third of the world's population lives in areas where water is scarce or extremely scarce. Moreover, by 2025, that number is expected to grow to two-thirds. Therefore, the second half of the twentieth century witnessed the increasing prominence of concerns over water management issues, which is now very present.

With regard to the water crises summarized above, the problem can be solved not only by implementing new technologies, but also through changes in water use practices and water resources management. In this sense, the primary reasons that water problems afflict developing countries are accepted to be of political and institutional nature, and not technical ones. In this respect, the Global Water Partnership concluded that "the water crisis is mainly a crisis of governance." It is accepted that sectorial regulation of water resources management leads to "splintered and uncoordinated" water use and hinders the organization of water protection mechanisms. One of the ways to find reasonable solutions to water-related problems in these countries is to implement the principles of integrated water resources management. There is also a need for ever-efficient water technologies, improving the situation in respect of excessive water use in agriculture. It should be noted that agriculture is the biggest water consumer worldwide. Apart from technological innovations aiming at "more crop for every drop," demand management tools is also another proposed solution for increased water efficiency which could also lead to increase in improved water productivity in the agricultural sector.

The issue of "sustainability" in terms of water use lies at the heart of this dynamic debate. Taking it more broadly, sustainability means not only seeking a balance between today's and tomorrow's needs, but also working towards a balanced view of water with consideration of intertwined relationships among all stakeholders, namely policymakers, water users, water service providers, and others, all competing water needs (of industry, energy sector, households, irrigation, recreation, ecological flows), and all relevant economic sectors (manufacturing, tourism, agriculture, water services sector, etc.). Reaching food security, particularly under the shadow of climate change, has become one of the utmost priorities for many countries adding further complications to existing equations of competition.

This book is located at the crossroads of two key phenomena: sustainability and water use. These themes should be taken in their width, meaning that the axis of sustainability and water use brings together academic research and discussions on water efficiency, new technologies, water-agriculture nexus, transboundary cooperation towards river-basin management, pricing issues, participatory water management, role of women in sustainable water use, and other themes. It is divided into two parts:

Part I deals with approaches in sustainable water use and management and offers users an overview of the theoretical basis and elements which have been guiding the implementation of sound approaches to use water resources.

Part II contains a set of case studies in sustainable water use and management, where ongoing projects and initiatives are demonstrated in practice.

Consistent with its editorial objectives, this publication aims to contribute to this growing debate with discussions of new approaches, methods, concepts, arguments, and findings. We hope that not only water experts but also readers from different backgrounds and disciplines will benefit from this volume.

In the process of preparing this edited volume, we, the editors, were supported both financially and logistically by our own respective institutions, namely Hamburg University of Applied Sciences, Manchester Metropolitan University, UNC—Global Research Institute, and Selcuk University, Konya, in Turkey. We would like to acknowledge their support. Special thanks are due to Erika Glazaciavoite, for helping to produce this book. And, last but not least, we would like to thank our families for their continuous support and patience all through the research and writing process.

Autumn 2015   Walter Leal Filho
                                                                                  Vakur Sümer

Contents

Part I
Approaches in Sustainable Water Use and Management

Ethics, Sustainability, and Water Management: A Canadian Case Study

Ingrid Leman Stefanovic

Abstract This paper argues that values, perceptions, and attitudes affect decision making in water management and that a better understanding of water ethics will ensure more reliable management practices. A Canadian case study, focusing on the City of Toronto's Biosolids and Residuals Master Plan (BRMP), illustrates the importance of values in water management practices. In 2007, the author served as one of a seven member expert peer review panel to evaluate the model used by consultants to recommend biosolids management upgrades at each of the city's four wastewater treatment plants. Both the decision-making model as well as community reactions to the model and master plan revealed value judgments that ultimately affected the management process and implementation of recommendations over recent years.

Keywords Ethics · Values in sustainability · Biosolids and water management · Perceptions and attitudes in decision making

1 Introduction

According to the United Nations (2012: 1), more than 50 % of the global population now resides in cities. Within these urban areas, sanitary sewage and stormwater drainage often constitute the biggest source of pollution to surface water. Given that the United Nations (2012: 1) projects a global population increase of more than 2 billion people from 2011 to 2050, the development and management of efficient and flexible wastewater treatment systems constitute a clear priority for city planners and politicians worldwide.

I.L. Stefanovic (✉)
Faculty of Environment, Simon Fraser University, 8888 University Dr., Burnaby, BC V5A 1S6, Canada
e-mail: ingrid.stefanovic@neimargroup.com

© Springer International Publishing Switzerland 2015
W. Leal Filho and V. Sümer (eds.), *Sustainable Water Use and Management*, Green Energy and Technology, DOI 10.1007/978-3-319-12394-3_1

In any advanced wastewater treatment plant, untreated solids that are removed from the sewage treatment process are referred to as "sludge." The biological treatment of sludge and wastewater produces a nutrient-rich material called "biosolids." A central element, therefore, of wastewater control includes a strategy for biosolids management as well. It is expected that over the coming years, "biosolids management is likely to become even more challenging due to external forces such as the need for energy conservation, increased regulations on greenhouse gas emissions, tighter regulations on contaminant emissions to water and air, higher national standards for trace inorganic and organic contaminants in the land application of biosolids, greater urbanization, and more competition for taxpayer dollars" (Ehl Harrison Consulting Inc and Genivar 2008).

This chapter draws upon a Canadian example of a planning effort for long-term wastewater management. More specifically, it describes how a number of values and assumptions drove the development of a Biosolids and Residuals Master Plan (BRMP) for Canada's largest metropolis, the City of Toronto. A description of the methodology employed within the plan will be followed by a discussion of how ethics and value systems affected both the drafting of the plan as well as community responses. The case will be made that water management decisions are hardly value free. The final section of the paper offers recommendations on how to enhance sustainable water management by addressing the impact of ethical judgments upon decision making.

2 Case Study: Managing Toronto's Wastewater Biosolids

While both provincial and federal governments in Canada have a number of supervisory functions, the majority of wastewater systems are municipally owned and operated. (Johns and Rasmussen 2008: 83). In Ontario's capital city, "Toronto Water" holds responsibility for providing high-quality drinking water, as well as for all phases of water transmission and distribution, wastewater and storm-water collection and treatment (AECOM 2009: 1). Together with a series of pumping stations and forcemains, a sewer system stretching over a length of 9,000 km conveys 1.3 million cubic meters of wastewater to four separate treatment plants daily. As much as 174,000 wet tonnes of wastewater biosolids are generated annually (City of Toronto 2013).

More than 2.7 million people reside in Toronto, the province of Ontario's capital city. In fact, over 30 % of all recent immigrants to Canada find their home here. (City of Toronto 2012). Ontario's population growth, through both immigration and births, is expected to be higher than the national average over the coming decades as the province absorbs an increasing proportion of the national population overall (Statistics Canada 2012).

Anticipating continued metropolitan growth, officials have recognized the need for long-term wastewater and biosolids management planning. Historically, disposal of biosolids occurred through incineration or landfills. While some land application has occurred in Ontario since the 1970s, a 1996 Great Lakes Water Quality Agreement caused the province to update regulations 2 years later. In the same year, 1998, the amalgamation of seven municipalities resulted in the creation of the new City of Toronto. Almost immediately, interest began to be expressed by councillors and planners in developing a long-term program of 100 % beneficial use of biosolids, in place of incineration or landfill disposal.

Today, there is a diversity of biosolids management options that the City of Toronto utilizes. On the one hand, "Beneficial Use Options" are said to profit from the soil-conditioning features of biosolids when they are applied as compost, pellets or dewatered cake to agricultural lands, tree farms, land rehabilitation needs, and other agricultural and horticultural locations. Other options, however, continue to be thermal reduction and incineration, landfill disposal, co-management with municipal solid waste, or green bin composting disposal, as well as market sales for use as a fuel product or proprietary fertilizer (City of Toronto 2009).

In order to plan ahead and navigate among these management options, the City's BRMP was developed in 2002 to provide guidance to the year 2025. The principal decision-making method utilized in the plan was a multi-criteria analysis (MCA) weighted scoring model, considered to be "the most common" approach used by engineers involved in significant biosolids management decisions (Osinga 2011).

It is also a method that aims to ensure that "rational, quantitative conclusions" are developed for large-scale planning decisions (Osinga 2011). Such a weighted scoring model is:

> a tool that provides a systematic process for selecting projects based on many criteria. The first step in the weighted scoring model is to identify the criteria important for the project selection process. The second step is to assign weights (percentages) to each criterion so that the total weights add up to 100 %. The next step is to assemble an evaluation team, and have each member evaluate and assign scores to each criterion for each project. In the last step, the scores are multiplied by the weights and the resulting products are summed to get the total weighted scores. Projects with higher weighted scores are the best options for selection, since "the higher the weighted score, the better" (Lessard and Lessard 2007: 27).

As was to be expected, the BRMP was developed in fulfillment of all provincial planning requirements stipulated in Ontario's Environmental Assessment Act as well as the Municipal Engineers Association Class Environmental Assessment process. Key components of this process (Osinga 2011) included:

- Stakeholder consultation
- Consideration of a "reasonable range" of alternatives
- An evaluation of the environmental effects of each alternative
- Systematic evaluation of each option
- Clear documentation and a transparent decision-making procedure

Despite this careful planning process, the issue of the draft Master Plan in September, 2004, generated serious public concern when released for a 30-day comment period. Approximately 200 responses were received, many of them from residents who objected to a recommendation that favored a fluidized bed incinerator in their neighborhood. Consequently, in March 2005, two city council-lors requested that a formal peer review be undertaken to evaluate the methodol-ogy utilized within the plan. Following a consultation process with other municipalities, industry, and scientific experts, "it was determined that the most objective way to undertake a peer review would be by forming an expert panel with selected, qualified, independent panel members whose expertise matched the specific needs of the project" (City of Toronto 2008: 3). The author of this chapter was one of the seven members selected for the peer review panel.[1]

3 The Peer Review Process and Its Findings

The panel was not charged with reviewing the biosolids management *technologies*. Instead, its task was to assess the appropriateness of the decision-making model, its criteria, and its scoring process. Overseen by Toronto Water and Toronto Public Health staff, the work of the peer review panel was coordinated and directed by Ehl Harrison Consulting Inc, together with Genivar, an environmental engineering firm specializing in integrated urban and environmental planning solutions. The peer review process included several meetings, public presentations, question and answer sessions, and preparation of a final written response to the draft Master Plan.

The panel concluded that the decision-making model utilized in developing the Master Plan was an example of those "commonly used" in generating both mas-ter plans and environmental assessments and, to that extent, it was "not unreason-able." Nevertheless, the panel did find "shortcomings in its implementation and suggested improvements, as well as additional tools that could be used to add rigor to the decision-making process" (Ehl Harrison Consulting Inc and Genivar 2008).

Specifically, five problem areas were flagged: (1) There was a lack of detail and clarity in the BRMP documentation; (2) there was "limited reach" of both the con-sultation and the tools that were utilized; (3) there was insufficient recognition and incorporation of public risk perceptions; (4) the process of weighting and scor-ing alternatives was unclear; and (5) a mediation agreement that was drawn up between one local community and the city to respond to concerns of the Master Plan was itself problematic. That agreement sought to allay concerns around the proposed incineration technology, and yet portions of the agreement were "ambig-uous" and indeed appeared to be "contradictory," implying that incineration might be an option even as the spirit of the document recommended against it.

---

[1] Other members of the Peer Review Panel were Dr. Ida Ferrara, York University; Mr. Paul Kadota, P.Eng., Greater Vancouver Regional District; Mr. Mark C. Meckes, United States Environmental Protection Agency; Dr. David Pengally, McMaster University; Dr. Lesbia Smith, University of Toronto; and Dr. Paul Voroney, University of Guelph. Ms. Tracey Ehl, MCIP, and Ms. Fredelle Brief of Ehl Harrison Consulting Inc., chaired the deliberations of the panel.

In the end, the following major recommendations for improvements to the Master Plan and decision-making process were presented by consensus of the panel to the city staff (Ehl Harrison Consulting Inc and Genivar 2008):

- *Enhance detail and overall clarity*: A number of elements in the decision-making model and mediation agreement were not readily understandable. The panel called for further "elaboration of definitions, and step-by-step descriptions of the calculations behind some of the outcomes" (Ehl Harrison Consulting Inc and Genivar 2008).
- *Broaden stakeholder consultation*: The panel felt that some members of the public—for instance, rural communities impacted by agricultural land application or landfill disposal—had not been properly consulted. Additionally, it felt that "the City engaged a relatively small number of individuals in the various stakeholder groups, who, for the most part, may not be statistically representative of their communities" (Ehl Harrison Consulting Inc and Genivar 2008). Consequently, it was suggested that additional tools be utilized to capture broader stakeholder input that was statistically valid.
- *Acknowledge the significance of public perceptions of risk*: While recognizing that no technology is risk free, the panel recommended that a risk assessment framework be a more explicit part of the Master Plan. The public's perception of health risks associated with incineration, for instance, was a primary factor behind many stakeholder responses to the plan. A diversity of risk assessments to address uncertainties and identify best practices was suggested (Osinga 2011).
- *Improve process for developing weighting criteria and scoring alternatives*: The Master Plan presented findings but did not provide clear explanation as to the reasoning behind the numbers in the weighted scoring model. The panel suggested the need for a review of the criteria and their weightings, together with clear documentation of the calculation process so that results could be easily replicated by others and the public could better understand elements of the decision-making process.
- *Consider additional, alternative decision-making models*: While a weighted scoring model was understood to be reasonable, the panel suggested that additional methods be utilized for decision-making purposes. Such methods could include risk assessments, public opinion surveys, and a triple-bottom-line decision-making model that focused on minimizing environmental, social, and economic impacts (Osinga 2011).
- *Re-assess scoring priorities*: Rather than privileging financial, technical, operational, and managerial elements, the panel suggested that higher values needed to be placed upon community concerns, public health, and environmental considerations (Osinga 2011).
- *Establish a longer term perspective on biosolids management*: Since there is a need to continually update the public about biosolids options, the panel suggested a long-term strategy and resource commitment to ensure public education programs. Additional quantitative surveys and qualitative research were

proposed in order to "help to set the planning context for future projects" to a 50—rather than 25 year—planning horizon (Ehl Harrison Consulting Inc and Genivar 2008).

The peer review panel's recommendations were presented to the City of Toronto in February, 2008. Following a number of public information sessions, the city initiated a Biosolids Master Plan (BMP) Update in 2008. AECOM—a consulting engineering firm—was hired to finalize the Master Plan which was approved by city council in 2010 and provides a blueprint for biosolids management to the year 2055.

A number of improvements to the original draft Master Plan were made, following the peer review process. Key changes reported by the City of Toronto (2009) included the following:

- Evaluation criteria and categories were revised in the weighted scoring model to ensure that they were more easily understood and legible to a lay audience.
- Quantitative surveys were conducted by telephone and focus groups organized to obtain statistically relevant public feedback about biosolids management options and decision-making criteria.
- Rather than providing a single, universal set of recommendations for such a large metropolis with a diversity of community expectations, options were evaluated with respect to the specific needs of each of the four wastewater treatment facilities, within the context of the city's overall needs.
- How each management option was scored was explained in greater detail, ensuring that information was provided about the meaning of each criterion and why it was used in the decision-making process.
- Information was updated with respect to developments in biosolids technologies and management opportunities.
- A more holistic accounting of impacts and opportunities was utilized, drawing from a "triple-bottom-line" approach that addressed environmental, social, and economic concerns of the city.
- While weightings are often evenly distributed in such cases of decision making, in this instance, the final plan weighed the environmental and social indices more heavily than cost indices, reflecting community values (AECOM 2009: 12).
- The overall strategy was now to maximize programs that encourage beneficial use of biosolids cake, relying upon landfill disposal purely as a "contingency measure" (AECOM 2009: 16).

Seven years of consultants' reports, peer review panel deliberations, focus groups, surveys, and public workshops have resulted in the final approval in December, 2009, by city council of a BMP for the City of Toronto. Certainly, the Master Plan management process has required a significant commitment to date, both financially as well as in terms of human resources.

One cannot help but wonder however: might the process have been more streamlined, had underlying values and judgment calls been more explicitly addressed? What were some of those values and ethical assumptions that affected the process of decision making? The following section looks at those questions specifically.

4 Values, Judgments, and Ethical Assumptions

It is common to perceive the role of ethics as a matter of clarifying universal moral principles to provide a theoretical framework for complexities of decision making. Through such a top-down model of justification, the expectation is that ethics consists simply of "applying a general rule (principle, ideal, right etc.) to a particular case that falls under the rule" (Beauchamp 2005: 7).

As appealing as such a model may be to some, others argue that ethics is more than a top-down intellectual exercise of applying theories and principles to specific situations. Rather, ethics is better understood as a bottom-up process of deciphering implicit values that underlie decision-making practices. Moral principles, on such a reading, are derivative, informed by the vagaries of each particular case, rather than intellectually conclusive, foundational, and resolved in advance of engaging with lived experience (Beauchamp 2005: 8).

To be sure, the fact is that "sometimes we do not *know* what our actual beliefs and values are" (Hinman 2013: 5). Values are often deeply embedded in our daily decisions and, in that respect, are implicit or even operate at a subconscious level (Stefanovic 2012). In that regard, the task for philosophers is perhaps less one of creating grand, speculative theories than of serving as "stand-in interpreters" who help communities to clarify and critically evaluate those values that impact in a significant way upon important decisions (Morito 2010).

When it comes to the case of biosolids management within the City of Toronto, values infused the decision-making process from the very start and on a number of different levels. Let me draw upon a few salient examples in order to then explore how they impacted upon the long process of evolving a master plan.

Consider the decision taken by engineers to base the original draft of the Master Plan on a quantitative, weighted scoring model. The 2004 report points out that, given the complexity of biosolids and residuals management processes, "experience in other communities has shown that developing a *systematic, step-wise method* for making decisions at the start of the project helps to focus and clarify decision making" (KMK Consultants 2004: 80. Italics added). Employing such a logical model is indeed common when it comes to large-scale planning projects, precisely because it is seen to set a framework "for a *systematic, rational and replicable* environmental planning process" (KMK Consultants 2004: 7 Italics added). Employing such an apparently "rational" and "replicable" model of decision making was intended to enable the identification of "actual benefits and impacts of the specific option" by way of "a quantitative comparison of one alternative to another" (KMK Consultants 2004: 83).

The language utilized here reflects a positivist paradigm that is characteristic of the mainstream western understanding of modern water management which begins, as some ethicists point out, "with humanity as the main focus of moral concern, separate from and generally understood to be superior to the rest of the world" (Brown and Schmidt 2010: 268). The decision-making model was intended to ensure a process that was intended to be *objective, quantitative, systematic,* and *methodical.* By virtue of presenting "actual" numerical *scores* for the various alternatives, the perceived value of *technical efficiency and control* was a primary driving force behind the model. Indeed, the way in which the numbers were presented was meant to indicate that the findings were not merely subjective but rather had the verity and scientific objectivity of mathematical calculation behind their truth value. The "right" way, on this reading, to undertake a comprehensive and rational decision-making process was to ensure that the value of quantification was taken seriously.

For instance, the plan noted that "value weights were applied to differentiate between those individual criteria which are very important, and those which are less important" (KMK Consultants 2004: 83). However, it is important to acknowledge that "value weights were applied" not in some absolutist manner but by actual *people*—human subjects who were engaged in the interpretation and prioritizing of criteria according to judgment calls that *were not always made explicit within the final plan.* To be sure, a sensitivity analysis was undertaken as part of the process and it was deemed significant that the same options were consistently identified as receiving the highest scores (KMK Consultants 2004: 157). Overall, the "value" of calculation was supreme, and it was assumed that such a rational approach would ensure the greatest distribution of good overall to the citizens of Toronto.

While the overt quantitative approach was meant to suggest *objectivity* of the final recommendations, the fact is, however, that the vocal reaction of the local community revealed that the scoring process was not as calculatively certain as it may have been meant to appear.

Moreover, the calculative paradigm of this model betrayed the common characteristic of many large-scale environmental planning processes, that is, it assumed the validity of a *utilitarian value system.* Utilitarianism is arguably one of two dominant schools of thought in the western ethics tradition, the other being deontology (Callicot 2005: 284). Arising from the writings of John Stuart Mill and Jeremy Bentham, utilitarianism aims to facilitate "the greatest good for the greatest number," usually of human beings, although often, animals are included in the formula (Mill 1863; Bentham 1970). Some interpret the "greatest good" in terms of "greatest happiness," while others refer to the significance of promoting the "greatest welfare" overall, but in any case, the utilitarian theory suggests that the morally superior decision is the one that advances the greatest good overall.

Cost-benefit analysis that seeks to weigh advantages and disadvantages in order to obtain an optimal result is a penultimate example of utilitarianism in action within the field of economics. But a utilitarian framework also emerges from other common decision-making models as well. The weighted scoring approach utilized in the City's BMP reveals a utilitarian value framework to the extent that the process was meant to deliver a set of recommendations that weighed alternatives in an objective manner and quantified mathematical scores to advance the greatest net benefit overall. As the writers of the plan explained, by way of systematic evaluation and weighing of the advantages vis-à-vis disadvantages of a particular alternative, the aim of evaluating each alternative was "to determine their *net* environmental effects" (KMK Consultants 2004: 7).

Needless to say, and as the peer review panel members themselves stated, such a utilitarian model of decision making that aims to advance the "greatest good for the greatest number" of citizens in the City of Toronto is hardly unreasonable. On the contrary, it is frequently utilized because it is deemed to be most efficient and fair, satisfying the demands of distributive justice, particularly when it comes to large-scale environmental decisions that affect a large population, such as biosolids management.

However, while deemed by many to be a "reasonable" approach, the City's Biosolids decision-making model left little room for stakeholder values that emerged later and that represented a second dominant western model of values, that is, a *deontological* rather than utilitarian moral framework. "Deon" is the Greek word for "duty," and so "deontological" approaches emphasize notions of duty and *individual rights*. Philosopher J. Baird Callicott offers the example of Roman gladiator contests: Quantitatively speaking, thousands of spectators received great satisfaction at the expense of the pain incurred upon five or ten gladiatorial contestants; nevertheless, each of those contestants had a *right* to human dignity and respect in principle that today we recognize must override the "repugnant outcome of the unbridled utilitarian welfare calculus" (Callicott 2005: 285).

Drawing upon similar arguments, residents of a single neighborhood were opposed *in principle* to the incineration option that was calculated within the original draft Master Plan as an option that promoted "the greater good" to citizens of Toronto as a whole. Those neighborhood residents argued that they had a *right* to refuse the incineration option, *no matter the overall welfare calculus.* Because they had longstanding concerns about impacts upon human and environmental health of a previous incineration unit within their community, their view was that the municipal government had a *duty* to respect their concerns and residents had a right to demand such a hearing, irrespective of the calculation of overall good to the city as a whole.

This underlying divergence between utilitarian efficiency of the greater good, on the one hand, and a deontological belief in principles of human rights—is commonly observed and often helps to explain what is at the root of many stakeholder conflicts (Stefanovic 2012). Had the engineers who drafted the original Master Plan recognized the deep significance of this community's rights-based objection to incineration, they might have identified different biosolids use options right from the start.

In that connection, it becomes important in any decision-making process to (a) make such divergent value systems explicit, early in the game and (b) encourage ways in which to communicate *across* the values divide. Philosopher Bruce Morito offers an example of how this strategy might be employed. He describes a forum where First Nations' people, industry representatives, and others came together to discuss resource management issues (Morito 2010). A resource manager approached him, frustrated that the Aboriginal communities were unwilling to allow the building of a dam on their territory, despite being offered "more than adequate" compensation. Morito turned to him and asked whether he would agree to sell his daughter into slavery for a "more than adequate" amount of money.

Clearly, the manager was unwilling to do so, but through the analogy, he began to better understand the First Nations' unwillingness to compromise their principles with respect to the land. Morito (2010: 110) concludes that the basic idea of bringing value systems to light is to "seek mutual understanding among stakeholders concerning their values and then allow this understanding to generate prescription principles."

Admittedly, identifying taken-for-granted value systems and interpreting conflicting moral paradigms is not a easy task. But the argument can be made that this is precisely the role that ethicists and philosophers should be undertaking. Otherwise, values will affect perceptions and attitudes of both experts and the broader public in ways that remain hidden, even as they exert a powerful influence upon decisions made.

For instance, let us consider another example of how values and attitudes affected the scoring of alternatives within the City of Toronto's Biosolids decision-making model. During the master plan peer review process, the team recognized that scoring criteria such as resource inputs to the biosolids management system were given more weight and importance over public health and environmental outputs. Financial, operational, managerial, technical performance and construction considerations, representing 50.9 % of the weight in the overall scoring, were found to be privileged by the engineering firm who prepared the initial draft plan, over community, public health and natural environment considerations which represented only 49.1 % of the weight of the overall scoring. "The consequences of affording so much weight on the input criteria," the panel reported, "is the potential of reduced sensitivity to concerns expressed by external stakeholders" (Ehl Harrison Consulting Inc and Genivar 2008: 34–35).

In fact, once those external stakeholder *were* taken into account, the decision-making model was redesigned to emphasize community values in a more meaningful way. As the Master Plan Update reports, "although in this type of model, weightings are usually evenly distributed between the three indices, for the BMP Update, the Environmental Index was weighed more heavily, followed by the Social and Cost Indices. This is to reflect the level of importance of each criteria group to the public and consulted stakeholders" (AECOM 2009: 12). In other words, while *technical and economic* concerns were more heavily weighted by engineers in the earlier drafts of the Master Plan, it gradually became evident, through a more sustained stakeholder communication process, that an emphasis on *environmental sustainability and health* considerations more accurately reflected the values of the community as a whole. Had such a meaningful consultation process occurred earlier, presumably time and money will have been saved by the city because the plan will have reflected the pervading community values from the start.

Another way in which values arise on water management projects such as this one relates to perceptions and attitudes regarding *risk*. The peer review panel recommended that "public perception of the risks related to both human health and other environmental impacts associated with various technologies should be addressed across all communities" (Ehl Harrison Consulting Inc and Genivar 2008: 43). There was a duty, the panel felt, of the City of Toronto to demonstrate that it was following best practices "to mitigate risks to the public's health and safety, so that no community bears a disproportionate amount of risk" (Ehl Harrison Consulting Inc and Genivar 2008: 43). For these reasons, the panel proposed that a risk assessment framework be added to the Master Plan.

Interestingly, the city disagreed. A staff report indicated that including such a risk assessment framework "would be costly, time consuming and, in this instance, would not add significantly to the decision making process" (City of Toronto 2008: 6). Yet, the fact is that excluding risk assessment in any project can itself be a risky move: Management professionals recognize that "addressing risks proactively will increase the chances of accomplishing the project objective. Waiting for unfavorable events to occur and then reacting to them can result in panic and costly responses" (Gido and Clements 2012: 284). In those instances where uncertainty exists and the stakes are high, risk management is particularly crucial. It is only by incorporating a risk framework that "surprises that become problems will be diminished, because emphasis will now be on proactive rather than reactive management" (Kerzner 2001: 904).

In the case of the Toronto BMP, engineers did not themselves adequately anticipate or plan for the risk of antagonistic community responses to their initial draft plan. Master Planners' neglect of perceived risks of incineration technologies by community members in one Toronto neighborhood eventually became a significant stumbling block and cause of delays in the overall planning process.

Other ongoing concerns of community members reflected important value judgments regarding risks, even with regard to the safety of "beneficial use" options such as land applications. An article in a leading Toronto newspaper expressed concern, for instance, that biosolids constitute a "disaster waiting to happen" (Vynak 2008). Certainly, environment ministry officials promote biosolids as a "safe" alternative to other commercial land applications, insisting that guidelines are "both up to date and adequate" (Vynak 2008). The feeling within government circles is that risks are thereby mitigated to a reasonable degree.

Others, however, are not convinced. Opponents argue that biosolids "may contain thousands of toxic chemicals, the effects of which we know little about. Regulatory guidelines for spreading biosolids on farmland are outdated and inadequate," having been updated as far back as 1998 (Vynak 2008). Stories abound about rural residents near sludged properties who complain about respiratory and stomach problems, headaches, nausea, rashes, and fatigue. Soil scientists express concerns about concoctions of pharmaceutical medications excreted in human waste or pathogens like *Escherichia coli* bacteria persisting through the water treatment process and affecting the health of the land and surrounding residents. "I don't know how (the Ministry of the Environment) can believe regulated heavy metals are the only contaminants in sludge we need to worry about," laments soil scientist Murray McBridge (Vynak 2008).

To be sure, there is no such thing as "no risk" in life. In the words of the peer review panel, it is always helpful to remember that "there are no biosolids management options that are totally risk free" (Ehl Harrison Consulting Inc and

Genivar 2008: 43). Nevertheless, "risk management is not done by machines or robots.... It requires human judgment" (Hillson and Murray-Webster 2005: 19). Different risk personalities assess risk differently. For instance, as has been shown in other instances, mothers are frequently unwilling to balance risks and benefits through a utilitarian calculus when it comes to the health of their children, arguing instead in favor of a precautionary approach to risks (Stefanovic 2012). To argue that a sustained pattern of risk management is either value free or not worth the investment is simply irresponsible in water management scenarios.

A range of other judgment calls can impact upon project decisions. How the problem is defined in the first place inevitably reflects attitudes regarding what ought to be included or excluded within the scope of a project. In the Toronto Biosolids example, choices about how to define and scope evaluation criteria, together with the decision to rely upon a particular scoring method, were seen by the peer review panel to have clearly influenced the outcome of the original bio-solids assessment (Osinga 2011: 7). That only urban residents were consulted may have seemed reasonable in the beginning inasmuch as all water treatment plants were geographically located within the urban core. However, the potential for rural applications of biosolids meant that rural municipalities should have also been consulted. The peer review panel, therefore, recommended expanding stakeholder consultation beyond the city limits. The takeaway lesson here is that an ethical stakeholder management process is one that ensures that less vocal contributions (in this case, the rural municipalities) are meaningfully represented.

Another example of how values affect project definition relates to how the project as a whole is perceived within the context of the broader community plan. While incineration was a management option that was scored third for one major wastewater facility, ultimately, it was not recommended within the final, Master Plan Update because of the city's "plans to make significant investment in a 20-year program to improve the waterfront" within the surrounding area (AECOM 2009: 17). In other words, when the incineration option was considered within the larger spatial and temporal city planning scales, it was no longer seen to be as viable a biosolids management option for this particular community, despite its apparent technical efficiency. The fact that a longer time frame—amended from 2025 to 2050—was proposed similarly contextualized options within a different and broader planning horizon. Both the spatial and temporal *contexts* influence the identification and assessment of water management options, as the case from Toronto clearly indicates.

5 Next Steps: Enhancing Water Management Practices with Ethics

As we have seen, human factors and judgment calls affect management decisions at many levels and at all stages of the decision-making process. Few decisions can be said to be meaningfully value free. In that regard, the job for ethicists is to help to identify and critically evaluate ethical dimensions of water management decisions. Doing so will help to anticipate and proactively address potential conflicts that might emerge as a result of value judgments that frequently operate implicitly within the decision-making process.

Sometimes, those value judgements emerge due to different theoretical beliefs, such as in cases where utilitarian and deontological values conflict. In other cases, they underlie our risk assessments of the "safety" of new technologies. In fact, how projects are scoped—which alternatives are deemed to be "reasonable" and how they are quantified within scoring systems—also reflect judgment calls regarding what ought to be included and/or excluded as a viable option in the decision-making process.

It is naïve to assume that value judgements do not matter. They can affect policies and politics: As the City of Toronto's Biosolids Management example shows, when a community's values and risk perceptions are not taken seriously by planners, a project can experience severe delays, particularly when a community elicits the voice of a powerful politician to represent their core values.

Water ethicists Peter Brown and Jeremy Schmidt summarize the point succinctly when they conclude that:

> from a decision making perspective, purely rational and technocratic management cannot go far enough…What we also need is a new narrative that positions scientific knowledge and technological know-how as part of the broader systems people seek to manage, and which include the cultural, religious and ethical values by which the managers and users are informed (2010: 274).

It is in the spirit of such a new narrative that this paper invites those involved in the water management process to reflect upon and to critically evaluate taken-for-granted values that affect decisions that are, ultimately, always more than merely technical.

References

AECOM Canada Ltd (2009) City of Toronto biosolids master plan update. Brampton, Ontario

Beauchamp T (2005) The nature of applied ethics. In: Frey RG, Wellman CH (eds) A companion to applied ethics. Blackwell, Malden

Bentham J (1970) Introduction to the principles of morals and legislation. Clarendon Press, Oxford

Brown P, Schmidt J (2010) An ethic of compassionate retreat. In: Brown P, Schmidt J (eds) Water ethics: foundational readings for students and professionals. Island Press, Washington

Calicott J (2005) The intrinsic value of nature in public policy: the case of the endangered species act. In: Cohen A, Wellman C (eds) Contemporary debates in applied ethics. Blackwell, Malden

City of Toronto (2013) Toronto's wastewater treatment plants. http://www.toronto.ca/water/wastewater_treatment/treatment_plants/index.htm. Accessed on 26 June 2013

City of Toronto (2009) Responsible choices. http://www.toronto.ca/wes/techservices/involved/wws/biosolids/pdf/newsletter/brmp_6_newsletter.pdf. Accessed on 25 June 2013

City of Toronto (2012) Toronto's racial diversity. http://www.toronto.ca/toronto_facts/diversity.htm. Accessed on 24 June 2013

City of Toronto (2008) Staff report on the biosolids and residuals master plan peer review panel report. Submitted by Toronto Water to the Public Works and Infrastructure Committee. http://www.toronto.ca/legdocs/mmis/2008/pw/bgrd/backgroundfile-13879.pdf. Accessed 27 June 2013

Ehl Harrison Consulting Inc (EHC), Genivar (2008) City of Toronto biosolids and residuals master plan decision making model peer review

Gido J, Clements J (2012) Successful project management. South-Western Cengage Learning, Mason

Hinman L (2013) Ethics: a pluralistic approach to moral theory. Wadsworth, Boston

Hillson D, Murray-Webster R (2005) Understanding and managing risk attitude. Gower Publishing Company, Burlington

Johns C, Rasmussen K (2008) Institutions for water resource management in Canada. In: Sproule-Jones M, Johns C, Heinmiller B (eds) Canadian water politics: conflicts and institutions. McGill-Queen's University Press, Montreal

Kahneman D (2011) Thinking fast and slow. Farrar, Straus and Giroux, New York

Kerzner H (2001) Project management: a systems approach to planning, scheduling and controlling, 7th edn. Wiley, New York

KMK Consultants Limited, in Association with Black and Veatch Canada and Cantox Environmental Inc (2004) City of Toronto biosolids and residual master plan. City of Toronto, Canada

Lessard C, Lessard J (2007) Project management for engineering design. Morgan and Claypool, San Rafael

Mill JS (1863) Utilitarianism. Longmans, London

Morito B (2010) Ethics of climate change: adopting an empirical approach to moral concern. Hum Ecol Rev 17(2):106–116

Osinga I (2011) City of Toronto biosolids and residuals master plan decision-making model peer review. In: Proceedings of the residuals and biosolids 2011 conference. Water Environment Federation (WEF), Alexandria, VA

Statistics Canada (2012) Population projections for Canada, the provinces and the territories. http://www.statcan.gc.ca/pub/91-520-x/2010001/aftertoc-aprestdm1-eng.htm. Accessed 24 June 2013

Stefanovic I (2012) To build or not to build: how do we honour the landscape through thoughtful decision making? Minding Nat 5(1):12–18

United Nations (2012) World urbanization prospects: the 2011 revision highlights. United Nations Department of Economic and Social Affairs, New York.http://esa.un.org/unup/pdf/WUP2011_Highlights.pdf. Accessed 24 June 2013

Vynak C (2008) Biosolids a 'disaster waiting to happen'. The Toronto star. http://www.thestar.com/news/gta/2008/07/13/biosolids_a_disaster_waiting_to_happen.html. Accessed 19 Sept 2013

Author Biography

Ingrid Leman Stefanovic is Dean of the Faculty of Environment at Simon Fraser University, Vancouver, Canada, and Professor Emeritus, Department of Philosophy, University of Toronto. She has served as Executive Co-Director of the International Association for Environmental Philosophy; Senior Scholar at the Center for Humans and Nature, Chicago and New York; and Academic Fellow of the Potomac Institute for Policy Studies, Arlington, Virginia. Her teaching and research center on how values and perceptions affect public policy, planning, and environmental decision making. Recent books include *Safeguarding Our Common Future: Rethinking Sustainable Development* and *The Natural City: Re-Envisioning the Built Environment*.

Water as an Element of Urban Design: Drawing Lessons from Four European Case Studies

Carlos Smaniotto Costa, Conor Norton, Elena Domene, Jacqueline Hoyer, Joan Marull and Outi Salminen

Abstract One of the most challenging problems that urban areas will face in the future is adaptation to the effects of climate change, particularly with regard to local problems of water management (e.g., flooding caused by heavy rain events, degradation of urban streams, and water scarcity). Sustainable local management of stormwater calls for approaches that connect technical and ecological solutions with urban design aspects and socioeconomic factors. This in turn opens up great opportunities to advance knowledge toward the application of water-sensitive urban design (WSUD), an approach that integrates the water cycle into urban design to simultaneously minimize environmental degradation, improve aesthetic and recreational appeal,

C. Smaniotto Costa (✉)
Department of Urban Planning, Universidade Lusófona de Humanidades e Tecnologias, Campo Grande, 376, Lisbon 1749-024, Portugal
e-mail: smaniotto.costa@ulusofona.pt

C. Norton
School of Spatial Planning, Dublin Institute of Technology, Bolton Street, Dublin 1, Ireland
e-mail: conor.norton@dit.ie

E. Domene · J. Marull
Barcelona Institute of Regional and Metropolitan Studies (IERMB), Edifici MRA, Autonomous University of Barcelona, 08193 Bellaterra, Barcelona, Spain
e-mail: elena.domene@uab.cat

J. Marull
e-mail: joan.marull@uab.cat

J. Hoyer
Sustainable Urban and Infrastructure Planning, HafenCity Universität Hamburg, Hebebrandstraße 1, 22297 Hamburg, Germany
e-mail: jacqueline.hoyer@hcu-hamburg.de

O. Salminen
Department of Forest Sciences, University of Helsinki, Latokartanonkaari 7, 00014 Helsinki, Finland
e-mail: outi.m.salminen@helsinki.fi

© Springer International Publishing Switzerland 2015
W. Leal Filho and V. Sümer (eds.), *Sustainable Water Use and Management*,
Green Energy and Technology, DOI 10.1007/978-3-319-12394-3_2

17

and support social cohesion. A comparative study of four case studies across Europe reveals some of the successes and limits of WSUD implemented so far and presents new considerations for future developments. Best practices on integrated management as well as concepts to re-establish natural water cycles in the urban system while ensuring water quality, river health, and sociocultural values are included. In the selected case studies, water takes a structuring role in urban development, which has been designed to serve diverse public functions and maximize environmental quality, urban renovation, resilience to change and sustainable growth.

Keywords Water sensitive urban design · Decentralized water management · Stormwater management · Open spaces · Urban development · Urban design

1 Introduction

Urban growth is a world phenomenon that implies a profound change in the natural environment. Urban development is a dynamic and very diverse process, and although with distinctions in terms of social and economic drivers and magnitudes of impact, it has a common feature: It is increasingly space intensive (UNFPA 2007). Considering space as a finite resource, the ever-increasing demand for land for urban purposes, e.g., for housing, workplaces, recreation, infrastructure, and transport networks, speeds up and intensifies processes of pressure on landscape and ecosystems.

More and more, the land-take affects the natural environment, both functionally and morphologically, with far-reaching effects also on the built environment, whose quality depends very much on the nourishing quality of the natural environment. Although these two aspects are part of the same process, they are usually not discussed and treated with the same concern and consideration. In numerous cases, urban growth results both in landscape fragmentation and in a substantial loss and degradation of open spaces. Open spaces, as components of the green infrastructure, are often treated as "potentially developable" land within the urban fabric. Rural zones, forests, and seminatural and natural water bodies are disappearing in favor of a high percentage of sealed-off ground. The land taken for urban purposes and related infrastructure affect biodiversity also, since it provokes degradation of habitats and reduction of the living space of species, along with the loss of landscape segments that support and connect the remaining habitats with each other (EEA 2010).

One of the most challenging problems that urban areas will face in the future is the adaptation to the effects of climate change, particularly with regard to water: in Southern Europe long drought periods followed by floods due to heavy rain events are expected, while increases in heavy rain events are expected in many parts of Northern Europe. Increasing problems with water quality in rivers and urban streams are among the challenges that Europe is facing in the future (IPCC 2012). Stormwater management policies in most European urban areas have traditionally removed water runoff from communities by burying water systems in underground impervious pipe and culvert infrastructures for the protection of human health and property, with a low priority in the conservation of natural water cycles.

In the context of this work, urban stormwater is to be understood under its broad meaning: rainfall and snowmelt that seeps into the ground or runs off the land into storm sewers, streams, and lakes. When functioning well, the conventional urban water infrastructure systems are effective. However, the increase of urban population, expanding land uptake and soil sealing, and climate change effects along with growing environmental concerns raise questions on the limits of its functionality. The practice of conveying (storm) water away from urban areas results in wasted economic resources and loss of opportunities for making use of alternatives for the sustainable development of cities. Sustainable alternatives include the increase of water supply via decentralized stormwater systems or harvesting, the improvement and strengthening of the urban environment by reducing the risk of natural events, as well as making cities more distinctive and attractive through the reduction of impervious surfaces and the enhancement of infiltration and detention. Moreover, alternative water "elements" enrich the urban landscape and create high landscape values, which makes cities more attractive with a well-defined identity.

To respond to future challenges as well as to seize the opportunities, stormwater management is best tackled by decentralized and locally appropriate approaches in a tight connection with urban development issues. This calls for approaches that break from the traditional pipe-bound systems and combine stormwater management, sustainable urban design, policy change, and capacity building. Such practice change can be logistically and technically more difficult than just discharging stormwater into the drains or watercourses, and change is still often perceived irrelevant by service managers and the local communities (Hoyer et al. 2011). More integrated and innovative alternatives to the conventional management of stormwater are emerging. Integrating practices is seldom problem free due to a limited culture of cooperation between stormwater managers and urban planners and often unexpected institutional barriers. Greatest public and political support emerges out of the process when examples of successful integration are shown (Barbosa et al. 2012). In fact, water-sensitive urban design (WSUD) measures apply to all aspects of change in the built environment and urban management. These aspects are extensively discussed with the help of the four practical cases analyzed: Dublin (Ireland), Santa Coloma de Gramenet (Spain), Cascais (Portugal), and Nummela (Finland). These cases are examples of an alternative approach to local stormwater management. Their motivation, achievements and barriers are discussed in the following sections.

2 The Undergoing Process of Urbanization, Land-Take, and Stormwater Management

Sustainability calls for effective ways for cities both to reverse the level of land-take in a socially and economically meaningful way and to include the urban landscape aspects in planning procedures and decision-making processes. Heading toward sustainable urban stormwater management, cities are faced with several challenges. Coping with these challenges will put urban areas increasingly under pressure on account of the sensitive problems of resource inefficiency and waste. Five issues are hereby paramount:

1. Growing urban population: More and more people will be living in urban areas in the future. In 2012, already around 41 % of the EU27 population lived in urban areas (Eurostat 2012). The growth of urban population puts enormous pressure on water infrastructure systems, as cities are one of the main water resource consumers, as well as one of the main polluters. With the rise of population, an expansion of land-take and soil sealing can be expected, which will negatively affect stormwater infiltration and storage capacity at local level. This leads to reduced groundwater recharge rates, increased surface runoff, deficient soil water storage, and not sufficient availability of water for vegetation. Moreover, due to the fact that urban streams are receiving waters for stormwater runoff from urban areas, there is an observed ecological degradation of streams draining urban land. Walsh et al. (2005) describe degradation as elevated concentrations of nutrients and contaminants, altered channel morphology, with the reduction of biotic richness along increased dominance of tolerant species.

2. Climate change: Some climate change models predict changes in the frequency and intensity of meteorological extreme events. Forecasts are uncertain; in Europe, an increase in long-term droughts in summer and an increase in heavy rain events are expected (IPCC 2012). Long-term droughts will lead to a shortage in available water suitable for urban vegetation and can lead to a shortage of drinking water as well. In Australia, a drought that lasted more than 10 years provoked the heavy decrease of drinking water resources, e.g., in the metropolitan area of Melbourne, 35 % reductions were experienced (MW, n.d.). Drought concerns particularly those countries in Southern Europe, such as Greece, Italy, Portugal, and Spain, where the climate is dry and hot. Moreover, increases in frequency of heavy rain events in central and northern parts of Europe (e.g., Nordic and Baltic countries, the Netherlands, Germany), as well as heavy rain events in the south (e.g., in Tuscany in November 2012), forecast greater intensity floods with high damage potential. Changes in winter freeze–thaw weather patterns in the northern cities and cities of high elevation may cause additional challenges such as rain on frozen grounds; ice cover on a lawn can act as a highly impervious surface for wintertime rain.

3. Inflexible and cost-intensive systems: Existing systems for stormwater management, which are based on pipes and culverts, are not flexible enough to be adapted to uncertain changing conditions from increased urban development or climate change. This leads to unmanageable stormwater runoff. Adapting the existing (mostly centralized) systems to current and future changes calls for higher running costs and investments which municipalities may not be able to afford in the near future. Therefore, there is a need for more flexible, decentralized systems (Cettner et al. 2013; Hoyer et al. 2011). As pipeless alternatives

and open stormwater management systems are not hidden under streets, they need more space in the urban fabric. When planned and implemented together and within the green infrastructure, such systems can result in both capital and maintenance cost reductions.

4. Stormwater is seen as having no value: Urban water management is historically driven by the engineering sector and based on pipes and sewers, from providing drinking water up to the collection and treatment of wastewater. In this system, stormwater is seen as having no value. Thus, stormwater, despite its inherent qualities, is still today discharged to public sewers and pipes, while there are many opportunities to use this water in urban environments. The SWITCH project evaluated several stormwater management experiences across the world, and this has shown that stormwater can be an important resource in urban areas (Hoyer et al. 2011). In particular, some urban subsystems may achieve water self-sufficiency by means of rainwater harvesting (Farreny et al. 2011a). Harvested stormwater can be used for devices, where drinking water quality is not necessary, such as for toilets, industry, landscape irrigation, and water features in public spaces. Using stormwater contributes to safe drinking water and aquifers as well as the improvement of livability in the city. To tap water-fed installations in the urban landscape, water features utilizing stormwater can be designed with gravity flow and dynamically vitalize the scenery year round even in cold climates. Hence, a paradigm shift valuing stormwater as a resource can be used to promote both water and energy conservation.

5. Lack of acceptance for sustainable systems: Some of the EU countries, particularly Germany and the UK, are already quite advanced in developing new approaches and techniques for sustainable stormwater management tackling all the problems mentioned above. The decentralized rainwater management (DRWM) and sustainable urban drainage systems (SUDS) in the UK can be cited as examples. However, acceptance of these systems is still low even in the respective countries (Domènech and Saurí 2012; Hoyer et al. 2011). Even when the positive effects of sustainable alternatives are well demonstrated and described, their incorporation as components of urban infrastructure is still seen as a "tree huggers" alternative and not as a technically sound standard. There is a potentially widening gap between what is known and what has been converting into reality. There are several issues that can be identified as reasons for the lacking of acceptance of decentralized stormwater management systems:

(a) Too strong technical focus and a need for change: Decentralized stormwater measures have been predominantly shaped by the technical sector, which resulted in the development of different techniques (e.g., swales, infiltration trenches, retention basins) with corresponding technical guidelines. However, and unfortunately, stormwater facilities have often been engineered just according to technical points of view without considering socioeconomic and urban design aspects. In consequence, too few projects have been applied in a manner that is appreciated by the community (Echols 2007). Furthermore, hydraulic engineering-driven design underestimates the importance of phytotechnologies; vegetation and associated microbes are the key to water quality improvement in sustainable stormwater design. Stormwater management should move from a regulated requirement into "a clear value added component of good design" (Hoyer et al. 2011: 16).

(b) Missing integration with urban design: Most of the technologies for decentralized stormwater management have been advanced with regard to technical design and functionality, but what has been disregarded so far is their potential to be integrated in urban areas and the establishment of appropriate standards and guidelines along with policy support. The results of the EU project SWITCH have shown that integration of water management into the urban development is critical. Therefore, identifying opportunities to adapt and integrate form and design of the urban water systems is essential. A look at a few existing projects shows that there is a potential waiting to be tapped. Among professionals, there is a perception of the legal requirements related to the provision of drainage services that inhibits the utilization of non-piped solutions (Cettner et al. 2012). There is specifically a need to explore possibilities that enable the development and easy application of locally adapted solutions, which contribute to a transition to more sustainable water management and ultimately the development of (water) smart communities. Current urban planning practice also commonly neglects watersheds as a basis for water sustainable design (Krebs et al. 2013). The high damage potential of the forecasted climate change floods is coupled with poor urban planning: in many locations dense urban development with little or no landscape structure continues near to flood and erosion sensitive areas. This gives rise to increased imperviousness and the potential for high levels of poor quality run-off to enter receiving and conveying streams in the event of heavy rainfall. Planning for the location and extent of urban structures (both the problems causing imperviousness and the sustainable management offering mitigation structures) in regard to natural waterways requires watershed-sensitive design across municipality borders and through collaborative actions within all landowners.

(c) Uncertainties in economic issues: As Chocat et al. (2007) note, sustainable stormwater management is often regarded as being more expensive than conventional solutions of stormwater drainage. In fact, the costs are the most frequent argument for not using more sustainable systems, although in several circumstances, they are in fact cheaper (Farreny et al. 2011a, b). Such calculations may be based on retrofitted designs on top of existing sites with much associated renovation works and, most importantly, commonly disregard the positive effects that sustainable stormwater management structures bring for increasing the amenity and livability of an urban area. Such values are not easy to be calculated in monetary means (Smaniotto 2007). With regard to the fact that urban areas need to be prepared for being adapted to the effects of intensification of land use, with the likelihood of additional ground sealing, and climate change, and both producing high extra costs for expanding existing stormwater pipe systems, such argument even becomes outdated. Project reviews and research results have shown that the earlier the engineers and urban designers cooperate in the planning process, the cheaper the implementing sustainable solutions become (Hoyer et al. 2011). Thus, in this way, synergies can be better used and complex solutions, which might cause high investment and maintenance costs can be avoided. An early and continuous collaboration can make proper use of synergies and complex solutions, which might cause high investment and maintenance costs, avoided. Each landscape requires its own solution, with respect to the natural environment and its interdependences. Sustainable solutions can usually be found for any given site and implemented with long-term cost efficiency.

(d) Fragmented responsibilities and strong competition around open spaces: As sustainable stormwater management techniques are mostly focused on infiltration, they often require to be installed on or above the surface, and therefore, they influence the design, usability, and aesthetics of the urban spaces to a large extent. This raises conflicts particularly in densely built-up areas, where different requirements and interests rule for the available open spaces, such as traffic, recreation, and greenery. This leads to high competition, where stormwater management is often not going to win (Smaniotto and Hoyer 2013). By using synergies and combinations of stormwater techniques beyond the techniques for infiltration and adapting these to the local conditions, the development of integrated solutions can be initiated, which in consequence can be used to create multifunctional spaces. Such spaces can thus, besides integrating stormwater management systems, give rise to recreation facilities and/or habitat conservation, enhancement and creation measures. Based on the principles of landscape ecology and an ecosystem approach to land use planning and management, such measures enable the creation of networks connected by water which simultaneously provide habitat and stormwater corridors.

(e) Lack of public awareness, public participation, and institutional capacity: Public awareness for urban surface water management issues is very limited, with the exception of severe events affecting communities such as drought or flooding. The majority of the urban population is not aware of the existence and functionality of methods, e.g., for rainwater retention, infiltration, and usage. In urban areas, there is often the problem that stormwater management facilities can just be used on publically owned property. Contrary to this, initiatives, e.g., Portland, USA, or Melbourne, Australia (SWITCH project), show that private owners can make important contributions toward implementing approaches for managing stormwater on site. Therefore, incentives need to be developed and public knowledge on the effects of sustainable stormwater management raised (Domenech and Sauri 2012). Citizen involvement and better collaboration allow the construction of social capital that revalues the right to a healthy urban environment. Urban sprawl can lead to areas without identity and associated social instability. Well-sited and fitted stormwater landscaping can provide the identity and sense of place to a completely new development, as well as a rehabilitated area, allowing dwellers to take pride and ownership of their neighborhood. According to Florida (2002), the human capital plays the central role in the level of happiness with the environment and personal well-being, outperforming every other variable, including income. In addition, urban planners and local councils need support and exchange experiences and knowledge in order to be involved in WSUD practices (Farreny et al. 2011c).

3 The Rationale of Water-Sensitive Urban Design and Water as an Element of Urban Design

WSUD is a design and planning approach, developed in Australia and adapted in other locations (Coutts et al. 2013; Dolman et al. 2013; Fryd et al. 2013; Roy et al. 2008). It connects technical solutions for water management with urban design and socioeconomic aspects. Water-sensitive solutions comprise concepts of stormwater reuse and recycling in urban areas as well as concepts to re-establish natural water cycles in the city and ensure water quality and river health. This approach is, in a whole, not new, but it puts individual measures into a conceptual framework with determined guidelines and objectives. It manages the transition and offers an alternative to the traditional conveyance approach to stormwater management.

WSUD aims to integrate the water cycle in urban settings. It centers urban design and landscape planning in the heart of water management in order to reuse water on site including its permanent or temporary storage. The possible measures are multifaceted and range from detention and retention basins to lower peak flows, grassed swales, and vegetation to facilitate water infiltration and treatment of pollutants, to ecological restoration of channeled watercourses. A concept for sustainable stormwater management under the WSUD approach includes also comprehensive and far-reaching measures such as approaches to reduce the impervious surfaces and embrace green infrastructure. Rooftops are part of the urban water cycle and can be designed as blue roofs for temporary detention or as green roofs with added amenities by the vegetative cover.

The relevant feature is that WSUD measures detain and filter stormwater where it falls and use the synergies while creating natural and environmental values to enhance urban livability. The housing and road layouts that preserve vegetated landscape and minimize imperviousness can be mentioned as a classical example for intervention. Integrating sustainable water management in housing, residential, and industrial developments includes more compact settlement layouts, space-saving buildings, and minimal sealed surfaces. The last is also applicable for road layout, an objective of which is to decrease the length and width of low-traffic local roads and design a shorter road network. The space-saving measures allow the conservation and installation of drainage corridors and retention and detention basins. Well designed, these elements not only protect the water quality but also act as multipurpose spaces. Besides the technical solution and safeguarding of all other water uses that serve public interest (recreational, sports, educational opportunities, etc.), using water as an element of urban design requires also the attention to visual amenity and aesthetic aspects. High aesthetic values enrich the urban landscape contributing to the character and identity of an area. Attractive urban landscape is often associated with the reinforcement of cultural or social identity and sense of place and belonging. Hence, WSUD offers good prospects for promotion or innovation as potentials for waterfront residential areas, retail, and businesses, and facilities for leisure and recreation.

In the future, it is essential that stormwater management measures in urban settings consider environmental protection and urban design as a whole, recognizing that they cannot be patched up by separate sectoral decisions. Therefore, (re-) integration of water bodies in urban areas calls for cooperation across disciplines, to enable the development and implementation of comprehensive approaches. Considering local conditions and circumstances is a key aspect in successful planning, design, and implementation of water smart communities that coexist within the natural water cycles.

4 Examples of Local Stormwater Management in Europe: Achievements and Challenges

Urban intensification is a key factor in the four case examples outlined here. They all address issues of poor urban quality (in social and/or environmental terms) and seek alternative means of stormwater management. As a positive side effect, they all seek to achieve an improvement of the urban landscape as a whole. All four cases have a pioneer character in their contexts and regions and are used to showcase how with few economic resources much can be reached, as river/stream degradation and their restoration/revitalization are central parts of stormwater management.

In terms of geographic situation, the cases represent diversity in Europe, as they are from Southern, Western, and Northern Europe. These exemplify very different climatic conditions which lead to different approaches to managing water in the urban environment. Looking at climate change scenarios, the sites vary in regard to estimated changes in rainfall. The potential for droughts is present in the south, whereas snowmelt is another issue in the north. Finland and Ireland are at the maximum and minimum range of increased rainfall, and Portugal and Spain are at the maximum range of decreased rainfall within Europe. Therefore, the need and requirements for stormwater management vary among the cases, while the water runoff is permanent in Besòs and Nummela; it is seasonal in Cascais and sporadic Dublin. Also considering the urban form, these cities are also quite different, while the cases in Dublin and the river Besòs represent very urban settings; Cascais and Nummela correspond to mid-sized cities inserted in metropolitan areas; these differences impact on governance issues.

The cases share the need to manage stormwater in a more efficient way and one in which the citizens can also benefit. The cases represent a local, plan-led approach using different instruments rather than an ad hoc project-by-project approach. They all seek multiple benefits for the citizen and the environment and seek to make a strong contribution toward sustainable urban development. They all challenge conventional approaches to managing water in urban settings—the solely engineering approach. Importantly, the case studies show an attempt for cities to return to their origins, where water was the driver and the life-giver for the settlement. This chapter looks briefly at how this was attempted by the four different case areas.

5 Case 1: Water-Sensitive Urban Design as Part of the Green Infrastructure of Local Area Plans— Administrative Area of Dublin City Council, Ireland

5.1 Developing Green Infrastructure Strategies, with Detailed Plans and Proposals for Water-Sensitive Urban Design for Local Spatial Plans in the City

Dublin City Council and the Heritage Council of Ireland commissioned consultants Norton UDP and Áit, to prepare a pilot Green Infrastructure Strategy to inform two statutory, local spatial plans (Local Area Plans) in Dublin City. One area (Georges Quay) is located in a densely urbanized area in the city center, while the other (Naas Road) is, in contrast, located in the middle suburbs to the west of the city center in an area dominated by large-scale industry and distribution uses and on a major road artery of the city. The pilot study was a follow-up to the then recent policy paper *Creating Green Infrastructure for Ireland* (Comhar, The Sustainability Council for Ireland, 2010).

The Georges Quay area comprises some 14 ha and it lies on the south bank of the river Liffey in what could be described as a transition zone between the recent redevelopment of the city docklands and the established core of the city center. Parts of the area, nearest the city center, witnessed the first clustering of modern office development in the city in the 1960s, while other areas have witnessed only scattered, incremental redevelopment of traditional mixed use functions. Today, the area could be defined as "gray" in character, being fragmented in terms of urban form and lacking in local green spaces and biodiversity resources.

The green infrastructure strategy for the area envisaged a central role for WSUD, based on a local spatial structure of hubs and corridors through the existing urban structure of the area and surrounding areas. The multipurpose green infrastructure is also intended to meet sustainable transport objectives, by incorporating more generous and attractive footpaths which reduce road widths and incorporate new cycle ways. Biodiversity is also promoted through the development of new green spaces and corridors and measures in new and existing private development (see Fig. 1). The strategy presents a number of concept projects, including an experimental tree line, constructed in load bearing soil and connected to the piped surface water drainage system. The purpose of this concept is twofold: to attenuate surface water from the street and to provide a filter for water before it reaches the piped system. The overall concepts of the final Local Area Plan for the area are strongly influenced by the green infrastructure strategy and spatial concepts.

The strategy for the Naas Road was quite different in terms of priorities and spatial structure. The area is located on a long-established artery of the city, the Naas Road. The area does not present any coherent or memorable urban character. It comprises extensive areas of active and obsolete light industry and distribution, some more recent office and retail warehouse uses, disconnected but extensive public and private open spaces, and scattered groups of housing (see Fig. 2). Despite extensive culverting and insensitive, surrounding development, the Camac River, which flows to the river Liffey, remains a unifier for the larger area.

As with Georges Quay, the green infrastructure for the area was based on identifying and connecting multipurpose hubs or green spaces of various sizes and scales and developing a network of new and improved green corridors. The green infrastructure was developed on the basis of a multilevel spatial strategy which was broadly structured on the Camac and its tributaries. Space will be provided for a connected biodiversity along with new pedestrian and cycle links.

Fig. 1 Concept proposal for a multipurpose urban tree line, Georges Quay area, Dublin. *Source* Dublin City Council (2012)

Fig. 2 Spatial structure for the Naas Road area. *Source* Norton UDP and Ait Urbanism and Landscape (2012)

The strategy for Naas Road presented a number of concept proposals, including: a new multipurpose, water attenuation swale in a boulevard design idiom along the major road artery, proposals for uncovering the river in the existing industrial areas and reconstructed wetland along the banks of the Camac to enhance the character of the river, to manage and filter water and to assist in the flood risk management measures for the city (see Fig. 3). The green infrastructure strategies were included in the community consultation for the Local Area Plan for the Naas Road. The community response was positive, and the final Local Area Plan incorporates the spatial structure and many of the objectives and strategies of the Green Infrastructure Strategy.

The Dublin case examples show the importance of introducing WSUD concepts into the early stage of urban planning process and the potential to do so using the multipurpose concepts in green infrastructure.

Fig. 3 Proposal for multipurpose swale for the Naas Road area. *Source* Norton UDP and Áit Urbanism and Landscape (2012)

6 Case 2: Besòs River—Metropolitan Area of Barcelona, Spain

6.1 Environmental Restoration and Flood Risk Protection in the Lower Course of the River Besòs

The Besòs River flows through the municipalities of Barcelona, Santa Coloma de Gramenet, Sant Adria de Besòs, and Montcada i Reixac, an urban area with a population of over two million people. The lower course of the Besòs River (9 km) has been profoundly altered by human action to support industrial and urban uses. The Besòs River has a clear Mediterranean character with irregular flows. Almost all of its water comes from reclaimed water and the transfer from other watersheds. The last stretch of the river was channeled in 1962, after a catastrophic flood. With the progressive reduction of the river channel, it lost hydraulic capacity and suffered from water pollution, ecosystem deterioration and marginalization during decades.

The environmental restoration project at the Besòs River started in 1996 with multiple aims: to recover the ecological and landscape quality of the river (then considered an open sewer), to improve the outflow from the Waste Water Treatment Plant of Montcada by implementing tertiary treatment based on the regeneration of wetlands, to increase the hydraulic capacity of the river, and finally, to use certain areas of the river for leisure purposes, with the development of the Besòs River Park (115 ha).

Broadly, the project defined two major areas of implementation, depending on the degree of urban development. In the first area, the main objective was to promote the ecological restoration of a river course with the design and development of constructed wetlands on the lateral margins of the riverbed. This area is 3.8 km long, restricted to the public and characterized by containing 60 plots (7.66 ha) of constructed wetlands that perform tertiary treatment to 30 % of the effluent from the Montcada wastewater treatment plant. Together with the 30 ha of meadows that surround the constructed wetlands, the site creates an attractive river ecosystem for birds and other species.

In the most urbanized part of the Besòs River Park, which passes along the municipalities of Santa Coloma and Sant Adria de Besòs and Barcelona, the main implementation actions included:

(a) Creating 13 ha of meadow landscape for public use;
(b) Improving public access to the river corridor landscape by building ramps;
(c) Extending the central channel width from 20 to 50 meters, and installing 5 inflatable dams to increase the hydraulic capacity of the river and to allow the creation of lagoons that favor the self-purification of water; and
(d) Establishing hydrological monitoring and alerting for rapid evacuation in case of flood risk, and emergency plan information (lighting, traffic lights, and signs).

The result of this regeneration process is the creation of the Besòs River Park (9 km and 115 ha). The project is the result of an initial agreement between the municipalities of Barcelona, Montcada i Reixac, Sant Adrià de Besòs, and Santa Coloma de Gramenet that has allowed to convert what has been considered an open sewer located within one of the most important green areas of the metropolitan Barcelona (Figs. 4 and 5). The Park has 22 public accesses (though ramps and stairs) to the green grassland area and a 5-km-long bike path. The area of public use is equipped with a flood warning system that guarantees user safety against flooding.

Fig. 4 Besòs River Park in Santa Coloma de Gramenet (*upstream views*). *Photograph* Pons-Sanvidal (IERMB) (2013)

Fig. 5 Besòs river park in Santa Coloma de Gramenet (*downstream views*). *Photograph* Pons-Sanvidal (IERMB) (2013)

Apart from the improvement of the environmental conditions and increased biodiversity, the Besòs River Park project has enhanced living conditions for the dwellers of municipalities on both sides of the river and it has attracted visitors from the metropolitan area to this water landscape (more than 500,000 visits every year). This, along with recent and planned changes to the new riverfront Besòs, will make the leap of scale necessary to meet the challenges of economic competitiveness, social cohesion, and environmental quality of the municipalities along the Besòs River.

7 Case 3: Ribeira de Sassoeiros, Cascais—Lisbon Metropolitan Area, Portugal

7.1 The Implementation of a Municipal Ecological Structure Along the Sassoeiros Stream (With the Collaboration of João Cardoso de Melo and Bernardo Cunha, EMAC, Cascais)

Ribeira de Sassoeiros is an ephemeral stream that flows through the Cascais municipality into the Atlantic Ocean. In order to enlarge the built-up areas and accelerate water runoff, the Sassoeiros bed has been narrowed and canalized over

Fig. 6 Concrete walls canalizing the Sassoeiros stream. Such monumental structures neither protect the residential areas from flooding nor offer a habitat for the riparian fauna and flora. *Photograph* Smaniotto (2013)

time. Due to uncontrolled urban expansion, the area along the Sassoeiros is suffering serious social, environmental, and flood problems. The result is the loss of permeable and productive soils and the complete destruction of riparian vegetation, as shown in Fig. 6.

The stream's canalization increased not only the runoff velocity but also the material flow. Habitat destruction, pollution, and soil erosion were detected as the main environmental problems. Floods became a recurring problem, causing economic damage and frequently homeless families.

Regarding the social fabric and mobility, the scarcity of bridges dictates the distance between the people on the two sides of the stream. The situation is aggravated by the absence of road hierarchy and the lack of equipped public open spaces.

In response to these problems, Cascais City Council started a project toward a comprehensive water management alongside a 1.6 km section of the stream in an area of 86,500 m^2 with a population of 4,029 people. However, the initial stages in 2009 followed classical engineering methods based on flow recovery and regularization of the stream. The Council recognized that the technical focus of the project was not solving the problems and moved its approach toward a more natural, biophysical engineering solution, in order to both stabilize the banks and reconstruct the riparian gallery. However, many of the technical plans were already implemented taking the space from innovative approaches available, as shown in Fig. 6.

The project is interesting in that while it primarily aimed at solving flooding problems, it also moved toward a more comprehensive approach for urban regeneration. The ongoing implementation of bioengineering-optimized measures not only provides a safe residential area but also aims to benefit from the ecological structure acting as an axis-supporting measure for urban revitalization allowing the creation of new social gathering places. The Council set as project objectives the following measures:

1. to manage the stream water flow through the construction of retention basins;
2. to stabilize the stream banks through the restoration of riparian galleries;
3. to improve riparian habitat; and
4. to create a pedestrian distribution axis throughout the stream allowing connectivity between the two sides of the stream, a measure that opens new opportunities for strengthening social cohesion.

This last objective is important, in that while the vehicular access, vehicle parking, and service areas are adequate and in some neighborhoods even oversized compared to the needs (see Fig. 7), the access for pedestrian and especially for people with disabilities is very poor or totally non-existent.

The ecological restoration of the Sassoeiros Stream opens the chance to use the stream as a backbone for integrating the implementation of ecological mitigation structures with the revitalization of social aspects within the nearby urban areas.

Fig. 7 The measures implemented 2013 for flood protection in the Sassoeiros stream. Without any other measures, e.g., settling adequate plants, the water flow will wash the banks away in the next rain. *Photograph* Smaniotto (2013)

With an extensive cultural heritage from centuries of historic influences, the City of Cascais has a variety of infrastructures (hotels, golf courses, roads, and public transportation) and a wide range of recreational facilities such as sand beaches, nature areas, and historical and urban parks. However, these are distributed unevenly across the municipality. The project along the Sassoeiros stream presents an opportunity to cope with these inequities. In Sassoeiros area, the City Council is tackling factors that functionally and unduly compromise other development aspects of the site or even the visual amenity of the whole area.

8 Case 4: Kilsoi Stream—Nummela Community, Municipality of Vihti, Finland

8.1 Making a Network of Water Quality Mitigating Wetland Parks Out of a Degraded Urban Stream

Nummela is an urban community of 11,000 inhabitants located by a shallow Lake Enäjärvi in the Municipality of Vihti. Due to its location within a commuter proximity to the capital Helsinki region, Nummela is urbanizing rapidly. Poor water quality in the lake raised concerns about stormwater quality. Fish deaths and blue green algal blooms reduce recreational amenities as well as habitat values provided by the lake.

Site analysis revealed that the majority of Nummela was located within a 550 ha subwatershed of the Lake Enäjärvi. No reference was made on local maps to the heavily altered stream draining to the lake. The clay stream was also disappearing from the landscape into storm sewers and culverts as a conventional measure to manage erosion by high and rapid stormwater pulses from the urbanized watershed during rain and snowmelt events, as shown in Fig. 8. Phosphorus, the limiting nutrient to the blue green algae, is transported to the lake bound in clay particles.

Actions toward sustainable stormwater management were initiated with the goal of mitigating the effects on the health of the receiving lake and stream, as well as the urban landscape as a whole. The municipality of Vihti, the Regional Environmental Institute (UUDELY), academic outreach (the Department of Forest Sciences of the University of Helsinki), a local Lake Enäjärvi protection association (VESY), and a regional watershed protection association (VHVSY) collaborate to carry out these goals. As the very first step, the straightened and eroded stream was restored by reintroducing its old local name, Kilsoi, into maps. Due to the significant land use changes within the watershed, no further stream restoration to any previous state in time was possible: Rather, natural processes respecting urban landscape design with stabilizing stream corridor vegetation and constructed wetlands were applied as water and urban landscape mitigation elements.

The most eroded 200 m stretch of the stream was stabilized. Widening of the stream was possible only to a very limited extent as preceding water-insensitive development had zoned private housing close to the bank of the stream. The sides

Fig. 8 Starting point
in Nummela was rapid
urbanization with a
major community stream
disappearing in culverts.
Photograph Salminen (2007)

owned by the municipality hosted underground utility lines, which are often hidden in the easily excavated stream sides. Land use changes and high imperviousness within the upstream 260 ha urban watershed, as shown in Fig. 9, required stabilization of the receiving stream banks with rocks at the erosive flow heights. Biodegradable coconut meadow seed mats and native trees were planted at higher elevations to establish a stable stream corridor. Seed mat installation and tree planting were conducted as an awareness-raising and collaboration-enhancing municipality staff and local dweller's joint volunteer event. The stream bed self-established rapidly with wetland plants.

The cost of stream stabilization by vegetation and rocks was only 20 % of the originally intended culvert extension. The municipality engaged in preserving the rest of the stream open and zoned a broad park network space with no underground utilities along the stream Kilsoi. Following the change in zoning practice, a 2 ha constructed stormwater management wetland was planned and implemented at the mouth of the Kilsoi in 2010.

The aim of this wetland park, which is now named the Gateway (located both at the mouth of Kilsoi and at the main commuter road entrance to Nummela) was to

KILSOI STREAM WATERSHED

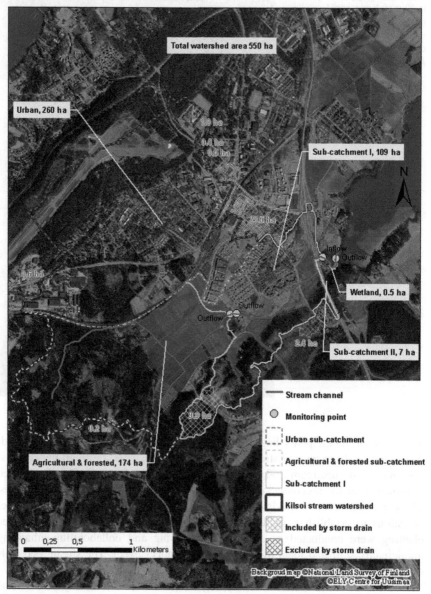

Fig. 9 The Kilsoi watershed includes a 260 ha stormwater sewer subwatershed. The yet remaining agricultural areas are undergoing urbanization. A wide wetland park area will be located along the entire length of the stream. Land use impact, as well as the impact of the constructed Gateway wetland park at the mouth of the Kilsoi stream, is monitored to improve design for water treatment landscapes. *Map by* Salminen and Jussila (2013)

treat water quality and manage flows and also to gain acceptance by urban dwellers for the treatment of wetland environments as urban green. Furthermore, habitat for sensitive species was attempted at the prior crop field site of clay soils. All drainage ditches were blocked to create amphibian habitat and wet meadows. A half-hectare of inundated wetland was excavated as wintertime dry construction, as shown in Fig. 10. Winter excavation is cost efficient and protects dormant vegetation. The stream was bypassed the construction site to avoid clay soils from entering the lake. Excavations were careful to avoid soil compaction. The excavated wetland rapidly self-established with over 100 native herbaceous species. Frogs, newts, and birds rapidly found the site. A nature trail with two bridges and a bird-watching tower was built. The cost of the two-hectare park establishment was minimal compared to construction costs of any conventional urban park. No impervious paving is present in the park.

Maintenance of the Gateway park includes moving the alignment of the one-kilometer-long nature trail by one meter each year. Meadow areas and the sediment-trapping pond at the beginning of the wetland area are monitored to establish a maintenance schedule for estimated every 5–10-year meadow cutting and sediment-trapping pond re-excavation. Water purification by the wetland, which only comprises 0.1 % of its watershed size, has reached up to 70 % turbidity (corresponding to clay-bound phosphorus) event reduction and 10% annual (over four seasons) reduction by the end of the fourth growth season (Salminen et al. 2012). Interviews with the local dwellers have revealed that the constructed water environment parks have provided with local pride and sense of place, yet instead of focusing on water treatment, locals view the parks as dynamically changing with diverse places of beauty with sights and sounds of nature. Listening to singing birds such as the nightingale and watching spawning fish and amphibians has created vivid nature experiences. Local elementary schools visit the park regularly.

The municipality viewed the wetland park as both a cost-efficient park amenity to dwellers and a cost-efficient water management facility. An interest in improving knowledge on multifunctional landscaping grew strong among the

Fig. 10 The Nummela Gateway wetland park was constructed in 2010 by wintertime dry excavation, with some native trees planted on the site by an engaging volunteer event in May (*left*). In June 2013, the first children's summer camp was held at the site (*right*). *Photograph* Salminen (2010, 2013)

collaborators. The EU Life + 11 ENV/FI/911 Urban Oases funding was granted for 2012–2017 to expand knowledge and implementation of the constructed urban wetland parks within the Kilsoi watershed. Thanks to the acceptance gained by the Gateway wetland park, a seven-hectare water environment mitigation park is under construction upstream in the middle of urban Nummela.

9 Water-Sensitive Urban Design in Action: Drawing Lessons from Experiences in Four Case Studies

To illustrate and confirm the experience discussed in this chapter, a series of case studies have been analyzed in different cities across Europe. This analysis encompassed cities with a long urban history driven by different policies, government regulations, and community expectations.

Although part of the same process, urban development, green infrastructure, and stormwater management are not usually discussed and treated with the same concern and consideration. Urban development impacts as a process on the design of landscapes and cities as a whole in functional, aesthetic, and symbolic ways. For this reason, it is very important to increase understanding and knowledge of water sensitiveness in urban settings as well as political awareness about its importance, because these are crucial elements in all cities.

The case studies explored here demonstrate the variety of approaches adopted under very different conditions. They cover only a few topics of WSUD—rainwater harvesting, stormwater management, floodplain design, and constructed wetlands—but the range of options is enormous. There are a number of elements that have led to stormwater management and WSUD as concepts to re-establish natural water cycles and ensure water quality and river health. These experiences are vast, varied, and difficult to compare. However, a number of broad conditions for success and leading practices can be identified. The success conditions are grouped under four headings; early stage consideration, integration with urban design, active listening, and proactive and continuing work.

(a) Consider WSUD at the early stages of the planning process, not added as an afterthought.

WSUD must be present at the early stages of the planning process, so that it can directly inform or influence local and city planning and plans. It must be part of the consultative process of planning process, where communities can engage with the concepts and accept advocate or reject them. Local communities are generally very responsive to concepts of water-sensitive urban design.

The negative aspect of an afterthought can be illustrated by the case of Cascais. Even if the paradigm shift from a traditional approach to stormwater management has to be considered an improvement, the later addressing WSUD issues have now to cope with already implemented structures (see Fig. 7). Many of them do not match the rationale behind WSUD and have anyhow to be integrated in the system. The project in Cascais is a kind of pioneer work, as in Portugal, there are no major experiences and national empirical evidences from bioengineering-optimized measures. Moving from a traditional approach needs time and effort. The positive aspect is that the Council is open to new experiences, developing ideas and taking the initiative. In Nummela, the combination of establishing sustainable stormwater structures while providing recreation areas for the inhabitants opens a new concept of an urban park. Furthermore, the cost efficiency, the added park amenities, and local knowledge gained from only a 200 m stream landscaping were enough to make a rapid shift in zoning to allow a broad chain of water mitigation park spaces along the thus revitalized urban stream. Such zoning was possible as the entire watershed is located within the same municipality which was also able to purchase the new park land.

(b) Water sensitiveness has to be an integral part of the urban design

WSUD will be most successful where multiple benefits, such as landscape or urban quality, recreation and amenity, walking and cycling, flood risk management, and biodiversity, can be achieved. It can also help to break down institutional silos (e.g., planners, engineers, landscape architects, architects, ecologists). Such actions might be used also to gain support and unlock funding, as the case of Besòs illustrates. Under the WSUD package, stormwater management should always create a little more scope for expansion and innovation. This can be a resource for local education, involvement, and empowerment, especially considering that planning of urban open space often involves a large number of stakeholders: local authorities, architects, developers, consultants, local inhabitants, and so forth.

Taking the case of Cascais, the ecological restoration of the stream enables not only the (re)integration of water into the urban landscape but also the creation of new connections, such as radial elements cutting through the mostly concentric urban fabric, opening new prospects for increasing environmental quality and recreation opportunities. In the case of Nummela, the importance of amenities such as the sounds of birds associated with the water treatment in wetland areas raised the issue of true multifunctionality and how the added values to the urban dwellers are much broader than any planning process currently considers.

There are also reservations about a dichotomy that inhibits WSUD actions—is stormwater an issue for the planning department or for the water department? Who is responsible for the construction and maintenance of the disciplines crossing structures? It is concluded that water professionals have unique opportunities to integrate stormwater management approaches within wider urban planning practice and hence are able to encourage the use of alternative systems that are more sustainable than using traditional pipes or sewers.

(c) Active listening to stakeholders and creating partnerships

The environmental degradation and river marginalization and the increasing environmental awareness and recreational and other demands from the population urge local and regional governments to find solutions. In the case of

Besòs, these issues resulted in 1995 in an institutional agreement between different municipalities. The municipalities of Barcelona, Santa Coloma de Gramenet, Sant Adria de Besòs, and Montcada i Reixac decided to start a series of programs and actions framed in a single project for the regeneration performance of the watershed, whose budget came mainly from European funds, and the establishment of the consortium for the defense of the river. On the other hand, political action was driven by a range of legislation and policy, in particular the EU Water Framework Directive (WFD) (2000/60/EC). All together, the Besòs case shows a great political commitment at different scales (local, subregional, and regional). Moreover, this project emerged from public concern, preliminary studies, and a high degree of cooperation between professionals of different fields, and the public stakeholder participation was very active in their preliminary stages.

Partnerships are crucial to sustainable delivery. Considering that the local community and their organization can be a valuable source of local knowledge and an incubator of ideas, they should be fully engaged in the process. Actions such as giving a name to a water treatment park or participating in the construction are both awareness raising and engaging to local dwellers. In the case of Nummela, these activities have also helped the decision of the municipality to acquire land from private owners who have learnt to view sustainable water management parks as a preferred future for their land.

(d) Proactive and continuing work

Actively listening to stakeholders and creating partnerships can be particularly difficult where interest in sustainable development and alternative stormwater management is low. WSUD cannot and should not place its reliance alone on positive stakeholder influence; councils must take their own initiatives to drive the idea forward. Proactive acting involves thinking, planning, and foreseeing in advance of a future situation, rather than just reacting on anticipated events.

Despite the improvements in the case studies, however, there is still room for improvement in the management of stormwater. In view of the future uncertainties from climate change and impacts from current legislation (especially the Water Framework Directive), stormwater management will need to take a more central role in all aspects of urban planning. For example, an integrated plan of stormwater management in the Metropolitan Region of Barcelona is needed, including extended measures of WSUD, to the occasional already existing experiences. The European Directive on the Assessment and Management of Flood Risk (2007/60/EC) can be a good framework to develop these plans. For a successful implementation of stormwater management, measures should be part of an overall strategy, with a strong political consensus, public acceptance, and stakeholder participation, and must be adapted to the particular (urban, socioeconomic, environmental, and climate) characteristics of each municipality.

Measures to re-establish healthy water systems, in particular preventing water bodies pollution and restoring their ecological balance, achieve standard levels that allow their integration into the urban fabric, once they do not pose any threat or risk to the population or environment. As the cases show the improvement in water quality, strengthening biodiversity and safe access opens up a new dimension in experiencing water and waterways in urban settings.

To benefit from all these aspects, it is explicitly necessary to work out a multidisciplinary approach, as the Dublin Strategy demonstrates—which included planners, engineers, landscape architects, heritage officers, and ecologists. In Nummela, both local and regional water protection associations complement the list of interdisciplinary municipal, academic, and regional government staff. A comprehensive scope is a necessity in the delivery of cost-effective and long-term sustainable solutions and this needs to span each phase of planning process, from goals setting to planning, design, construction and management.

The structural change with the shift from agrarian to industrialized and service economies, the deindustrialization, and the economic cycles (growth, decline, and regrowth) is a cumulative process involving almost all sectors of the urban development. These changes in the location patterns of industries have a spatial consequence: a growing number of brownfields. Such abandoned (urban) properties are often environmentally contaminated and therefore unusable. As in the case of Besòs, heavy industries are often strategically located on very attractive areas by the river banks. In many cases, there is no pressure to reuse such areas, leaving behind white spots in the urban fabric. These white spots, however, make land available with the possibility to develop a comprehensive approach to transform watercourses into linear structural elements.

The cases show that improved and well-managed water resources in the urban environment can provide a wide range of benefits to communities. And this despite the current trend of seeing stormwater in urban settings as more of a threat than an opportunity for improving the urban environment in different ways. Water is a vital asset, so there is nothing more fundamental than protecting it. A healthy environment has many benefits for the health of people and cities. Ensuring that waters are clean costs an enormous amount of resources and takes a huge amount of management effort but on the other side offers many opportunities to create sustainable and livable cities. Sustainable water management requires integrated, collaborative, and far-reaching approaches. While the paradigm shift requires resources for activation, the added benefits are numerous and long-term cost savings become enormous. In simple words, measures for a successful implementation of stormwater management should be part of an overall strategy, based on strong political will and social consensus, Which is backed by public acceptance and stakeholder participation and tailored to the particular urban, socioeconomic, environmental, and climate characteristics of each individual location.

References

Barbosa AE, Fernandes JN, David LM (2012) Key issues for sustainable urban stormwater management. Water Res 46(20):6787–6798

Cettner A, Söderholm K, Viklander M (2012) An adaptive stormwater culture? Historical perspectives on the status of stormwater within the Swedish urban water system. J Urban Technol 19(3):25–40

Cettner A, Ashley R, Hedström A, Viklander M (2013) Sustainable development and urban stormwater practice. Urban Water J:1–13

Chocat B, Ashley R, Marsalek J, Matos MR, Rauch W, Schilling W, Urbonas B (2007) Toward sustainable management of urban storm water. Indoor Built Environ 16 (3):273-285

Coutts AM, Tapper NJ, Beringer J, Loughnan M, Demuzere M (2013) Watering our cities: the capacity for water sensitive urban design to support urban cooling and improve human thermal comfort in the Australian context. Prog Phys Geogr 37(1):2–28

Dolman N, Savage A, Ogunyoye F (2013) Water-sensitive urban design: learning from experience. Proc Inst Civil Eng-Municipal Eng 166(2):86–97

Domenech L, Sauri D (2012) A comparative appraisal of the use of rainwater harvesting in single and multi-family buildings of the metropolitan area of Barcelona (Spain): social experience, drinking water savings and economic costs. J Clean Prod 19(6–7):598–608

Dublin City Council and The Heritage Council (2012). Green infrastructure strategies for local area plans. Dublin

Echols S (2007) Artful rainwater design in the urban landscape. J Green Build 2(4):1–19

EEA—European Environment Agency (2010) The European environment state and outlook. Copenhagen

Eurostat (2012) Around 40 % of the EU27 population live in urban regions and almost a quarter in rural regions. http://epp.eurostat.ec.europa.eu/cache/ITY_PUBLIC/1-30032012-BP/EN/1-30032012-BP-EN.PDF Accessed on 15 Jan 2013

Farreny R, Morales-Pinzón T, Guisasola A, Taya C, Rieradevall J, Gabarrell X (2011a) Roof selection for rainwater harvesting: quantity and quality assessments in Spain. Water Res 45(10):3245–3254

Farreny R, Oliver-Solà J, Montlleó M, Escribà E, Gabarrell X, Rieradevall J (2011b) Transition towards sustainable cities: opportunities, constraints, and strategies in planning. A neighbourhood ecodesign case study in Barcelona. Landscape Urban Plann A 43:1118–1134

Farreny R, Gabarrell X, Rieradevall J (2011c) Cost-efficiency of rainwater harvesting strategies in dense Mediterranean neighbourhoods. Resour Conserv Recycl 55(7):686–694

Florida R (2002) The rise of the creative class. Basic Books, New York

Fryd O, Backhaus A, Birch H, Fratini CF, Ingvertsen ST, Jeppesen J et al (2013) Water sensitive urban design retrofits in Copenhagen-40 % to the sewer, 60 % to the city. Water Sci Technol 67(9):1945–1952

Hoyer J, Dickhaut W, Kronawitter L, Weber B (2011) Water sensitive urban design—principles and inspiration for sustainable stormwater management in the city of the future. Jovis, Berlin

IPCC (2012) Summary for policymakers. In: Managing the risks of extreme events and disasters to advance climate change adaptation. A special report of working groups I and II of the intergovernmental panel on climate change. Cambridge University Press, Cambridge, pp 1–19. http://www.ipcc-wg2.gov/SREX. Accessed on 04 Nov 2013

Krebs G, Rimpiläinen UM, Salminen O (2013) How does imperviousness develop and affect runoff generation in an urbanizing watershed? Fennia 191(2):143–159

MW—Melbourne Water. n.d. City of Melbourne WSUD guidelines. pp 10–11

Roy A, Wenger S, Fletcher T, Walsh C, Ladson A, Shuster W, Thurston H, Brown R (2008) Impediments and solutions to sustainable, watershed-scale urban stormwater management: lessons from Australia and the United States. Environ Manage 42(2):344–359

Salminen O, Ahponen H, Valkama P, Vessman T, Rantakokko K, Vaahtera E, Taylor A, Vasander H, Nikinmaa E (2012) Benefits of green infrastructure – socioeconomic importance of constructed urban wetlands (Nummela, Finland). In Kettunen et al (eds) Socio-economic importance of ecosystem services in the Nordic Countries – Synthesis in the context of The Economics of Ecosystems and Biodiversity (TEEB). Nordic Council of Ministers, Copenhagen pp 247–254

Smaniotto Costa C, Hoyer J (2013) Why invest in urban landscape? The importance of green spaces and water for urban sustainable development. In: Wagner I, Zalewski M, Butterworth J (eds) Ecohydrology in urban areas: experiences of the SWITCH demonstration, City of Lodz, Poland

Smaniotto Costa C (2007) Ökonomische Argumente für Grünflächenentwicklung. Stadt und Grün 2:13–19

SWITCH project—managing water for the city of the future. www.switchurbanwater. Accessed on 16 Nov 2012

UNFPA (UN Population Fund) (2007) The state of the World population 2007—unleashing the potential of urban growth. http://www.unfpa.org Accessed on 16 Nov 2012

Walsh C, Roy A, Feminella J, Cottingham P, Groffman P, Morgan R (2005) The urban stream syndrome: current knowledge and the search for a cure. J North Am Benthol Soc 24(3):706–723

Authors Biography

Carlos Smaniotto Costa (Ph.D.) is a Landscape Architect and Environmental Planner, graduated at the University of Hanover, Germany. He works in the fields of design of urban environment, open space planning, and urban development in Germany, Italy, and Brazil. His Ph.D. focused in landscape planning as directive for sustainable urban development. He is professor of Urban Ecology and Landscape Design at Universidade Lusófona, Lisbon, and the head of its Experimental Laboratory on Public Spaces. His research activities deal with issues of sustainable urban development strategies for integrating open spaces and nature conservation in an urban context.

Dr. Conor Norton is a Head of Department at the School of Spatial Planning, Dublin Institute of Technology in Ireland

Elena Domene is a researcher at the Barcelona Institute of Regional and Metropolitan Studies (IERMB) in Spain.

Jacqueline Hoyer is a researcher at Hafen City University in Hamburg, Germany.

Joan Marull is a researcher at the Barcelona Institute of Regional and Metropolitan Studies (IERMB) in Spain.

Outi Salminen is a researcher at the Department of Forest Sciences, University of Helsinki, Finland.

Water Consumption in Dormitories: Insight from an Analysis in the USA

Umberto Berardi and Nakisa Alborzfard

Abstract Worldwide depletion of resources has brought many sustainability issues to the forefront including the consumption of water use for indoor purposes. Based on various studies, the third largest consumption of water occurs in buildings, mainly for flushing and personal hygiene. The United States Department of Energy and European Commission places domestic indoor water use at more than 250 L per person per day. This chapter examines the water consumption in Leadership in Energy and Environmental Design (LEED) and non-LEED-certified dormitories. LEED is a sustainability rating system providing guidance on incorporating sustainable design strategies in the design of buildings. LEED offers various rating levels including certified, silver, gold, and platinum out of a possible 100 base points. The varying levels are associated with target points achieved. Three LEED and six non-LEED dormitories, located in the northeast, serving over 2,000 students, were selected for this comparative study. Different categorization of dormitories by varied agencies and the inconsistency in water-use studies make isolating water consumption in dormitories problematic. Considering the fact that the International and Uniform Plumbing Codes do not require to calculate the water consumption in buildings, and engineers' calculations have been used to create baseline water use for the nine dormitories. The perception of water consumption behavior of occupants has also been investigated through users' surveys. Finally, a comparison among the design evaluation, actual water consumption and subjectively evaluated consumption allows highlighting water consumption in dormitories.

Keywords Dormitories · LEED · Sustainable buildings · Water consumption

U. Berardi (✉)
Faculty of Engineering and Architectural Science, Ryerson University,
Toronto, Canada
e-mail: uberardi@ryerson.ca

N. Alborzfard
Department of Civil and Environmental Engineering, Worcester Polytechnic Institute,
Worcester, MA, USA
e-mail: nalborzfard@wpi.edu

© Springer International Publishing Switzerland 2015
W. Leal Filho and V. Sümer (eds.), *Sustainable Water Use and Management*,
Green Energy and Technology, DOI 10.1007/978-3-319-12394-3_3

Abbreviations

AIA   American Institute of Architects
AWWA   American Water Works Association
BREEAM   Building Research Establishment Environmental Assessment Method
CASBEE   Comprehensive Assessment System for Built Environment Efficiency
CNT   Center for Neighborhood Technology
EEA   European Environment Agency
EC   European Commission
EPA   Environmental Protection Agency
EU   European Union
HE   Higher education
IAMPO   International Association of Plumbing and Mechanical Officials
ICC   International Code Council
ILFI   International Living Future Institute
LBC   Living Building Challenge
LEED   Leadership in Energy and Environmental Design
LPD   Liters per person per day
LPF   Liters per flush
LPM   Liters per minute
NWS   National Weather Service
OECD   Organization for Economic Co-operation and Development
POE   Post Occupancy Evaluation
RIBA   Royal Institute of British Architects
SIU   Southern Illinois University
USGS   United States Geological Survey
US   United States
US-DOE   United States Department of Energy
WE   Water efficiency

1 Introduction

In 2050, global population, water demand, and global gross domestic product should increase by 30, 55, and 100 %, respectively (OECD 2012). Moreover, by 2050, almost 70 % of the world population is projected to live in cities, relying on public water supply (OECD 2012). As a result, future urban developments will further stress public water supply infrastructures.

Less than 1 % of the world water is freshwater and can be adapted for human use (ILFI 2011). Given this already limited resource, current and future challenges of sustainable water consumption and recharge have become ever more pressing. The current state of water extraction from groundwater and freshwater sources has resulted in dramatic negative environmental impacts, such as water depletion, quality reduction, waterlogging, salinization, annual discharge reduction, and contamination of potable water sources (OECD 2012; EEA 2012). Excessive diversion of river waters has also led to lowering of groundwater tables and saltwater infiltration in coastal areas (EEA 2012). These considerations impose to promote more sustainable water management and use. In particular, this chapter will focus on the opportunities available in a particular typology of buildings in the USA, which is the dormitory.

In the USA, the United States Geological Survey (USGS) works in collaboration with local, state, and federal agencies to collect water-use data. USGS has several goals including: (1) analyzing source, use, and disposition of water resources at local, state, and national levels; (2) replying to water-use information requests from the public; (3) documenting water-use trends; (4) cooperating with state and local agencies on projects of special interest; (5) developing water-use databases; and (6) publishing water-use data reports outlining domestic (residential) water consumption from self-supplied (i.e., wells) and public-supplied (i.e., state agencies) sources. Domestic (residential) water use typically includes drinking, food preparation, washing clothes and dishes, flushing toilets, and outdoor applications include watering lawns and washing cars (USGS 2013).

In the last decade, almost every region in the USA has experienced water shortages, and at least 36 US states have recently anticipated local, regional, or statewide water shortages under non-drought conditions (Shi et al. 2013). Researches show that due to increases in water demand and droughts, water has not been recharged at sustainable rates (Shi et al. 2013). This points to the need to promote sustainable pathways, which consider population growth, climate change, and water-use habits to decrease risks of future water shortages and challenges in our ability to source water (The National Academies 2008; Shi et al. 2013).

From the total water withdrawn for all uses in the USA, domestic water use has an estimated value of 111.3 billion L per day (LPD) (USGS 2009). The consumptions differ from 193 L per person per day in Maine to 715 LPD in Nevada, with the national average at 375 LPD (USGS 2009). The Environmental Protection Agency (EPA) WaterSense program reports a similar average value of 379 LPD, of which 70 % (265 LPD) is assumed for indoor purposes (EPA 2013).

Since this chapter focuses on water consumption in dormitories, it may be a misrepresentation to compare the residential case studies to dormitories, as they include outdoor water consumption values. A lack of uniformity in USA water-use study methods and variables results in the inability to use available reports for comparisons (SIU 2002). Categorical disparities of dormitories (commercial or domestic) by USGS and United States Department of Energy (US-DOE) further complicate isolating water use in dormitories (USGS 2009; US-DOE 2013a, b). USGS does not explicitly categorize building types, resulting in ambiguity on whether dormitories fall under the commercial or residential data set. Commercial water-use data were not collected by USGS in the 2000 and 2005 reports (USGS

2000, 2009). However, in the 1995 report, below the commercial category, the following building typologies were identified: hotels, motels, restaurants, office buildings, other commercial facilities, and civilian and military institutions (USGS 1995). These building types are very different from dormitories, and their water consumption values do not reflect the indoor water-use purposes in dormitories. Residential values suggested by USGS seem more applicable to dormitories, although they include outdoor applications (watering lawns, gardening, and washing cars).

US-DOE categorizes dormitories under lodging, a commercial category. However, the US-DOE relies on the USGS datasets for water use reporting per sector. Given the inconsistency between the USGS and US-DOE building categorization, no explicit US data on indoor water consumption of dormitories exist.

Examining water consumption in the European Union (EU) between 60 and 80 % of public supply water is used for domestic applications, of which personal hygiene and flushing account for 60 % (Mudgal and Lauranson 2009). Case studies from different member states showed domestic water consumption of 168 LPD on average (Mudgal and Lauranson 2009).

The overall withdrawals in the EU are projected to decrease by almost 11 % in 2020 (Floerke and Alcamo 2004). However, a major unknown variable of water use in EU is the domestic water consumption (Floerke and Alcamo 2004). Given the current increase in water consumption in urban area and the increasing effects of climate changes, the Mediterranean river basins are continuing to face water stress (EEA 2012). These stresses pose threats to the availability of clean potable water and might increase the need for more sophisticated wastewater treatment methods. The Environment Directorate-General European Commission (EC) carried out a water performance of buildings study (Mudgal and Lauranson 2009), which does not explicitly categorize dormitories. However, EC identifies educational buildings in the non-residential public sector, although a lack of water consumption data exists for this category.

Differences between EU and US study methodologies and building categorizations compound problems of isolating dormitory water consumption. To address the lack of available water consumption data in dormitories, this chapter assesses and compares the water consumption in some US dormitories. Different uses of water, such as washing dishes and clothing, flushing toilets, and showering, are taken into account (Vickers 2001; Schleich and Hillenbrand 2009).

Many factors influence water consumption such as geographical location, climate, culture, gender, and occupant behavior (Vickers 2001; Balling et al. 2008; Randolph and Troy 2008; Schleich and Hillenbrand 2009; Vinz 2009; Elliott 2013; Berardi 2013a). To mitigate the effect of these variables, the present study considers water-related practices in several dormitories over the last 10 years.

This chapter is structured in the following way: Sect. 2 focuses on water efficiency (WE) strategies in sustainability rating systems, Sect. 3 presents the methodology of the case study research, Sect. 4 presents the case study results, and Sect. 5 highlights main conclusions.

2 Sustainability Rating Systems and Water Efficiency Strategies

Voluntary sustainability rating systems including LEED (USGBC 2009), BRE Environmental Assessment Method (BREEAM 2008), Comprehensive Assessment System for Built Environment Efficiency (CASBEE 2010), and Living Building Challenge (LBC 2012; Green Globes 2012) recommend use of water-efficient flow fixtures to minimize water demand. Guidance is also provided for the minimization of wastewater effluent into existing treatment infrastructures by implementing onsite treatment strategies.

Some of the shared water-saving strategies recommended by the rating systems and professional associations such as the American Institute of Architects (AIA) and the Royal Institute of British Architects (RIBA) include low-flow fixtures, dual-flush toilets, ultra-low-flow or waterless urinals, infrared sensors, timed automatic shutoff faucets, low water-use washing machines and dishwashers, rainwater catchment, gray water use, and onsite wastewater treatment. Gray water is untreated wastewater which has not come in contact with toilet water, and it includes water from bathroom washbasins or laundry tubs (USGBC 2009). Onsite wastewater treatment can reduce the quantity of effluent treated in the public treatment infrastructures, reducing overall energy demands to treat and transport effluent (AIA 2007; USGBC 2009; LBC 2012). Onsite-treated water can be reused within the building for non-potable purposes such as toilet flushing, minimizing demand from public water supply infrastructures. Various strategies might be implemented to accomplish secondary- or tertiary-level treatment of wastewater including anaerobic septic tanks, anoxic reactors, closed aerobic tanks with plants to filter gases, open aerobic tanks with snails, shrimp and fish, redirection of sludge to septic tanks or composting of sludge, and redirection of polluted water to indoor wetlands for filtration (AIA 2007; ILFI 2011; LBC 2012).

Table 1 provides an overview of recommended water-saving flow fixture efficiencies in liters per flush (LPF) for toilets and liters per minute (LPM) for showerheads, lavatory, and kitchen faucets. As can be seen, difference between recommended efficiencies by rating system exists. In cases such as CASBEE and LBC, a prescriptive value is missing, and it is at the discretion of designers to select and specify the appropriate fixture technology to meet water-saving target goals.

However, water-efficient fixtures and treatment strategies alone may be insufficient to reduce consumption, as users' behavior is critical in lowering overall water consumption (Stevenson and Leaman 2010). The collection of users' feedback about WE strategies in the buildings and the education on consuming less water plays a key role in supporting WE strategies. Active participation of users and post occupancy evaluations (POEs) are significant to uphold sustainability in practice. Various researchers highlight the need to adopt education campaigns to promote more sustainable users' behaviors (Stevenson and Leaman 2010; Sterling et al. 2013; Berardi 2013a).

Table 1 Efficiencies of water-saving flow fixtures

Rating system and professional best practices	Toilet efficiency targets (LPF)	Shower efficiency targets (LPM)	Lavatory faucet targets (LPM)	Kitchen faucet targets (LPM)
LEED	≤6 LPF	≤9.5 LPM	≤8.5 LPM	≤8.5 LPM
BREEAM	Dual flush: ≤3 LPF (low) to ≤4.5 LPF (full)		≤6 LPM	Two-stage faucets with low flow for rinsing and higher flow for filling objects
CASBEE	Specific target values not provided			
LBC	Specific target values not provided			
Green Globes	≤6 LPF	≤9 LPM	≤7.5 LPM	≤7.5 LPM
AIA	≤4.9 LPF	≤6.6 LPM	≤3.8 LPM	≤7.6 LPM
	Dual flush: ≤3.6 LPF (low) to ≤5.7 LPF (full)			
RIBA	Specific target values not provided, suggested referring to other reference sources including BREEAM			

Some examples of organizations, agencies, and programs that are promoting sustainable water practices in the USA include the following: the Center for Neighborhood Technology (CNT), Nature's Voice-Our Choice, Water Use it Wisely, Save our Water, Stop the water while using me, and EPA WaterSense. These agencies and programs provide suggestions on water conservation and water-saving strategies and promote WE through behavioral changes.

Design strategies and users are hence strongly linked in the process of making sustainability a reality (Stevenson and Leaman 2010; Berardi 2013a; GhaffarianHoseini et al. 2013). A bridge between modeled design and actual outcome is represented by POEs. POEs ensure users are satisfied with their current conditions and inform future designs (Bordass et al. 2006, 2010; Stevenson and Leaman 2010; Berardi 2012). Design strategy labeling can also be developed through the collection of user feedback, further identifying which sustainable strategies to avoid and promote in practice (Bordass et al. 2006; Berardi 2013b).

3 Case Study Overview and Methodology

Three LEED and six non-LEED dormitories, varying from 3 to 62 years of age, comprise the studied dataset. The research methodology involved the collection of various specifications including number and gender split of students served, flow fixture efficiencies, actual water meter readings, and LEED documentation pertaining to WE credits in LEED-certified dormitories.

Data were gathered from designers, facilities departments, and residential life offices of the various higher education (HE) institutions. All dormitories are

Table 2 Overview of dormitories

Bldg.	Rating	Age years	No. of users	Gender split (% of female)	Location	Building zone[a]
EH	LEED-Gold	5	232	F = 31	Northeast	Cold
CSC	LEED-Gold	3	450	F = 53	Northeast	Mixed-humid
PS	LEED-Silver	3	622	F = 44	West Coast	Hot-dry
WT	Non-LEED	11	475	F = 18	Northeast	Cold
MH1	Non-LEED	62	284	F = 60	Northeast	Cold
MH2	Non-LEED	52	190	F = 49	Northeast	Cold
MH3	Non-LEED	47	190	F = 60	Northeast	Cold
HH	Non-LEED	54	163	F = 50	Northeast	Cold
KH	Non-LEED	52	191	F = 53	Northeast	Cold

[a]Based on US-DOE (2013a, b)

located in the USA with eight in the northeast and one on the West Coast. For the purposes of anonymity, acronyms designate the dormitories. Table 2 provides main building data of the selected dormitories.

Monthly actual water meter readings were collected for EH, PS, WT, MH1, MH2, MH3, HH, and KH and quarterly actual water meter readings for CSC. The average number of students served per year allowed calculating the liters per person per day (LPD) metrics and comparing water performance. Dormitories EH, CSC, WT, MH1, MH2, MH3, HH, and KH are located in the northeast, experiencing cold to mixed-humid climates, whereas dormitory PS is located on the West Coast, experiencing a hot-dry climate.

Typically, the peak water consumption occurs in summer (AWWA 1999). The weather in the USA followed typical patterns in the years from 2002 to 2009 and in 2011 and 2013; reversely, in 2010, the coldest winter was experienced, and in 2012, record summer heat and mildest winter was recorded (NWS 2013).

Flow fixture efficiency values were collected to highlight differences in technologies used in dormitories. Non-LEED flow fixture data were collected from the HE facilities departments and walkthroughs, while WE documentation was collected from designers for LEED dormitories. Dormitory age was also recorded as newer dormitories are less likely to experience plumbing leakages and may have implemented higher efficiency fixtures.

3.1 Engineer's Metrics

The International and Uniform Plumbing Codes do not require designers to calculate total water consumption of buildings (ICC 2009; IAMPO 2009); hence, engineer's metrics were calculated based on the EC report, providing European metrics (Mudgal and Lauranson 2009), and the AWWA report, providing guidance on US metrics (AWWA 1999).

The AWWA report values are based on data from over 1,000 households in 12 study sites around the USA. The data include historic billing records and detailed mail surveys, broken into two sets to capture winter and summer indoor water consumption. The AWWA water end-use findings are as follows: 70 LPD for toilet use, 57 LPD for clothes washer, 44 LPD for shower use, 41 LPD for faucet use, 36 LPD for leaks, 5 LPD for baths, 4 LPD for dishwasher, and 6 LPD for other domestic use (AWWA 1999). In calculating the comparative AWWA metric, the value applicable to dormitories was assessed to be 212 LPD (including toilet use, clothes washer, shower use, and faucet).

The EC report values are based on information collected from local case studies in different European member states, feedback from stakeholders, and a literature search. Findings in water using products of residential buildings are 41 LPD toilet use, 26 LPD clothes washer, 37 LPD showers, 29 LPD faucet use, 10 LPD dishwasher, and 11 LPD outdoor use (Mudgal and Lauranson 2009). Calculating the comparative EC metric, the value applicable to dormitories is 143 LPD.

4 Results and Discussion

4.1 Average Overall (LEED and Non-LEED) Actual Water Consumption

As indicated in Table 3, the overall range of actual LEED and non-LEED dormitory water consumption fell between 85 and 175 LPD, with an average of 144 LPD and a standard deviation of 34 LPD. Comparing the average consumption to the EC and AWWA engineer's metrics, the consumption was higher by almost 1 and 32 %, respectively.

Figure 1 depicts the actual water consumption of the nine dormitories in LPD: LEED dormitory EH is the top performer followed by non-LEED dormitories WT and HH, while LEED dormitory PS performed slightly better than the poorest performer non-LEED dormitory MH1.

4.2 Non-LEED Dormitories

The average water consumption of non-LEED dormitories was 146 LPD with a standard deviation of 30 LPD. Figure 2 provides a profile of the water consumption of the six non-LEED dormitories over the years. In the dormitory WT, the averaged consumption resulted 107 LPD with a 3 % increase in consumption over the 12 years. Although the increasing consumptions, these are lower than to the engineer's metrics by 25 and 45 %, respectively. Excluding WT from the non-LEED dataset, the average consumption resulted 154 LPD. Comparing this average to the engineer's metrics, the consumption is higher by 8 % and lower by

Table 3 Average overall water consumption results in liters per person per day (LPD)

Bldg.	Data range dates	Sample size 'N'	Actual average consumption (LPD)	Standard deviation of Bldg. Dataset (LPD)	Comparison of actual to EC engineer's metric (143 LPD) (%)	Comparison of actual to US engineer's metric (212 LPD) (%)
EH	September '08–June '12	46	85	52	−41	−60
WT	January '02–June '13	138	107	37	−25	−50
HH	July '07–May '12	59	110	74	−23	−48
MH2	July '07–June '12	60	160	104	+12	−25
KH	July '07–June '12	60	162	114	+13	−24
CSC	May '11–April '13	24	163	82	+14	−23
MH3	July '07–June '12	60	164	98	+15	−23
PS	July '11–May '13	23	172	107	+20	−19
MH1	July '07–June '12	60	175	101	+22	−18

38 %, respectively. In dormitories MH1, MH2, MH3, KH, and HH, the percent net change over the 5 years was 3 % indicating an uptick. Dormitories HH and KH showed the highest variation over the years versus steadier consumption in MH1, MH2, MH3, and WT (Fig. 2).

Factors specific to dormitories that affect the vary consumption include institutional academic schedules together with water technologies the other factors

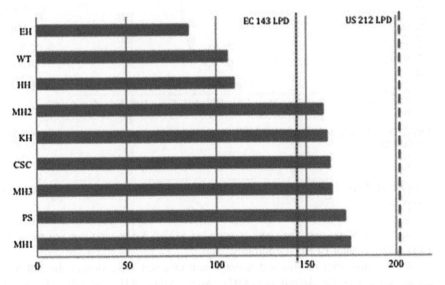

Fig. 1 Actual water consumption of the nine dormitories in LPD (compared to engineer's metrics)

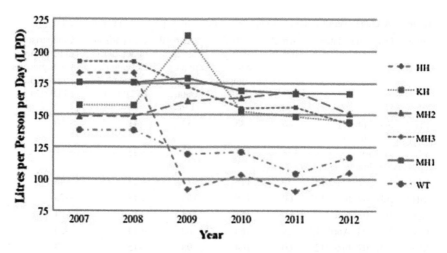

Fig. 2 Actual average yearly water consumption of non-LEED dormitories in LPD

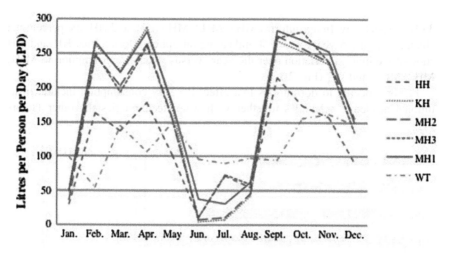

Fig. 3 Actual average (from 2007 to 2012) monthly water consumption of non-LEED dormitories in LPD

outlined in Sect. 1 (geographical location, climate, culture, gender, and occupant behavior).

In an effort to investigate the high variations, an exploration of the monthly consumption values over the years is provided in Fig. 3, showing average monthly LPD of the six non-LEED dormitories.

The months with the highest average consumption were during the fall and spring semesters for dormitories MH1, MH2, MH3, KH, and HH. The water consumption for the summer months (June, July, and August) was the lowest, followed by January winter recess. The highest consumption periods were attributed to periods of high occupancy (returning students) and the warmer months within those periods. Dormitory WT also experienced consumption during the summer months (June, July, and August) as it operates year round due to academic requirements in the summer. Reversely, dormitories MH1, MH2, MH3, KH, and HH do not have summer sessions and showed minimal summer consumption.

4.3 LEED Dormitories

4.3.1 Dormitory EH

In calculating the LEED green case, designers assume a specific number of days the dormitory will be in operation. The assumed operational days play an important part over the water performance calculation. The assumption is generally based on the information provided by owner's facilities departments according to academic schedules.

Designers of dormitory EH used 305 days and estimated a green case consumption of 89 LPD. Using the 305-day assumption, dormitory EH resulted in the lowest average water consumption when compared to all the dormitories (LEED and non-LEED). The average yearly consumption values from 2008 to 2012 were 133, 62, 68, 78, and 82 LPD, respectively. Although dormitory EH outperformed its counterparts in further dissecting the water consumption over the years, an increase resulted. If the consumption of the first (commissioning) and last years (hottest summer) is excluded, the average consumption is 69 LPD. This value is 29 % lower than the 'green' case. EH actual consumption was less than modeled consumption by 22 % over the 3-year period (2009–2011), but only by 4 % over the 5-year period (2008–2012).

To further explore the discrepancy between actual and LEED case consumption values, an online user survey was distributed to EH occupants. Since 44 % of indoor residential water end use is related to shower and toilet use (AWWA 1999), questions were developed on the shared assumptions used in LEED (USGBC 2009) and AWWA (1999) about shower duration (8 min), shower frequency (1/day/occupant), and toilet flushes (5 flushes/occupant/day). Sixty occupants answered the questionnaire in the 2 weeks following the survey distribution (November 2010), a value corresponding to 26 % of students living in the dormitory at the time. Figure 4 provides the percent breakdown of responses to the LEED and AWWA assumptions posed in the user survey.

The responses indicate shower frequency and daily toilet flushes fall within shared thresholds of AWWA and LEED design assumptions. However, the shower duration assumptions of 8 min dramatically fell short. Over 87 % of respondents indicated taking longer than 15-min showers. Such variations in actual practice versus modeled assumptions can result in large differences in water estimations and performance evaluations. These results confirm that highlight

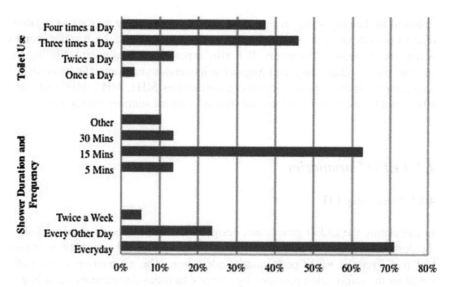

Fig. 4 Occupant responses on toilet use, shower duration, and shower frequency in dormitory EH

occupants' attitudes and behaviors have substantial impacts on promoting sustainability in practice (Barr 2003; Bamberg 2003; Hand et al. 2003; Hurlimann 2006; Alshuwaikhat and Abubakar 2008; Randolph and Troy 2008).

4.3.2 Dormitory CSC

CSC designers assumed 360 operational days, with a LEED 'green' case of 88 LPD. CSC exceeded modeled consumption by an average of 85 % over the 3-year period (2011–2013). The yearly consumption values for 2011, 2012, and 2013 were 147, 170, and 172 LPD, respectively, resulting in drastic percent increase in consumption as compared to the modeled case of 67 % higher, 93 % higher, and 95 % consumption in 2011, 2012, and 2013, respectively. As previously mentioned, part of the increase may be due to record heat in 2012. However, drastic percent increases in consumption over the years, echo the findings of other dormitories which behaved less sustainably over time.

4.3.3 Dormitory PS

PS designers assumed 250 operational days with a LEED 'green' case of 87 LPD. The yearly consumption values for 2011, 2012, and 2013 were 198, 146, and 171 LPD, respectively, resulting in differences in consumption as compared to the modeled case of 128 % higher, 68 % higher, and 97 % higher in 2011, 2012, and 2103, respectively. Dormitory PS actual consumption exceeded modeled consumption by an average of 98 % over the three-year period. It must be noted given the dormitories location that its occupants may have been better equipped to handle the heat of 2012, as consumption of PS in that year was lower than in any other year.

4.3.4 Comparison of LEED and Non-LEED Dormitories

Exploring the age and technologies employed among the dormitories, the average age of non-LEED dormitories is 46 years, while the average age of LEED dormitories is 4 years. Dormitories EH, WT, CSC, and PS were built in 2008, 2002, 2011, and 2011, respectively, where the 1992 and 2005 Federal Energy Policy Act (FEPA) were already in place. This act includes maximum consumption for fixtures of 9.5 LPM and 6.0 LPF. MH1, HH, MH2, KH, and MH3 were built in 1951, 1959, 1961, 1961 and 1966, respectively, and do not comply with the 1992 or 2005 Federal Energy Policy Act.

All non-LEED dormitories and dormitory CSC used full flush toilets, while EH and PS used dual-flush toilets (low/full). Figure 5 represents the average and standard deviation of flow fixture rates in LEED and non-LEED dormitories in LPM for lavatory, kitchen sink, and shower fixtures and in LPF for toilets.

Non-LEED dormitories used flow fixtures with 6.4, 7.9, and 7.9 LPM for shower, lavatory, and kitchen sink, respectively, with toilets using 10.9 LPF, whereas LEED dormitories used flow fixtures with 5.9, 1.9, and 8.1 LPM for shower, lavatory, and kitchen sink, respectively, with toilets using 3.6 and 5.7 LPF for low and full flush, respectively.

Even though non-LEED flow fixtures were higher on average, the dormitories outperformed LEED ones in terms of total LPD. This finding indicates sole reliance on technology to lower overall consumption which might not be the answer.

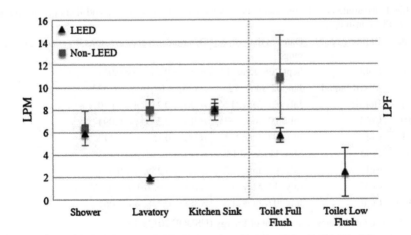

Fig. 5 Average flow fixture rates in LPM and LPF for LEED and non-LEED dormitories

Attention must be given to occupant expectations and behaviors. For example, some respondents in the EH survey commented about their frustrations with low-flow fixtures and declared they replaced low-flow showerheads with higher flow fixtures, while others indicated taking longer showers. Similar comments were provided for low-flow toilets, where respondents indicated often double and triple flushing as the toilet low flush was simply not sufficient. These results confirm the role of users as critical factors for sustainability.

To further highlight how climate impacted the consumption of the dataset, bivariate correlation analysis was carried out. The analysis tested the relationship between average monthly temperature and consumption in LPD. The bivariate correlation analysis was done using IBM SPSS Statistics software version 19. The analysis excluded summer months of all dormitories, except in the case of WT, which has summer semesters.

The results indicate a positive correlation between average monthly temperature and LPD consumption in all dormitories except PS; however, the correlations are not significant (95 % or above). It must be noted in dormitory EH, HH, MH2, MH3, and MH1 the significance surpass 90 %, supporting the work of previous researchers. Table 4 provides the bivariate correlation results per dormitory.

In the case of PS, the number of observations in the dataset was only 18; therefore, the negative correlation result may be attributed to the small sample size. In the case of WT with over 10 years of data and inclusion of the warmest months, the correlation between average monthly temperature and LPD consumption was positive yet weak. This indicates that temperature has a negligible impact on the consumption patterns. In order to dissect this weak correlation, the 12-month average monthly temperature moving average was compared to highlight variations due to seasonality. It can be seen in Fig. 6 that no variations due to seasonality exist, and average temperatures were relatively steady over the 10-year period.

Table 4 Bivariate correlation results of average monthly temperature and liters per person per day (LPD) consumption

Bldg.	Building zone[a]	Dates of data range[b]	Bivariate correlation results R $_{d.f.}$ $(N - 2) = r, \rho$
EH	Cold	Sept. '08–June '12	$r(30) = 0.284, \rho < 0.057$
WT	Mixed-humid	Jan '02–June '13	$r(136) = 0.015, \rho < 0.432$
HH	Hot-dry	July '07–May '12	$r(38) = 0.237, \rho < 0.070$
MH2	Cold	July '07–June '12	$r(38) = 0.213, \rho < 0.094$
KH	Cold	July '07–June '12	$r(38) = 0.150, \rho < 0.177$
CSC	Cold	May'11–April 13	$r(16) = 0.217, \rho < 0.193$
MH3	Cold	July '07–June '12	$r(38) = 0.259, \rho < 0.053$
PS	Cold	July '11–May '13	$r(16) = -0.079, \rho < 0.378$
MH1	Cold	July '07–June '12	$r(38) = 0.248, \rho < 0.061$

[a]Based on United States Department of Energy (USDOE 2013a, b)
[b]Excludes summer months when students are not on campus except in the case of WT, since summer semesters are required as part of the academic program

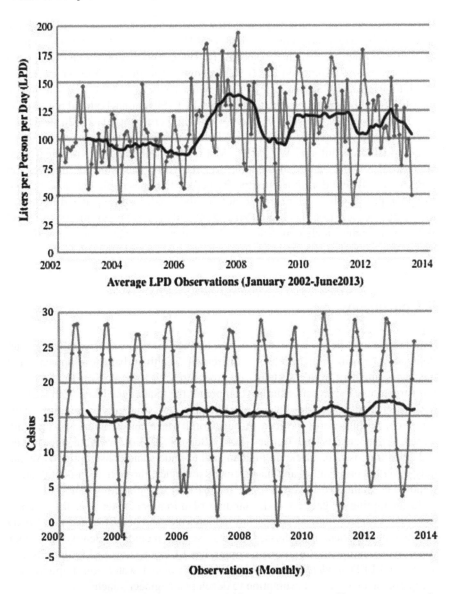

Fig. 6 Twelve-month LPD moving average and 12-month average monthly temperature moving average (January 2002–June 2013)

This indicates that other variables, such as user consumption behavior, might be the driving force behind consumption variations. Figure 6 also provides a plot of the 12-month LPD moving average over the 10-year period. As can be seen, the consumption patterns are not uniform and vary substantially from year to year.

Examining the average water consumption of LEED dormitories between years, building EH, CSC, and PS consumed 10 % more, 9 % more, and 5 % less, respectively, between yearly readings. However, compared to their LEED 'green' cases, the average yearly consumptions of EH, CSC, and PS were 4 % lower, 85 % higher, and 98 % higher, respectively. These values result in an overall percent increase in consumption of 60 % as compared to their LEED 'green' cases. Dormitory EH and CSC are LEED-Gold, while PS is LEED-Silver. Even though the LEED-Gold dormitory outperformed the LEED-Silver one, both did not provide the expected savings (Kats 2010). Moreover, LEED dormitory data indicate diminished consumption savings over time, rendering them less sustainable every year.

Non-LEED dormitories WT, MH1, MH2, MH3, KH, and HH resulted in an increase of 3 % in water consumption over the years. Based on the findings, on average, non-LEED dormitories outperformed LEED ones depicting steadier consumption profiles. It is interesting to note as the gender split equalized in dormitories, the consumption increased (Vinz 2009; Elliott 2013). Dormitories EH and WT had the highest male populations at 75 % on average, while dormitories MH1, MH2, MH3, KH, HH, PS, and CSC had average male populations of 47 %.

5 Conclusions

Water-related studies suggest we are consuming water at an unsustainable rate. Population growth, climate change, increased wealth, urban development, and mismanagement of water systems are over stressing our already fragile water infrastructures. These issues further compound the challenges faced with sustaining this necessity. As a result, we must engage new strategies to minimize consumption, pushing forth the idea of behavioral water conservation and not only fixture WE (Bennetts and Bordass 2007; Berardi 2013a). Tracking, measuring, and collecting user feedback are fundamental to understand consumptions. We can only develop conservation and management strategies, through an in-depth understanding of qualitative and quantitative feedback by implementing POEs.

In attempting to gain an understanding of dormitory water use, this chapter focused on identifying and comparing indoor water use of LEED and non-LEED certified dormitories. It addressed several scopes including identifying indoor water consumption in dormitories, comparing LEED to non-LEED dormitories, assessing LEED modeled case projections with actual water consumption, and comparing actual water consumption to developed engineer's metrics.

Evidently isolating water consumption of dormitories using US-DOE, USGS, AWWA, and EC data is problematic due to differences in the categorization of dormitories between water-use studies and a lack of available data. Different classifications of residential customers by utility companies also compound the problems in collecting published data on water consumption in dormitories.

To address this gap, actual consumption data were collected from nine dormitories, indicating indoor water ranges between 85 and 175 LPD. Overall average actual dormitory consumption was lower than values found in US-DOE (375 LPD), EPA (265 LPD), EC (168 LPD), AWWA US (212 LPD), and EC (143 LPD) engineer's metric. On average, non-LEED dormitories consumed 4 % more than LEED ones; however, the LEED buildings resulted in contrasting results with a high standard deviation of values.

On a yearly and monthly basis, non-LEED dormitories depicted steadier consumption values with an overall 3 % uptick for which the entire time data were collected. On the other hand, LEED dormitories showed an increase of 5 % over the years and, on average, higher variations in consumption patterns. The average water consumption of EH, CSC, and PS was 60 % higher when compared to the LEED 'green' cases. The data showed decreases in savings yearly, making LEED dormitories less sustainable every year. These results highlight the possibility that LEED labeling does not fully capture actual user behavior and might result in unrealistic savings expectations.

Examining assumptions of LEED and AWWA, over 87 % of respondents indicated longer than 15-min showers. Such vast differences in assumptions (8 min) and actual practice (over 15 min) must be ameliorated to ensure performance gaps are minimized. It is interesting to note as the gender differential equalized the consumption in the dormitories increased, tying to arguments made by researchers on the inequality of gender consumption. The best performing dormitories had 75 % males on average, while the poorer performing dormitories held 47 % males.

Finally, it is important to highlight technology alone may not guarantee water saving. Many factors impact water use including: geography, weather, socioeconomic factors, gender, and occupant behaviors. Larger reductions in water consumption need improved user attitudes and changes in occupant behaviors.

Further examination about the influence of previous variables on actual water consumption is needed. An in-depth understanding of users interact with designed building components is important to ensure sustainability in practice. Also, research about preferred water temperature is ongoing.

References

Alshuwaikhat HM, Abubakar I (2008) An integrated approach to achieving campus sustainability: assessment of the current campus environmental management practices. J Clean Prod 16(16):1777–1785

American Institute of Architects (AIA) (2007) 50 to 50: 50 strategies toward 50 percent fossil fuel reduction in buildings

American Water Works Association (AWWA) (1999) Residential end uses of water

Bamberg S (2003) How does environmental concern influence specific environmentally related behaviors? A new answer to an old question. J Environ Psychol 23(1):21–32

Barr S (2003) Strategies for sustainability: Citizens and responsible environmental behavior. Area 35(3):227–240

Balling RC, Gober P, Jones N (2008) Sensitivity of residential water consumption to variations in climate: an intra-urban analysis of Phoenix, Arizona. Water Resour Res 44(10):1–11

Bennetts R, Bordass W (2007) Keep it simple and do it well, sustainability supplement to building magazine. Digging beneath the greenwash, pp 8–11

Berardi U (2012) Sustainability assessment in the construction sector: rating systems and rated buildings. Sustain Dev 20(6):411–424

Berardi U (2013a) Clarifying the new interpretations of the concept of sustainable building. Sustain Cities Soc 8:72–78

Berardi U (2013b) Moving to sustainable buildings: paths to adopt green innovations in developed countries. Versita, DeGruyter. ISBN: 978-83-7656-010-6

Bordass W, Leaman A, Eley J (2006) A guide to feedback and post-occupancy evaluation. Usable Buildings Trust

Bordass B, Leaman A, Stevenson F (2010) Building evaluation: practice and principles. Build Res Inf 38(5):564–577

BREEAM Rating System (2008) Multi-residential scheme document

CASBEE (2010) Technical manual for new construction

Elliott R (2013) The taste for green: the possibilities and dynamics of status differentiation through "green" consumption. Poetics 41(3):294–322

Environmental Protection Agency (EPA) (2013) WaterSense Program

European Environment Agency (EEA) (2012) European waters-assessment of status and pressures. EEA report No. 8/2012

Floerke M, Alcamo J (2004) European outlook on water use. Center for Environmental Systems Research,University of Kassel. Final report, EEA/RNC/03/007

GhaffarianHoseini A, Dahlan N, Berardi U, GhaffarianHoseini A, Makaremi N, GhaffarianHoseini M (2013) Sustainable energy performances of green buildings: a review of current theories, implementations and challenges. Renew Sustain Energy Rev 25:1–17

Green Globes (2012) New construction, criteria and point allocation

Hand M, Southerton D, Shove E (2003) Explaining daily showering: a discussion of policy and practice, Economic and Social Science Research Council. Sustainable Technologies Programme, UK

Hurlimann A (2006) Water, water, everywhere—which drop should be drunk? Urban Policy Res 24(3):303–305

International Association of Plumbing and Mechanical Officials (IAPMO) (2009) Uniform plumbing code

International Code Council (ICC) (2009) International Plumbing Code (IPC,) Country Clubs Hills, IL

International Living Future Institute (ILFI) (2011) Cascadia green building council toward net zero water: best management practices for decentralized sourcing and treatment

Kats G (2010) Greening our built world: costs, benefits, and strategies. Island Press, Washington, DC

Living Building Challenge (LBC) (2012) International Living Future Institute

Mudgal S, Lauranson R (2009) Study on water performance of buildings. European Commission (DG ENV), Reference Number: 070307/2008/520703/ETU/D2

National Weather Service (NWS) (2013) Indices and forecasts daily arctic oscillation index. National Oceanic and Atmospheric Administration (NOAA)

Organization for Economic Co-operation and Development (OECD) (2012) OECD environmental outlook to 2050: the consequences of inaction. OECD Publishing, Paris

Randolph B, Troy P (2008) Attitudes to conservation and water consumption. Environ Sci Policy 11(5):441–455

Schleich J, Hillenbrand T (2009) Determinants of residential water demand in Germany. Ecol Econ 68(6):1756–1769

Shi D, Devineni N, Lall U, Pinero E (2013) America's water risk: water stress and climate variability. Earth Institute, Columbia University, New York

Southern Illinois University (SIU) (2002) Predictive models of water use: an analytical bibliography, Carbondale Illinois

Sterling S, Maxey L, Luna H (2013) The sustainable university: progress and prospects. Routledge, Taylor and Francis, London

Stevenson F, Leaman A (2010) Evaluating housing performance in relation to human behavior: new challenges. Build Res Inf 38(5):437–441

The National Academies (2008) Drinking water: understanding the science and policy behind a critical resource

United States Department of Energy (US-DOE) (2013a) Water, buildings energy data book (Chapter 8)

United States Department of Energy (US-DOE) (2013b) Building technologies office. Residential Buildings, Climate Zone Designations

United States Geological Survey (USGS) (1995) Water use in the United States report 1995

United States Geological Survey (USGS) (2000) Water use in the United States report 2000

United States Geological Survey (USGS) (2009) Water use in the United States report 2005

United States Geological Survey (USGS) (2013) General water use in the United States

United States Green Building Council (USGBC) (2009) LEED new construction design guide, vol 3

Vickers A (2001) Handbook of water use and conservation: homes, landscapes, businesses, industries, farms. Waterplow Press, Amherst

Vinz D (2009) Gender and sustainable consumption: a German environmental perspective. Eur J Women's Stud 16:159–179

Authors Biography

Umberto Berardi is an Assistant Professor in Building Science at Ryerson University. His researches concern the application of physics and sustainability principles to the built environment. His areas of expertise include green buildings, energy saving technologies, and architectural acoustics. He took an MSc degree in Building Engineering "summa cum laude" at the Politecnico di Bari (I), a MSc in Sound and Vibration at the ISVR of the University of Southampton (UK), and a PhD in "Product Development & Innovation Management" and in "Building Engineering" at the Scuola Interpolitecnica (I). Before joining Ryerson, he was an assistant professor at the Worcester Polytechnic Institute (MA, USA). He also had research experiences in the Faculty of Engineering of the University of Syracuse (US) and at the International Centre for Integrated assessment and Sustainable development of the University of Maastricht (NL). Umberto has an extensive publication records, which include 2 books, 40 peer-reviewed journal papers and 30 conference papers. Umberto acts as a member of the advisory boards of several journals, such as "Intelligent Building International", "International Journal of Sustainable Construction", and "Energy Technology & Policy". For his merits, Umberto has been awarded several grants and awards, including the best faculty paper at the American Society for Engineering Education (ASEE) Conference 2013, Norwich (US).

Nakisa Alborz is an Assistant Professor in the Civil Engineering Department at Wentworth Institute of Technology (WIT) in Boston, MA. She holds a Ph.D. in Civil Engineering from the Civil and Environmental Engineering Department at Worcester Polytechnic Institute (WPI). Her Ph.D. research is on the post-occupancy evaluation of LEED and sustainable buildings and Master's research investigated Life Cycle Cost Analysis of Sustainability Features in Buildings. Her work experience entails a decade of cost estimating and civil design on residential, commercial and heavy civil projects for public and private entities.

Water Resource Management in Larisa: A "Tragedy of the Commons?"

Paschalis A. Arvanitidis, Fotini Nasioka and Sofia Dimogianni

Abstract The commons are natural or man-made resources that due to non-excludability and subtractability face serious risks of overexploitation, mismanagement, or even destruction, the so-called "tragedy of the commons". Groundwater is a typical example of such a resource. Drawing on the framework developed by the 2009 Nobel laureate Elinor Ostrom, this research explores issues of collective management of groundwater using Larissa area, one of the most important agricultural areas of Greece, as a case study. More specifically, the paper assesses empirically the possibility of user-based management of groundwater used for irrigation purposes. This is done through a survey which explores, *inter alia*, the views of local stakeholders on the intensity of the water problem, the irrigation practices, and the existence of trust-based social relations between the farmers, which are seen as essential for the development of successful, long-enduring, user-based governance solutions. The research finds that farmers are rather reserved toward the possibility of groundwater self-management, which may be due to lack of trust both among them and toward the other players in the field. On these grounds, it seems that the most appropriate solution would be to create an independent coordinative body with multiple responsibilities and powers.

Keywords Groundwater · Common pool resources · Tragedy of the commons · Larisa

P.A. Arvanitidis (✉) · F. Nasioka
Department of Economics, University of Thessaly,
43 Korai Street, Volos, Greece
e-mail: parvanit@uth.gr

S. Dimogianni
Department of Planning and Regional Development,
University of Thessaly, Volos, Greece

© Springer International Publishing Switzerland 2015
W. Leal Filho and V. Sümer (eds.), *Sustainable Water Use and Management*,
Green Energy and Technology, DOI 10.1007/978-3-319-12394-3_4

1 Introduction

The common pool resource (CPRs), or simply commons, is a special category of natural or man-made resources characterized by non-excludability, meaning that it is too difficult (i.e., too costly) to exclude someone from using them, and sub-tractability, meaning that use by someone reduces the level of the resource available to others. These features of commons enable rational individuals (acting in their immediate self-interest) to use as much of the resource as they like, without taking full responsibility for their actions. As a result, the resource is gradually depleted and eventually led to degradation and destruction, a situation known as the "tragedy of the commons" (Hardin 1968; Feeny et al. 1990).

The reasons behind the "tragedy" are twofold. On the one hand, there is the economic rational behavior of the users (which seek to maximize their individual immediate benefit, disregarding the social/collective long-term costs of their actions), and on the other hand, there is a lack of a proper institutional structure for the sustainable "governance" of the CPRs, that is a framework which enables property rights on the resource to be properly defined, allocated, and enforced to all actors. On these grounds, possible solutions to the commons' tragedy could be to infuse stewardship ethic among users[1] and to enhance moral and altruistic behavior toward sustainability (Barclay 2004), and/or, as Hardin (1968) and others (e.g., Libecap 2009) have argued, to attribute clearly defined property rights, either to individuals (privatization) or to the state (nationalization), giving the owner the incentives and authority to enforce the sustainability of the resource.

However, the 2009 Nobel laureate in economics, Elinor Ostrom, has revisited Hardin's work and drawing on a number of empirical studies across the world demonstrated that communities can successfully manage commons even in the absence of private property rights and a strong regulatory authority. In particular, Ostrom (1990, 1992, 1999, 2000, 2008, 2010, as well as Stern et al. 2002 and Dietz et al. 2003) made clear that local users are able to overcome collective action problems and to develop indigenous, self-organized, and long-enduring institutions for the sustainable management of the CPRs. These institutions are particularly social arrangements (rules, norms, routines, customs, etc.) which define and allocate rights and obligations among users and provide the mechanisms for policing and enforcing them.

Combining field and experimental research on the commons, Ostrom (1990, 2006, as well as Ostrom et al. 1999) and other scholars (Wade 1987, 1988; Baland and Platteau 1996; Agrawal 2001, 2003) identified a number of characteristics that are common to all successful management structures. These can be organized under five headings (Briasouli 2003). The first group of elements regards the

---

[1] In its modern conception, stewardship ethic refers to the "responsible use (including conservation) of natural resources in a way that takes full and balanced account of the interests of society, future generations, and other species, as well as of private needs, and accepts significant answerability to society" (Worrell and Appleby 2000: 269).

resource itself; resources of smaller sizes with definable boundaries, for example, can be preserved much more easily. A second group concerns the characteristics of the user community; small and homogeneous populations with a thick social network[2] based on trust, with experience in self-regulation and with social values promoting conservation (e.g., stewardship ethic), do better. The third group of conditions has to do with users' dependence on the resource; there must be a perceptible threat of resource depletion, and the community (current and future generations) should depend to a high degree on the resource for its living. The fourth group refers to the governance structure, that is, the institutional arrangements that should be developed to manage the CPR; locally emerged, user-based, simple rules with simple, internal, and low-cost policing and enforcement procedures are preferable. Finally, the last group concerns the external environment; clear and supportive state regulations (with formal incentives and sanctions) and accommodating and collaborative local/regional authorities do help to a great extent.

Groundwater constitutes a typical example of CPRs (Easter et al. 1997; Theesfeld 2010). It is subject to rivalry in consumption, in the sense that there is a specific amount available (finite in the case of non-renewable, e.g., fossil groundwater), which must be shared over a variety of users/uses and geographical areas. In addition, the change of the climate of the planet, with the rise of the world temperatures and the reduction of the annual rainfalls, and the increase of the environmental degradation and the water demand (for agricultural, industrial, and residential uses) have made groundwater a valuable resource in scarcity (Mariolakos 2007). In this sense, academics (Starr 1991; Klare 2001; Bolton 2010), journalists (de Villiers 2003; Annin 2006; Solomon 2011), technocrats (Serageldin 2009), and politicians[3] alike have called into attention that disputes over freshwater would be the source of conflicts and wars in the near future (Mostert 2003). So, the "tragedy" might be even worse.

Drawing on the analytical framework of commons developed by Ostrom, this research explores issues of collective management of the groundwater resource using Larissa area, one of the most important agricultural areas of Greece, as a case study. In particular, the paper explores empirically the possibility of user-based management of groundwater used for irrigation purposes. This is done through a survey which sets out the views of local stakeholders on the intensity of the water problem (in terms of both quantity and quality), the irrigation practices and degree of dependence of the local farmers on the resource, and the existence of trust-based social relations between the users, which, as mentioned, are seen as essential for the development of successful, long-enduring, community-organized governance solutions.

---

[2] That is a network of strong personal relationships and social interactions between members of a community (individuals, groups, or organizations).

[3] In response to the water supply threat posed by an Ethiopian dam, Egyptian President Mohamed Morsi declared in a television speech to his people on June 10, 2013: "Egypt's water security cannot be violated at all. As president of the state, I confirm to you that all options are open" (BBC News 2013).

The paper is structured as follows. The following section assesses the formal regulatory framework that prescribes (ground) water use in Greece, whereas the next one moves to outline the condition of water resources in Thessaly, which is the region where Larisa is located. Section four presents the analysis and results of the case study, and section five concludes the chapter.

2 The Legal Framework of Water Management in Greece

As discussed, facilitative to sustainable CPR management is the provision of a formal institutional (legal) framework that clearly and credibly defines (property) rights and responsibilities and enforces compliance with those involved, providing incentives for proper consumption, management, and conservation of the resource.

As far as Greece is concerned, until three decades ago, there was a serious lack of legal provisions regarding the protection and management of the water resources.[4] Despite the several efforts to overcome problems and to provide a comprehensive institutional framework that deals with these issues,[5] the legal instruments available by the mid-1980s were multiple in number, limited in scope, and piecemeal in character, with weak policing and enforcement mechanisms and poor control and implementation powers (Kampa 2007; Kampa and Bressers 2008).

The Framework Laws 1650/1986, for the "Protection of the Environment," and 1739/1987, for the "Management of Water Resources," constituted the first serious attempts for the provision of an integrated legal frame able to support sustainable water management in Greece. Although they were only partially implemented,[6] mainly due to public sector inability to put into effect some of their provisions (CSEH 2003; Kampa 2007), the 16-year experience that they endowed to all relevant parties provided a valuable background for the transposition of Water Framework Directive 2000/60/EC into the Greek national legal context—see below (NTUA 2008).

The Water Framework Directive (WFD) provided a wider frame for European Union (EU) member states in the field of water policy in order to achieve good qualitative and quantitative status of all water bodies by 2015. To do so, it established a number of common objectives, principles, definitions, and measures for

---

[4] In practice, there were no restrictions in the abstraction of groundwater to both private and public users (Kampa 2007).

[5] Such as the Civil Code of 1940, the Law 481/1943 on the management and administration of waters used for irrigation (complemented with further acts in 1948, 1949, 1952, 1957), the Law 1988/1952 on wells, the Decree 3881/1958 on land reclamation works, the Code 420/1970 for the protection of the aquatic ecosystem, and the New Constitution of 1975 (which introduced environmental protection as an obligation).

[6] As a result of their weak implementation, water management continued in a piecemeal and opportunistic manner throughout the 1990s (Kampa 2007). In practice, this meant that water users could abstract, at their will, uncontrolled, large amounts of water with the tolerance of the local authorities (Delithanasi 2004).

the sustainable management of the water resources throughout EU and prescribed the steps that member states need to follow in order to reach the common goal, taking in due account not only environmental but economic and social considerations as well. Interestingly, in contrast to past mentality, the WFD has, *inter alia*, urged states to encourage the participation of all interested parties in the water management process (Article 14) and recommended the establishment of economic instruments to ensure incentive pricing to water savings and full cost recovery based on the polluter pays principle (Article 9).

Despite the shortcomings of the WFD (see *inter alia* Kallis and Butler 2001; Baltas and Mimikou 2006), Greece has been relatively prompt to incorporate it into the national legal context through the adoption of the 3199/2003 "Water Protection and Management" Framework Law (Kanakoudis and Tsitsifli 2010). This Law introduced most of the new definitions and notions of the Directive and determined the competent authorities and the analytical procedures that they should follow for each individual issue, but did not go through a number of important provisions specified by the WFD (left to be regulated in future time). This partial harmonization with the Directive brought Greece in front of the European Court of Justice for a couple of times (in 2006, 2008, 2011), giving rise to the 51/2007 Presidential Decree, which literally transposed (word by word) all the provisions left out from the Framework Law. This delay in transposition, however, has brought further delays to a number of implementation actions (Sofios et al. 2008), posing a serious threat to the overall process. In addition, the recent financial crisis that afflicted the country with the hard austerity measures that imposed to local bodies has put into question the financial feasibility and necessity of the program and made its requirements to be somewhat neglected (Kalampouka et al. 2011). This, nevertheless, brought new impetus to bottom-up user-based initiatives, aiming to the sustainable governance of the water resource.

3 The Characteristics of Water Resources in the Region

The Thessaly Water District (WD)[7] virtually coincides with the corresponding regional territory incorporating almost the whole prefecture of Larisa and large parts of the prefectures of Magnesia, Trikala, and Karditsa (see Fig. 1). Its total area is 13,136 km^2 (with population, as measured in 2011, of 746,714 residents), which is divided into three sections: the eastern costal and mountainous area with Mediterranean climate, the central flat area with continental climate, and the western mountainous area with mountainous climate (Baltas and Mimikou 2006). The Thessaly WD comprises the basins of Pineios River (and its tributaries) and the

---

[7] The 3199/2003 Framework Law adopted the existing division of Greece into 14 WDs (already defined by the 1739/1987). A WD is considered to be the entity of all runoff basins of as similar as possible hydrological–hydrogeological conditions, which constitute the regional level in the field of water management (NTUA 2008).

Fig. 1 The 14 WD in Greece (*source* NTUA 2008)

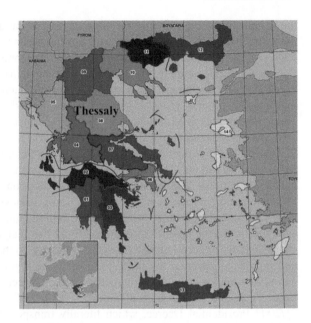

lake of Karla as well as two self-contained aquifers, the western and the eastern, covering 4,520 km², or 35 % of the region's area. The average annual temperature ranges from 16 to 17 °C (WMC 2005). The rainy season lasts from October to January and the dry one from July and August, giving an average annual precipitation of about 678 mm, which is one of the lowest in the country (WMC 2005). This provides a first indication of the water condition that the area exhibits.

Extended to an area of 14,000 km² (about 11 % of the whole country), the Thessaly region incorporates the highly fertile plain of Larisa, providing 14.2 % of the national agricultural product (40 % of cotton) and making it one of the most important agricultural areas of Greece. Agriculture is the main consumer of Thessaly's water resources (87 % of the total demand). The 2,500 km² of irrigated farmland requires about 1,550 million m³ of water annually, whereas the sustainable supply is about 750 million m³ (or which the 550 are groundwater) (Goumas 2006). This gives an annual deficit of roughly 800 million m³ of water (see Fig. 2), which is usually extracted through illegal borewells (count to be more than 30,000, according to some estimates, see Lialios 2011) depleting the groundwater resource and leading to 'tragedy.'

The dropping levels of the water table of the eastern aquifer, where our case study is located, provide another indication of the extent of the problem (see Fig. 3). As can be seen, from 1985 onward, there is a steady decrease of the groundwater level, apart from the years 2002–2003, when the area has experienced frequent and heavy rainfalls (Goumas 2012). In addition to the quantitative depletion of the resource, there is also qualitative deterioration, which comes from two main sources (Polyzos et al. 2006; Goumas 2012). First is saltwater intrusion (since the area is close to the coast and there is a hydraulic connection between

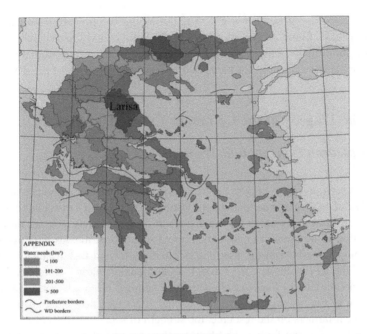

Fig. 2 Water deficit per prefecture (*source* NTUA 2008)

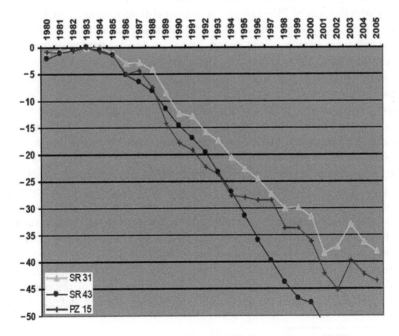

Fig. 3 Water table levels, eastern aquifer of Thessaly WD (*source* Goumas 2012)

the two bodies), and second is nitrate pollution due to crop overfertilization (as a result of lack of both proper education of the farmers and supervision by the regulatory authorities) both of which cause contamination to the groundwater, with catastrophic consequences for the agriculture and the economy of the area.

4 Is the Tragedy of the Commons Unavoidable?

4.1 Research Concept and Methodology

Previous sections made evident the extent of the groundwater degradation in Larisa (mainly due to illegal water extraction) leading to a tragedy of the commons and the deficiencies of the formal–legal framework to deal effectively with all these issues (at least up to the present point). The current section investigates the possibility of developing some bottom-up, user-based initiatives toward the sustainable management of the CPR. This is done through a questionnaire survey which explored the views and attitudes of local stakeholders on a number of relevant issues, such as the condition of the resource and the factors that affect this, the degree of dependence of the local farmers on the resource, the strength of social relations of users and the level of trust (among farmers and between farmers and other players), the willingness to contribute financially toward the maintenance of the resource ("willingness to pay"), and the institutional arrangements which are necessary toward sustainability.

Two groups of people have been surveyed. The first is local farmers (i.e., the users) from the area of Platykampos (a municipality located at about 10 km southeast of Larisa city), and the second is "informed technocrats," i.e., high-ranked public officials, scientists, and experts, who are involved in water management issues (affiliated to the regional authority, the local authorities, the Local Organization of Land Reclamation—TOEB, the local universities, the local branch of the Geotechnical Chamber of Greece, and the Geoponic Association of Larisa). Survey questions were pretested in a pilot study enabling fine-tuning of the instrument and improvement of its clarity. The final questionnaire consists of six parts containing 35 questions of all types: measurement, dichotomous, ordinal, as well as Likert-scale and semantic-differential ones scaled from 0 (denoting strong disagreement, negative opinion, etc.) to 10 (denoting strong agreement, positive opinion, etc.). The first part informs the respondent on the purpose of the research and ensures the anonymity of participation. The second part records views regarding the adequacy and quality of the groundwater (at present and in the near future) and the factors that affect its condition. The third part contains questions about the farming practices, their water consumption, and the willingness to pay for water conservation.[8] The fourth part assesses which institutional arrangements are conducive to sustainable water management. The final part of the questionnaire gathers information about the respondents, such as age, gender, and education.

---

[8] This part was included only in the questionnaire distributed to the farmers.

Table 1 Composition of respondents

			N	M	SD	Mdn	Percentiles		
							25	50	75
Group	Farmers	81.10 %	133						
	Technocrats	18.90 %	31						
Gender	Male	86.60 %	142						
	Female	13.40 %	22						
Age	<30	1.20 %	164	49.7	11.1	50	41	50	58
	30–50	51.80 %							
	>50	47.00 %							
Education	Primary or less	15.20 %	164	3	1.2	3	2	3	4
	Secondary	17.10 %							
	Post-secondary	33.50 %							
	Tertiary(university)	20.70 %							
	Postgraduate	13.40 %							

The survey was held during the first quarter of 2010. Questionnaires were distributed in person by the members of the research team and asked to be completed on the spot.[9] In order to increase the response rate and quality, participants were given the choice of having the questions read to them and responses recorded by the researcher, or, should they wish, to complete them on their own time and be picked up in a week. Questionnaires were collected, validated, and then coded and analyzed to generate a number of statistics illustrating the respondents' views on the issues raised. Data analyses were conducted with SPSS release 19.0. Since none of the developed variables satisfied the Shapiro–Wilk test for normality, nonparametric analysis was employed. Thus, correlations were assessed using the Spearman's rho correlation coefficient.

4.2 Response Rate and Composition of Respondents

A total of 250 distributed questionnaires yielded 164 properly completed responses (a response rate of about 66 %). The respondents were principally men (86.6 %), reflecting male dominance in both the agricultural sector (89.5 %) and high-ranked officer positions (74.2 %) (see Table 1). The 30–50 age bracket was the main group (51.8 %), followed by those over 50 (47 %) and those below 30 (1.2 %). The average age of the sample was about 50 years. Farmers comprised the majority of the sample (81.1 %). Most respondents (36.4 %) have completed post-secondary studies (33.5 %), followed by those holding a university degree (20.7 %), which are mainly technocrats. 81.2 % of the farmers had acquired only compulsory education.

---

[9] It should be noted that due to difficulties in defining with precision the statistical population, the choice of the sample was made by simple random sampling.

4.3 The Condition of Groundwater

When asked to assess the adequacy of groundwater for irrigation purposes, respondents almost unanimously acknowledged the problem (see Table 2). The majority of the sample (31.1 %) replied that there is a water shortage (scored 2, on a scale of 0: shortage to 10: abundance) and the average score was 2.6 (see Table 2). Similar, if not gloomier, was their response regarding the situation over the next decade. More than 75 % of the people said that ceteris paribus, the resource will diminish, whereas most respondents (28 %) gave the lowest score—zero (average score 1.3). Interestingly, a 3 % of the sample replied that there will be some increase in the groundwater reserves, all of which were farmers. In line was the next question, asking whether the resource faces a tragedy condition. Over 80 % of the respondents agreed that the amount of water extracted is not replenished, whereas more than 50 % gave the highest scores (indicating the severity of the problem).

Turning to the factors that held responsible for this situation, out of the five put forward: climate change, agricultural consumption, non-agricultural consumption, wasteful use, and bad management (by official authorities), the last scored higher (mean value of 7.6, with more than 25 % of the respondents giving the highest score), followed by climate change (mean 6.3) and by agricultural consumption and wasteful use (both scored 6.1). When asked to assess the percentage of illegally extracted water, technocrats indicated on average that this should be 32.1 % of that totally consumed, whereas the respective figure given by the farmers was 19.8 %. Similarly, technocrats deemed that 27.9 % of the farmers extract water illegally, while farmers provided a much lower figure (16.3 %).

What becomes evident from the above is that farmers (as well as technocrats) are fully aware of the intensity and causes of the groundwater problem in their area. This is good news, because realization of the problem constitutes the first step toward its solution.

4.4 Irrigation Practices and Attitudes

As regards to irrigation practices, the vast majority of the farmers (76.7 %) admit that they use as much water as there is available, with aim to maximize crop production. After all, they confess, even if they do not do so, someone else will. Of the rest 23.3 % who care for water conservation, 22.5 % do this due to concerns of water availability in the future, and only a 0.8 % act on purely altruistic motives (i.e., for water to be available to the others). Overall, using economics jargon, it becomes evident that economic rationality (utility maximization) drives to a large extent farmers' behavior, which, due to non-excludability of the groundwater resource, gives rise to a free-rider situation.

The above finding is also supported by the next question which explores whether farmers would be willing to slim down their water extraction levels as part of a program for the maintenance of the resource. Interestingly, 29.3 % of the

Table 2 Perceptions on the condition of groundwater

		0 (%)	1 (%)	2 (%)	3 (%)	4 (%)	5 (%)	6 (%)	7 (%)	8 (%)	9 (%)	10 (%)	N	M	SD	Mdn	Percentiles		
																	25	50	75
Groundwater condition	Water adequacy (0: shortage, 10: abundance)	10.4	20.7	31.1	18.3	5.5	3.0	3.7	1.8	4.3	1.2	0	164	2.6	2.0	2	1	2	3
	Quantity in next decade (0: deteriorate, 10: improve)	28.0	23.2	25.6	7.9	7.9	4.3	0.6	1.8	0.3	0.3	0.0	164	1.3	1.6	1	0	1	2
	"Tragedy of the commons" (0: disagree, 10: agree)	3.7	0.6	2.4	1.2	3.7	3.7	3.7	11.0	18.3	23	28.7	164	7.8	2.5	9	7	9	10
Factors affecting groundwater (0: not important, 10: very important)	Climate change	2.4	3	6.7	5.5	3.7	11.6	14.6	18.3	14.6	3.0	16.5	164	6.3	2.6	7	5	7	8
	Agricultural consumption	5.5	1.2	3.7	4.3	7.9	13.4	12.2	19.5	17.1	6.7	8.5	164	6.1	2.6	7	5	7	8
	Non-agricultural consumption.	3.0	9.8	8.5	25.0	14.0	17.1	14.6	3.7	1.8	0.6	1.8	164	4.0	2.0	4	3	4	5
	Wasteful use	7.9	1.8	4.3	6.1	7.9	10.4	8.5	15.2	15.9	8.5	13.4	164	6.1	3.0	7	4	7	8
	Bad management	1.2	0.6	1.8	1.2	3.7	11.0	8.5	10.4	17.1	18.9	25.6	164	7.6	2.3	8	6	8	10

Table 3 Willingness to pay (€)

0	1–100	101–500	501–1,000	1,001–2,000	>2,000	N	M	SD	Mdn	Percentiles		
										25	50	75
36.8 %	9.8 %	24.8 %	17.3 %	10.5 %	0.8 %	133	474.8	598.1	200	0	200	1,000

respondents have a rather non-positive stance, and of the rest 70.7 % who agrees to do so, most (39.8 %) seem willing only if sound economic incentives are given, whereas the others (20.3 %) if there would be additional measures for compliance by all farmers.[10]

4.5 Willingness to Pay

Given the above findings, the gloomy condition of the groundwater resource, and the provisions of the WFD for water service cost recovery through pricing policies, the next question is of particular interest. It is set as follows: "assuming that under current conditions the groundwater reserves will be run out in 10 years, what amount of money are you willing to pay on an annual basis in order successful corrective measures to be taken?". Table 3 presents the results. As can be seen, 36.8 % of the respondents were not willing to provide any financial support, on the basis (as subsequent conversations with the farmers revealed) that the water is a public good and the onus is on the state to ensure its adequate provision and maintenance. The amount of money that the rest of the farmers (63.2 %) were willing to contribute varied substantially, ranging from €50 to €3,000, with the mean value of €474.8.

4.6 Groundwater as Commons

The current section explores the need for, previous experience of, and willingness of the stakeholders to be engaged in some form of bottom-up, user-based initiatives toward the sustainable management of the groundwater. The specific issues examined are the degree of user dependence on the resource, the preferred allocation of ownership rights on groundwater, the kind of institutional arrangements regarded as conducive to sustainable management, the strength of trust-based social relations among users, and their past experience and willingness to cooperate with each other toward the aforementioned end.

Three questions were set to assess the degree of user dependence on groundwater and on agriculture in general. First, farmers were asked to estimate the change in their crop production capacity and resulted income if there was no groundwater

---

[10] These figures indicate that farmers were generally skeptical of the success of such an endeavor, especially given the acute economic conditions of the country and its population.

available. Though replies were varied considerably, on average, a 71.1 % reduction in production and a 67.9 % reduction in incomes were reported. The second question explored whether farmers would consider changing their occupation. Though 26.4 % of respondents were rather negative, the majority (46.7 %) were quite positive (and the rest 23.3 % were indecisive) (see Table 4). To assess the long-term intergenerational dependence, farmers were next asked whether they believe their offsprings would take over their family business. The results were overwhelming: 57.2 % of respondents deemed that their children will not continue farming, 20.4 % was not sure, and 22.6 % thought that rather they will. Overall, it became evident that although farmers and their families depend highly on groundwater for their living, this situation could be rather impermanent and short-termed. On these grounds, it is doubtful whether they would be willing to engage themselves and invest in long-standing relations regarding the management and maintenance of the resource.

The above findings explain relatively well the assignment of property rights that stakeholders seems to prefer, which is examined by the next question. In particular, respondents were asked to choose who should have the ownership of groundwater in order for sustainability to be achieved; should this be the central or local government (i.e., nationalization), a specialized management organization, formal associations/cooperatives of farmers, all farmers collectively, each farmer individually, private investors (i.e., privatization), or none of the above? The group of farmers showed a degree of divergence. Not unexpectedly (on the basis of the low trust among farmers—see below—and the low intergenerational commitment to farming and to self-governance of the resource), almost half of the respondents (49.1 %) opted for the specialized management organization, 20.5 % upheld the central state, 13.7 % argued that ownership should be split between farmers, and only 11.0 % endorsed a form of user-based ownership (i.e., 5.5 % voted for farmer associations and 5.5 % for collective ownership). Interestingly, a tiny 1.7 % chose privatization (i.e., ownership given to private investors) as the preferable solution. Perhaps, equally interesting was the outcome of the technocrats' group. Only two options were selected: the specialized management organization (getting the high 71.0 % of votes) and the central state, indicating that neither privatization nor any form of community ownership was deemed capable to ensure proper use and longevity of the resource.[11]

Next, respondents were asked to assess a number of institutional arrangements in terms of their significance for sustainable management (see Table 5). With the mean value of 8.0, first scored "rule enforcement," which, as seen, is the major deficiency of the Greek institutional framework. Next came the "specification of rules for use," the "specification of sanctions for violations," and the "monitoring of rule compliance" (with mean scores of 7.8). Last were placed arrangements

---

[11] Although further investigation is required, we could argue at this point that such a stance might be due to lack of confidence toward farmers' capacity for self-organization, fueled by relevant previous experience (e.g., the limited success of agricultural cooperatives in Greece—see, *inter alia*, Iliopoulos and Valentinov 2012).

Table 4 Farmer dependence on groundwater

	0 (%)	1 (%)	2 (%)	3 (%)	4 (%)	5 (%)	6 (%)	7 (%)	8 (%)	9 (%)	10 (%)	N	M	SD	Mdn	Percentiles		
																25	50	75
Occupation change	21.1	3.0	2.3	3.8	5.3	12.0	6.0	7.5	9.8	2.3	27.1	133	5.6	3.8	6	2	6	10
Offsprings continue farming	39.8	5.3	8.3	3.8	3.8	12.8	3.8	6.8	3.8	1.5	10.5	133	3.3	3.5	2	0	2	6

Table 5 Institutional arrangement significance

	0 (%)	1 (%)	2 (%)	3 (%)	4 (%)	5 (%)	6 (%)	7 (%)	8 (%)	9 (%)	10 (%)	N	M	SD	Mdn	Percentiles		
	0: Not important						10: Very important									25	50	75
Specification of users	7.0	1.3	5.0	4.4	1.9	7.6	4.4	15.2	19.0	13.9	20.3	158	6.8	3.0	8	5	8	9
Specification of rules for use	3.8	0	2.5	1.3	1.9	7.0	6.3	12.7	12.7	17	34.8	158	7.8	2.6	9	7	9	10
Specification of sanctions for violations	2.5	1.3	0.6	3.2	2.5	7.0	7.6	10.8	14.6	13.9	36	158	7.8	2.5	8.5	6.8	8.5	10
Monitoring of rule compliance	3.8	1.9	1.9	1.9	1.3	5.0	5.7	9.5	20.9	13.3	34.8	158	7.8	2.6	8	7	8	10
Rule enforcement	1.3	0.6	3.8	1.3	2.5	4.4	4.4	10.8	18.4	15.8	36.7	158	8.0	2.4	9	7	9	10
User coordination and conflict management	6.3	0	3.2	2.5	3.2	11.4	10.1	17.1	22.8	6.3	17.1	158	6.7	2.7	7	5	7	8
User participation in management	8.9	0.6	2.5	3.2	1.3	12.7	11.4	12.0	14.6	10.0	22.8	158	6.7	3.0	7	5	7	9

Table 6 Institutional arrangement significance by respondent group

	M (SD)		
	Farmers	Technocrats	Difference
	(a)	(b)	(b − a)
Specification of users	6.7 (3.2)	7.3 (2.0)	0.6
Specification of rules for use	7.5 (2.7)	9.0 (1.0)	1.5
Specification of sanctions for violations	7.7 (2.7)	8.1 (1.6)	0.4
Monitoring of rule compliance	7.5 (2.8)	8.9 (1.4)	1.4
Rule enforcement	7.9 (2.5)	8.5 (1.7)	0.6
User coordination and conflict management	6.6 (2.9)	7.5 (1.2)	0.9
User participation in management	7.0 (3.0)	5.4 (2.5)	−1.6

regarding the "precise specification of users" (6.8), "user coordination and conflict management" (6.7), and "user participation in management" (6.7). It is interesting to note that technocrats, as compared to farmers, valued higher all aforementioned institutional arrangements (see Table 6), apart from the one, the "user participation in management," which not only was placed at the bottom of the rank but also was regarded as having neutral significance.

The next set of two questions attempted to assess the strength of trust-based relations of users (a form of social capital). First, the trusting attitude of farmers was measured using a semantic-differential question with the following contrasting options: "I do not trust someone until there is clear evidence that (s)he can be trusted," indicating low trusting behavior (scored 0), and "I trust someone until there is clear evidence that (s)he cannot be trusted," indicating high trusting behavior (scored 10). Table 7 presents the results making apparent the low degree of trusting that characterizes farmers in Larisa. In particular, 58 % of respondents described themselves as rather reserved and suspicious (interestingly, 36.1 % picked the lowest scope), 13.6 % placed themselves on the middle of the scale, and a low 28.7 % put themselves on the high end of the trusting spectrum. Since interpersonal trust is a relative concept, depending on who it is directed at, the next question tried to assess the degree of trust farmers have on various people/entities: relatives, friends, fellow-villagers, other farmers, farmer associations/cooperatives, technocrats/scientists, specialized bodies, local authorities, and the central state. As Table 7 reveals, friends are the most trustworthy group (mean of 6.6), followed by technocrats (6.5) and relatives (6.0). Respondents were reserved against farmer associations (mean score of 5.6) and specialized bodies (4.9), and they distrusted local authorities (score of 3.8), other farmers (3.7), fellow-villagers (3.5), and the central state, which got the lowest score (3.0).

Finally, it has been examined whether farmers had previous cooperative experience and how willing they would be to cooperate with other farmers toward self-governance of the groundwater as commons. As regards the former, the majority of respondents (69.2 %) reported that they do participate in associations, cooperatives, clubs, etc. Of them, 46.2 % take part in one such organization, 37.4 % in two, and the rest in three or more, with average experience greater than 20 years of involvement. As concerns their attitude toward cooperation for self-governance

Table 7 Strength of social relations and trust

	0 (%)	1 (%)	2 (%)	3 (%)	4 (%)	5 (%)	6 (%)	7 (%)	8 (%)	9 (%)	10 (%)	N	M	SD	Mdn	Percentiles 25	Percentiles 50	Percentiles 75
	0: Not trust								10: Trust									
Trusting attitude	36.1	5.3	12.8	3.8	2.3	6.8	4.5	9.8	6.8	3.8	8.3	133	3.5	3.6	2	0	2	7
Trust on Relatives	6.1	1.5	2.3	6.9	6.1	18.3	9.9	16.0	16.0	9.2	7.6	131	6.0	2.6	6	5	6	8
Friends	2.3	2.3	2.3	6.1	3.8	11.5	14.5	14.5	18.3	13.0	11.5	131	6.6	2.5	7	5	7	8
Fellow-villagers	14.5	12.2	13.7	11.5	7.6	13.7	12.2	11.5	2.3	0.8	0	131	3.5	2.5	3	1	3	6
Other farmers	10.7	7.6	17.6	9.9	14.5	18.3	15.3	2.3	0.8	0.8	2.3	131	3.7	2.3	4	2	4	5
Farmer associations	5.3	3.1	6.1	6.1	11.5	15.3	12.2	12.2	15.3	3.1	9.9	131	5.6	2.7	6	4	6	8
Technocrats/scientists	8.4	0.8	0.8	5.3	2.3	13.0	10.7	14.5	19.8	13.7	10.7	131	6.5	2.8	7	5	7	8
Specialized bodies	11.5	4.6	5.3	6.1	9.9	16.8	12.2	16.0	12.2	2.3	3.1	131	4.9	2.7	5	3	5	7
Local authorities	13.7	8.4	10.7	16.8	11.5	12.2	10.7	8.4	3.8	2.3	1.5	131	3.8	2.6	4	2	4	6
Central state	24.4	11.5	11.5	15.3	8.4	9.9	6.9	4.6	4.6	0.8	2.3	131	3.0	2.7	3	1	3	5

of the commons, 59.3 % of the farmers were rather positive to work with farmers they know quite well (whereas 24.1 % were reserved) and 63.9 % were positive to join forces with organized groups (associations, cooperatives, etc.) of farmers (whereas 21.1 % were skeptical), but only 15.9 % were happy to work together with all interested farmers, in contrast to 58.6 % who were unwilling (see Table 8), indicating, one more time, the low level of trust among farmers in general.

4.7 Perceptions, Views, and Stakehold Characteristics

This section explores the degree to which the characteristics of the respondents, i.e., age, gender, education, and position/affiliation (viz. farmers or technocrats), affect their perceptions and attitudes with regard to examined groundwater issues. To do so, the Spearman's rho correlation coefficient is used, which measures the association between characteristics and perceptions/attitudes toward groundwater.[12] Table 9 presents the results of such statistically significant correlations.

As already mentioned, both farmers and technocrats have the same, gloomy perception of the groundwater conditions in Larisa. This does not seem to be affected by the age, gender, or education level of the respondents. As regards the factors that play a role in the depletion of the resource, positive correlations were detected between them and the gender, education level, and position of the respondents. On these grounds, it can be asserted that women, more educated people, and scientists–experts (as compared to men, less educated, and farmers) ascribe higher significance to agricultural and non-agricultural consumption, wasteful use, and poor management, as sources of groundwater degradation.

Turning to the irrigation practices of the farmers, it appears that water usage manners and care for groundwater conservation are not related to the age, gender, or educational differences between the respondents. On these grounds, explanations of both utility maximization or altruistic behavior and water conservation sensitivity of the farmers should be sought on other factors, related, perhaps, to socioeconomic or cultural characteristics. The same seems to be the case for the willingness of the farmers to pay for the groundwater services. On the other hand, farmer willingness to change occupation seems to be negatively related to their education background: The more educated people are more reluctant to change job, indicating probably how conscious and deliberate such decisions have been on their part.

As far as ownership on groundwater is concerned, once again, views are not differentiated by age, gender, or education level of the respondents. However, technocrats seem to draw apart from farmers, favoring allocation of property rights to a specialized management organization and discriminating against ownership by farmers (positive and negative coefficients, respectively).

---

[12] Correlations were also checked with the Sommer's d coefficient, giving the same results. For reasons of space efficiency, these have not been included in the paper.

Table 8 Attitude toward self-governance of the groundwater as commons

	0 (%)	1 (%)	2 (%)	3 (%)	4 (%)	5 (%)	6 (%)	7 (%)	8 (%)	9 (%)	10 (%)	N	M	SD	Mdn	Percentiles		
																25	50	75
Cooperation with	0: No							10: Yes										
…farmers I know well	15.8	1.5	3.0	3.8	1.5	9.8	5.3	10.5	12.0	7.5	29.3	133	6.4	3.6	7	4	7	10
…organized farmer groups	15.8	3.0	0	2.3	3.0	3.0	9.0	4.5	14.3	17.3	27.8	133	6.7	3.6	8	5	8	10
…all farmers	35.3	9.8	7.5	6.0	8.3	7.5	9.8	5.3	5.3	1.5	3.8	133	3.0	3.1	2	0	2	6

Table 9 Correlation coefficients between respondent characteristics and groundwater variables (Spearman's rho)

		Age	Gender	Education	Position
Water conditions	Water adequacy	–	–	–	–
	Quantity in next decade	–	–	–	–
	"Tragedy of the commons"	–	–	–	–
Degradation factors	Climate change	–	–	–	–
	Agricultural consumption	–	0.220[a]	0.201[a]	0.388[a]
	Non-agricultural consumption	–	0.194[b]	–	–
	Wasteful use	–	0.244[a]	0.261[a]	0.414[a]
	Bad management	–	–	0.162[b]	0.217[a]
Irrigation practice	Use as much as needed	–	–	–	
	Conserve for future	–	–	–	
	Leave for others	–	–	–	
Willingness to pay		–	–	–	
Farmer dependence	Occupation change	–	–	−0.161[b]	
	Offsprings continue farming	–	–	–	
Property rights	Central state	–	–	–	–
	Local authorities	–	–	–	–
	Specialized management organization	–	–	–	0.173[b]
	Farmer associations/ cooperatives	–	–	–	–
	All farmers collectively	–	–	–	–
	Each farmers individually	–	–	–	−0.170[b]
	Private investors	–	–	–	–
Institutional arrangements	Specification of users	–	–	–	–
	Specification of rules for use	–	–	0.161[b]	0.195[b]
	Specification of sanctions for violations	–	–	–	–
	Monitoring of rule compliance	–	–	–	0.194[b]
	Rule enforcement	–	–	–	–
	User coordination and conflict management	–	–	–	–
	User participation in management	–	–	–	−0.279[a]
Trust	Trusting attitude (general)	–	–	–	
	Relatives	–	–	–	
	Friends	–	−0.216[b]	–	
	Fellow-villagers	–	–	–	
	Other farmers	–	–	–	
	Farmer associations	–	–	–	
	Technocrats/scientists	–	–	–	
	Specialized bodies	–	–	–	
	Local authorities	–	–	–	
	Central state	–	–	–	

(continued)

Table 9 (continued)

		Age	Gender	Education	Position
Cooperation	With organized farmer groups	–	–	−0.190[b]	
	With farmers known well	–	–	–	
	With all farmers	–	–	–	

[a]Correlation is significant at the 0.01 level (two-tailed)
[b]Correlation is significant at the 0.05 level (two-tailed)
–Correlation is not statistically significant at the above levels

Perceptions regarding the significance of institutional arrangements for the sustainability of the groundwater resource differ according to the education background and position of the respondents. In particular, the more educated people seem to ascribe higher significance on the specification of credible rules for appropriate usage of groundwater. So do technocrats, which in addition set apart from farmers to highlight the importance of monitoring and policing procedures for rule compliance. Moreover, and in accordance with previous findings, technocrats are "significantly" skeptical on whether farmers should have an active role in the management of the groundwater resource (negative correlation coefficient).

A particularly interesting and valuable conclusion that this analysis yields relates to the trust issue: Not only is the lack of trust a characteristic of the farmer community examined, but this seems to be a pervasive phenomenon extending to all ages, sexes, and educational backgrounds.[13] In addition, the negative correlation coefficient between gender and the variable indicating trust to friends affirms other pieces of research (see *inter alia* Chaudhuri et al. 2013) that find women (as compared to men) to show lower levels of trust.

Finally, the schooling level seems to affect also the farmers' attitudes toward cooperation for the self-governance of the resource. In particular, the negative correlation between education and willingness to cooperate with organized groups of farmers implies that the less educated people are more prone to get involved in such relations, compared to the more educated farmers who are significantly reluctant.

5 Conclusions

Groundwater as a typical example of a common pool resource is subject to serious risk of overexploitation, pollution, degradation (in terms of both quantity and quality), and even total destruction (the so-called tragedy of the commons).

---

[13] Several other pieces of research report similar findings, that is, low and declining levels of social trust in Greece (see *inter alia* Paraskevopoulos 2006; Jones et al. 2008; Roumeliotou and Rontos 2009), offering a number of possible explanations: increasing levels of individualistic mentality and utilitarian political culture, increasing income disparities, strong clientelistic relations, increasing disappointment and distrust to political institutions, and a long tradition of authoritarian statism along with a problematic transition to democracy during the first post-dictatorship period (1974–mid-1990s).

The conventional literature prescribed either privatization or full nationalization of the resource as appropriate solutions to the problem. However, countries may exhibit a number of characteristics (e.g., weak property rights, deficient policing and enforcement mechanisms, rigid and bureaucratic institutions, lack of privatization experience) which preclude successful implementation of such top-down approaches. In turn, as the 2009 Nobel laureate in economics, Elinor Ostrom, has established, the users themselves can develop collective institutional arrangements that provide solutions to the commons problems which are more socially acceptable, more durable and sustainable, and with lower implementation costs.

Drawing on the analytical framework on commons that Ostrom and other scholars have developed, the current paper has examined issues of collective management of the groundwater resource using Larisa area (one of the most important agricultural regions in Greece) as a case study. Issues examined include the overall institutional/legal framework available for groundwater management, the irrigation practices in the area, the condition of Larisa's groundwater (and the perception stakeholders have about it), the institutional and other arrangements that local players deem as significant for the maintenance of the resource, and the capability of farmers to join forces toward the self-governance of commons. A number of emerged points should be highlighted.

First, adverse climate conditions, poor resource management, and overexploitation practices (e.g., illegal water extraction) have over the years depleted and downgraded the groundwater resource of Larisa, putting into great danger the agriculture industry and the whole economy of the region. Second, despite significant legal developments undertaken under the WFD, the existing regulatory framework lags behind in terms of ability to deal effectively with the tragedy condition that the groundwater of the area faces. Third, users (and stakeholders in general) are fully aware of the severity of the problem, but deficient policing and enforcement mechanisms on the part of the state and opportunistic, free-riding behavior on the part of the farmers (fed by the low intergenerational dependence on the resource and the subsequent short-term exploitation horizon) have intensified the condition and precluded the exploration of more innovative solutions to the case. Fourth, an additional and serious obstacle toward the development of community-emerged user-based governance arrangements has been the lack of trust both among farmers and between farmers and the state, both local and central (which in a sense constitutes a social capital deficit), hurting the confidence of technocrats that user participation can indeed be a key element of successful solutions. Fifth, given the reluctance of the farmers to engage themselves and invest in long-standing relations regarding the management and maintenance of the resource, the most pragmatic solution (acknowledged by all parties) would be the development of an independent coordinative body with multiple responsibilities and powers.

Though further research is necessary in order to specify the most acceptable form and structure of such an organization, some hints could be gained from the above findings and conclusions. Given the low trust both among users and between them and the state authorities on the one hand, and the high respect that technocrats enjoy on the other, it should be the latter to take the leading role in coordinating the whole initiative, bringing together all interested parties in a collaborative and participatory fashion. In such a scheme, state authorities could contribute legal credibility (formalizing successful practices) and, perhaps, financial support, whereas local users would infuse social validation, grassroots reinforcement, and safeguard.

References

Agrawal A (2001) Common property institutions and sustainable governance of resources. World Dev 29(10):1649–1672

Agrawal A (2003) Sustainable governance of common-pool resources: context, methods, and politics. Annu Rev Anthropol 32:243–262

Annin P (2006) The Great Lakes Water Wars. Island Press, Washington DC

Baland J-M, Platteau J-P (1996) Halting degradation of natural resources: is there a role for rural communities. Clarendon Press, Oxford

Baltas EA, Mimikou MA (2006) The water framework directive for the determination of new hydrologic prefectures in Greece. New Medit 5:59–64

Barclay P (2004) Trustworthiness and competitive altruism can also solve the "tragedy of the commons". Evol Hum Behav 25:209–220

BBC News (2013) Egyptian warning over Ethiopia Nile dam, 10 June 2013. http://www.bbc.co.uk/news/world-africa-22850124. Accessed on 2 Nov 2013

Bolton KR (2010) Water wars: rivalry over water resources. World Aff 14(1):52–83

Briasouli E (2003) The commons—resources of collective ownership and collective responsibility: concepts, problems and the issue of their management. Aeihoros 2(1):36–57 (in Greek)

Chaudhuri A, Paichayontvijit T, Shen L (2013) Gender differences in trust and trustworthiness: individuals, single sex and mixed sex groups. J Econ Psychol 34:181–194

CSEH—Committee for Social and Economic Issues of Hellas (2003) Opinion of CSEH on 'Protection and management of water resources—harmonisation with directive 2000/60 of the EU'. http://www.oke.gr/greek/gnomi96.htm. Accessed on 20 April 2013, (in Greek)

De Villiers M (2003) Water: the fate of our most precious resource. McClelland and Stewart, Toronto

Delithanassi, M (2004) Drill a well, you can! Kathimerini Press, 31-10-2004 Athens(in Greeks)

Dietz T, Ostrom E, Stern PC (2003) The struggle to govern the commons. Science 302(5652):1907–1912

Easter KW, Becker N, Tsur Y (1997) Economic mechanisms for managing water resources: Pricing, permits and markets. In: Biswas A (ed) Water resources: environmental planning, management and development. McGraw-Hill, New York, pp 579–622

Feeny D, Berkes F, McCay B, Acheson J (1990) The tragedy of the commons: twenty-two years later. Hum Ecol 18(1):1–19

Goumas K (2006) The irrigation of Thessaly plain: consequences for surface and ground water. In: Proceedings of the national conference of the Hellenic Hydrotechnical association on water resources and agriculture, Thessaloniki, pp 39–53 (in Greek)

Goumas K (2012) Strategic plans for the management of the water resources in Thessaly: opportunities and threats for the environment and the development. Paper presented at the water resources conference of the Technical Chamber of Greece, Larisa, Jun 2012 (in Greek)

Hardin G (1968) The tragedy of the commons. Science 162(3859):1243–1248

Iliopoulos C, Valentinov V (2012) Opportunism in agricultural cooperatives in Greece. Outlook Agric 41(1):15–19

Jones N, Malesios C, Iosifides T, Sophoulis CM (2008) Social capital in Greece: measurement and comparative perspectives. South Eur Soc Politics 13(2):175–193

Kalampouka K, Zaimes GN, Emmanouloudis D (2011) Harmonizing member state water policies to the EU water directive 2000/60/EU: the case of Greece. Int J Geol 2(5):29–33

Kallis G, Butler D (2001) The EU water framework directive: measures and implications. Water Policy 3(2):125–142

Kampa E (2007) Integrated institutional water regimes: realisation in Greece. Logos, Berlin

Kampa E, Bressers H (2008) Evolution of the Greek national regime for water resources. Water Policy 10(5):481–500

Kanakoudis V, Tsitsifli S (2010) On-going evaluation of the WFD 2000/60/EC implementation process in the European Union, seven years after its launch: are we behind schedule? Water Policy 12(1):70–91

Klare MT (2001) Resource wars: the new landscape of global conflict. Metropolitan Books, New York

Lialios G (2011) Thousands of unregistered borewells. Kathimerini Press, Athens (in Greeks)

Libecap G (2009) The tragedy of the commons: property rights and markets as solutions to resource and environmental problems. Aust J Agric Resour Econ 53(1):129–144

Mariolakos I (2007) Water resources management in the framework of sustainable development. Desalination 213(1):147–151

Mostert E (2003) Conflict and co-operation in International freshwater management: a global review. Int J River Basin Manag 1(3):1–12

National Technical University of Athens (NTUA) (2008) National program for water resource management and protection. NTUA, Athens (in Greek)

Ostrom E (1990) Governing the commons: the evolution of institutions for collective action. Cambridge University Press, New York

Ostrom E (1992) Community and the endogenous solution of commons problems. J Theor Polit 4(3):343–351

Ostrom E (1999) Coping with tragedies of the commons. Annu Rev Polit Sci 2:493–535

Ostrom E (2000) Reformulating the commons. Swiss Polit Sci Rev 6(1):29–52

Ostrom E (2006) The value-added of laboratory experiments for the study of institutions and common-pool resources. J Econ Behav Organ 61(2):149–163

Ostrom E (2008) The challenge of common-pool resources. Environ Sci Policy Sustain Dev 50(4):8–20

Ostrom E (2010) Analyzing collective action. Agric Econ 41(S1):155–166

Ostrom E, Burger J, Field C, Norgaard R, Policansky D (1999) Revisiting the commons: local lessons global challenges. Science 284(5412):278–282

Paraskevopoulos CJ (2006) Social capital and public policy in Greece. Sci Soc 16:69–105 (in Greek)

Polyzos S, Sofios S, Goumas K (2006) Temporal changes in Thessaly's groundwater and their implications on economy and the environment. In: Proceedings of the 9th Panhellenic conference of rural economics, Athens, pp 410–427 (in Greek)

Roumeliotou M, Rontos K (2009) Social trust in local communities and its demographic, socioeconomic predictors: the case of Kalloni, Lesvos, Greece. Int J Criminol Sociol Theor 2(1):230–250

Serageldin I (2009) Water: conflicts set to arise within as well as between states. Nature 459(7244):163

Sofios S, Arabatzis G, Baltas E (2008) Policy for management of water resources in Greece. Environmentalist 28(3):185–194

Solomon S (2011) Water: the epic struggle for wealth, power, and civilization. Harper Perennial, New York

Starr JR (1991) Water wars. Foreign Policy 82:17–36

Stern PC, Dietz T, Ostrom E (2002) Research on the commons: lessons for environmental resource managers. Environ Pract 4(2):61–64

Theesfeld I (2010) Institutional challenges for national groundwater governance: policies and issues. Ground Water 48(1):131–142

Wade R (1987) The management of common property resources: collective action as an alternative to privatisation or state regulation. Camb J Econ 11(2):95–106

Wade R (1988) Village republics: economics conditions for collective action in South India. Institute for Contemporary Studies, Oakland

Water Management Corporation of Central (WMC), Western Greece (2005) Developing systems and tools for the management for the water resources of the water districts of Western Mainland Greece. Eastern Mainland Greece, Epirus, Thessaly and Attica, phase A: Thessaly water district. Report 08-A-II-1, Athens (in Greek)

Worrell R, Appleby MC (2000) Stewardship of natural resources: definition, ethical and practical aspects. J Agric Environ Ethics 12(3):263–277

Authors Biography

Paschalis A. Arvanitidis (MEng., MLE, Ph.D.) is an assistant professor of institutional economics at the Department of Economics (University of Thessaly) and a visiting professor of urban economics and development at the Department of Planning and Regional Development (of the same institution). He is an engineer and an economist with specialization on institutional economics, spatial economics, and real estate markets. His recent research interests include institutional economics with emphasis on methodology and the analysis of the commons, and economic development at local, regional, and national levels.

Fotini Nasioka (B.Sc., M.Sc.) is a Ph.D. candidate at the Department of Economics (University of Thessaly). She is an economist with specialization on applied economics and local development. Her research interests include institutional economics with emphasis on the analysis of the common pool resources.

Sofia Dimogianni (MEng., M.Sc.) is a Ph.D. candidate at the Department of Planning and Regional Development (University of Thessaly). She is an architect with specialization on urban and regional planning. Her research interests include local economic development and the management of the commons.

Collective Versus Household Iron Removal from Groundwater at Villages in Lithuania

Linas Kliučininkas, Viktoras Račys, Inga Radžiūnienė
and Dalia Jankūnaitė

Abstract The Water Framework Directive (WFD) provides a framework to integrate high environmental standards for water quality and sustainable water resource management. Hydro-geological conditions typical for southwest part of Lithuania determine high concentrations of iron in the groundwater. Untreated groundwater is commonly used for every day needs by local inhabitants living in a villages (water consumption <100 m^3/day). Seasonal measurements indicated high variations of total iron concentrations in groundwater. The detected annual concentration of total iron in the water wells was 3.3 mg/L. The concentrations of total iron in the tap water were some 40 % lower compared to those in the groundwater. Iron removal from the ground drinking water yields advantages with the comfort of consumers; however, it entails environmental impacts and additional costs. A comparative analysis of collective and individual household iron removal systems for the selected village has been performed to estimate possible environmental impacts and costs. For assessment of costs and environmental impacts, authors applied input–output analysis. The chosen technique for collective iron removal was non-reagent method implying oxidation of contaminants in the drinking water and their containment in the filters. For individual households, reverse osmosis filtration method was selected. The environmental benefits of using central iron removal system result in formation of almost 70 % less of solid waste, 13 % less of wastewater, and 97 % less consumption of electric energy compared to the individual iron removal facility at each household. Estimated overall cost, including purchase, installation, and operational costs, for central iron removal system is 390 Euro/year per household, the respective cost for individual household iron removal facility—1,335 Euro/year. The analysis revealed that central iron removal system has advantages in comparison with iron removal facilities at each individual household.

Keywords Iron · Groundwater · Central and household iron removal

L. Kliučininkas (✉) · V. Račys · I. Radžiūnienė · D. Jankūnaitė
Department of Environmental Technology, Kaunas University of Technology,
Kaunas, Lithuania
e-mail: linas.kliucininkas@ktu.lt

© Springer International Publishing Switzerland 2015   91
W. Leal Filho and V. Sümer (eds.), *Sustainable Water Use and Management*,
Green Energy and Technology, DOI 10.1007/978-3-319-12394-3_5

1 Introduction

Mismanagement of groundwater may result in a cycle of unsustainable socioeconomic development—including risk of poverty, social distress, energy, and food security. Lack of good quality drinking water also limits processing of agricultural production, creation, and development of small businesses, as well as attraction of investments. To ensure the sustainable management of groundwater resources for domestic use, authors provide input–output analysis approach regarding selected village in Lithuania. The analysis covers innovative assessment of environmental impacts and cost of iron removal from groundwater.

During the last 40–50 years, groundwater use for drinking purposes has increased in many countries, especially in the developing countries and countries in transition (Shah 2005). However, considerable differences in the availability and quality of groundwater result in varying overall use of groundwater in individual countries. Groundwater part in a general balance of drinking water supply exceeds 70 % in Austria, Armenia, Belarus, Belgium, Hungary, Georgia, Denmark, Lithuania, Switzerland, and Germany. In most of these countries, groundwater is the primary source for water supply in rural areas (Zekster and Everett 2004). Since groundwater often needs some kind of pretreatment, local communities express willingness to have a possibility to use good quality drinking water. Surveys show that people living in those areas are prepared to pay for improved drinking water quality (Genius et al. 2008). In order to offer cost-efficient water pretreatment technologies, thorough analysis of possible alternatives of drinking water preparation systems is required (Lindhe et al. 2011).

Lithuania is one of the characteristic countries, generally using groundwater for drinking water needs. Iron removal from groundwater in cities and towns is no longer a matter of great concern in urban Lithuania, whereas rural areas face problems with drinking water quality. The quality of water is often the issue in small settlements. Residents of villages (with a drinking water consumption <100 m³/day) often extract water from the water wells, most of which are physically and technologically outdated and do not meet consumers' needs. Since for the time of the study Lithuania has an economy in transition, it was relevant to estimate the environmental impacts and costs of drinking water preparation.

2 The EU Water Resource Management

The Water Framework Directive (WFD, 2000/60/EC), providing a framework to integrate high environmental standards for water quality and sustainable water resource management, is a new approach to environmental policymaking from a European perspective. The main purpose is to improve the quality of all types of water bodies across the EU. Different instruments are used to obtain the objective, involving different level of organization—from public participation to national or European goals. The integration of water policy with other EU directives and sector policies as well as with spatial planning is also emphasized. WFD for the first time at European level provides a framework for integrated management of groundwater and surface water. The components of the WFD dealing with ground-water cover a number of different steps for achieving good quantitative and chemical status of groundwater by 2015.

The Drinking Water Directive (DWD, 98/83/EC) concerns the quality of water intended for human consumption. Its objective is to protect human health from adverse effects of any contamination of water intended for human consumption by ensuring that it is wholesome and clean. It sets standards at EU level for the most common substances (so-called parameters) that can be found in drinking water. According to the DWD, a total of 48 microbiological and chemical and indicator parameters must be monitored and tested regularly.

3 Characterization of Iron-rich Groundwater for Public Supply Purposes

The basic parameters that characterize and predetermine iron concentrations in the groundwater are pH and oxidation reduction potential (ORP) (Diliunas et al. 2006). Iron removal process is more effective in low-acidity environments (Bong-Yeon 2005) with high oxidation potential (Tekerlekopoulou et al. 2006). Water temperature and intensive aeration have no significant effect on iron removal process. The presence of ammonium is undesirable because it causes taste and odor problems, reduces disinfection possibilities, and also undergoes oxidation process, converting to nitrate (Katsoyiannis et al. 2008). Waters, containing high concentrations of chlorides, stimulate formation of iron corrosion products—green rusts (green-blue iron hydroxide compounds formed under reduction and weakly acid or weakly alkaline conditions as intermediate phases in the formation of FE oxides—goethite, lepidocrocite, magnetite). Water salinity can also influence iron release to the groundwater (Pezzetta et al. 2011). If water is containing organic substances, iron practically does not form the flocks or particles suitable for filtration or sedimentation. This problem is caused by the presence of stable iron colloids or iron complex compounds with dissolved organic substances (Serikov et al. 2009).

The DWD sets 200 $\mu g/L$ (98/83/EC) concentration as the threshold limit value for iron in drinking water, and the same level is set as the Specific Limit Value (SLV) in the national Hygiene Norm of Lithuania (HN 24:2003). Use of untreated water, containing high concentrations of iron, has no significant effect on human health. However, the reddish-brown color of water can cause discomfort when taking a bath, it can stain clothing, and it requires additional detergents for washing. It also has negative effect on the sanitary wear, mainly caused by the corrosion of metal components. If water containing high concentration of iron is used, negative impact on the environment is caused by use of chemical products for the daily living needs (comparatively more detergents, bleach, sanitary cleaning, and dish washing chemicals are needed to perform daily cleaning procedures); it also raises consumption of energy (in order to obtain the desired water quality, additional procedures of water boiling, laundry, etc., are used). If bottled drinking water is purchased, it results in additional amount of plastic waste. In summary, the possible additional costs of untreated water use are faster sanitary wear and additional electric energy consumption.

4 Methods for Iron Removal from Groundwater

Iron removal from groundwater is based on oxidation of soluble ferrous compounds to insoluble ones. Katsoyiannis and Zouboulis (2004) and Katsoyiannis et al. (2008) denoted that oxidation methods for iron removal can be divided into physical (without using chemical reagents), chemical (chemical reagent based), and biological. Generally, physical methods involve aeration–filtration technology and are advantageous for small- and medium-size applications. Other methods used for iron removal from drinking water are as follows: ion exchange (Vaaramaa and Lehto 2003); use of activated carbon or other adsorbing materials (Munter et al. 2005; Das et al. 2007); adsorption based on electro-coagulation processes (Vasudevan et al. 2009); oxidation–microfiltration systems (Ellis et al. 2000); subsurface treatment, involving aerated water injection into aquifer (van Halem et al. 2010); and other. Biological iron removal methods utilize microorganisms as oxidation catalysts (Munter et al. 2005).

In this study, for collective removal of iron from groundwater, authors selected the non-reagent method. This method was selected as the most economically feasible and effective iron removal technology, as the water met the following requirements: Iron concentration is below 3 mg/l, pH is less than 6, and permanganate index is below 7 mgO_2/L. The method implies oxidation of contaminants in the drinking water and their containment in the filters. The contaminants are removed by washing the filter. Filters are washed successively with water from the towers. It is proposed to install two parallel lines of filtering devices. This will reduce the instantaneous flow rate required for washing and will guarantee good operation of the system. The maximal capacity of the system for iron removal in the selected settlement—9.0 m^3/h. For iron removal from groundwater at the individual household, the reverse osmosis filtration method was selected. This method is widely used for removal of many types of large molecules and ions from groundwater. It is the reliable and effective technology for drinking water preparation. Since it is fully automated, the technology is not requiring the complex process control and is often chosen by the individual consumers.

The aim of this study was to analyze iron concentrations and related processes in the groundwater and tap water of the selected village, as well as to estimate collective versus individual household iron removal systems in respect to environmental impacts and costs. The estimates of cost for housing-related activities were based on current cost of living in Lithuania.

5 Methods and Materials

5.1 Study Area

Detailed studies of iron concentrations in the groundwater and tap water, as well as other water properties, were analyzed at Barzdai village, which is located 22 km southeast from the district municipality center Sakiai, Lithuania (Fig. 1). The 1,280 residents of the village use drinking water from upper cretaceous aquifer. The extracted water is directed to water tower and provided (without treatment) to local residents. Currently, there are three drinking water wells in Barzdai, two of them belong to the community of the village.

Hydro-geological conditions typical for southwest part of Lithuania determine high concentrations of iron in the groundwater. Chemistry of fresh groundwater is determined by rock composition. Water inflow from overlaying Quaternary intermorainic aquifers causes higher iron content (1–3 mg/L) in the groundwater. The distribution of iron concentrations is weakly related to the horizontal flow of groundwater; the uplift of more mineralized water from deeper aquifers via hydrogeological "windows" and tectonic faults is observed. The most important factor determining the iron content is the CO_2 regime, affected significantly by the conditions of inflow from deeper aquifers.

The closed groundwater system belongs to lower cretaceous aquifer, which is rich in ammonium compounds and organic matter. In the upper cretaceous aquifer, increased concentrations of ammonium, chlorides, and iron are recorded. Iron occurs in groundwater under reduction conditions (i.e., where dissolved oxygen is lacking and carbon dioxide content is high). Soluble form of iron (Fe^{2+}) is typically chemically bound with organic matter.

Fig. 1 Šakiai district municipality

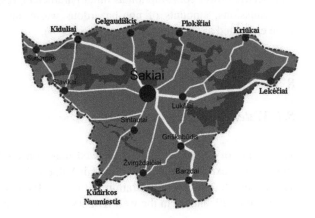

6 Groundwater and Tap Water Measurements

Seasonal concentrations of total iron as well as ammonium nitrogen, chlorides, which potentially have effect on iron levels in the groundwater or influence efficiency of iron removal process, were performed from March 2010 to February 2011. In addition, measurements of iron concentrations in the untreated tap water at different distances from the groundwater wells as well as measurements of pH, temperature, permanganate index (PI), and ORP were performed. The campaign involved in situ measurements as well as chemical analyses at the laboratories of Department of Environmental Engineering, Kaunas University of Technology. Triplicate samples were taken from drinking water wells and tap water. Sampling was carried out in accordance to the ISO 5667-5:2006 and ISO 5667-11:2009 standards. Concentrations of total iron, ammonium nitrogen, and chlorides were measured in conformity with the respective ISO 6332:1988, ISO 7150-1:1984, and ISO 9297:1998 standards; PI was determined in accordance with the ISO 8467-1993 standard. Multimeter WTW pH/Cond 340i/SET was used to determine temperature, pH, and ORP (WTW pH Electrode SenTix ORP). In order to assess deterioration/improvement of water quality in the water supply systems, additional analysis of tap water in the selected households was performed. The selected households were situated in 400–500 m, 500–600 m, and 600–800 m distances from the water well ID 34932.

7 Assessment of Alternative Iron Removal Systems

In order to estimate environmental impacts and costs of collective versus individual household iron removal, authors applied input–output analysis. The estimation of costs involved purchase, installation, and operational costs. Iron removal facilities were assessed with respect to commercial prices offered by local providers. The estimation of environmental impacts involved demand of filter load material and electric power consumption as input parameters, respectively, the CO_2 equivalent emissions; amounts of non-hazardous waste and discharges of wastewater were estimated as output parameters (see Fig. 4).

8 Results and Discussion

8.1 Water Quality Assessment

The results of chemical analysis showed that iron concentrations in all samples significantly exceeded the Specific Limit Value ($SLV_{Fe} = 0.2$ mg/L) (see Fig. 2). The average total iron concentration in the well ID 34932 has exceeded SLV_{Fe} by the factor of 20, and 29 in the well ID 26047, respective concentrations in the well

ID 38890 outreached SLV_{Fe} by the factor of 9. The highest concentration of total iron was observed in the well ID 26047 during summer measurement campaign and reached 6.89 mg/L.

Concentrations of ammonium nitrogen in the samples taken in the summer and the autumn sampling periods have not exceeded the Specific Limit Value ($SLV_{NH_4-N} = 0.5$ mg/L); however, average concentrations of ammonium nitrogen during the spring and the winter sampling campaigns in wells ID 34932 and ID 26047 exceeded the SLV_{NH_4-N} by 60 %, respectively, and in well ID 38890, average concentration was higher by factor 2 compared to SLV_{NH_4-N} (Fig. 2). The highest concentration of ammonium nitrogen was observed in the well ID 26047 during winter and reached 1.97 mg/L value.

Because of presence of ammonium in the groundwater, the required amount of oxygen during iron removal process would be higher. Ammonium forms the nitrates; therefore, it's presence in the water should be taken into account during the technological project preparation phase (Katsoyiannis et al. 2008).

Concentrations of chlorides showed high variation between the seasons and the wells (Fig. 2). The highest concentration (333.1 mg/L) was observed in the well ID 26047 during summer, and this was the only case when the Specific Limit Value ($SLV_{Cl} = 250.0$ mg/L) of chlorides was exceeded. The presence of chlorides in the groundwater is caused by the intrusion of these compounds into the groundwater from the Lower Cretaceous aquifer layer.

The water temperature in the wells analyzed varied from 7.5 to 11.0 °C. The pH values ranged from 7.3 to 9.5 ($SLV_{pH} = 6.5$–9.5); the highest values were observed during the summer sampling campaign and varied from 9.0 to 9.5. The high water pH values indicate faster oxidation of bivalent iron and manganese ions.

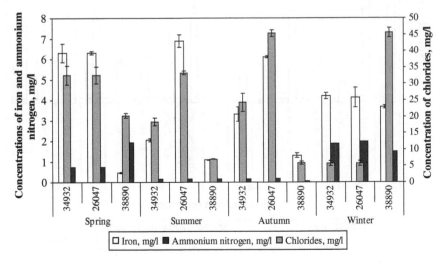

Fig. 2 Concentrations of iron, ammonium nitrogen, and chlorides in the groundwater, mg/L

The ORP of the groundwater usually varies between −480 and 550 mV. In our case, ORP measurements showed negative values and ranged from −250 to −80 mV. The ORP values confirm the reductive conditions in the groundwater. These conditions are usually caused by reducing agents, such as ammonium and bivalent iron.

The PI indicates water contamination by oxidizing organic and inorganic matters, and at the same, it is an important indicator of iron removal process. Each sampling campaign was followed by PI measurements. The PI values ranged from 0.5 to 2.1 mg/L O_2 and did not exceed the Specific Limit Value ($SLV_{PI} = 5.0$ mg/L O_2). Low PI values indicate that iron compounds in the water are of inorganic origin and their oxidation is easier.

In order to assess deterioration/improvement of water quality in the water supply systems, analysis of tap water in the selected households was performed. Observed iron and ammonium nitrogen concentrations in the tap water are presented in the Fig. 3.

In general, it could be stated that the observed iron concentrations in the tap water were lower than those in the water wells by some 40 %. The explanations of this phenomenon could be that the bivalent iron ions are oxidized and precipitated in the pipes of the water supply system. Sedimentation of iron oxide in the pipelines reduces water flow and creates conditions for biofilm formation. This also could increase microbiological contamination of drinking water. Ammonium nitrogen concentrations in the tap water were slightly lower compared to those measured in the water wells. This could be explained by the specific conditions of

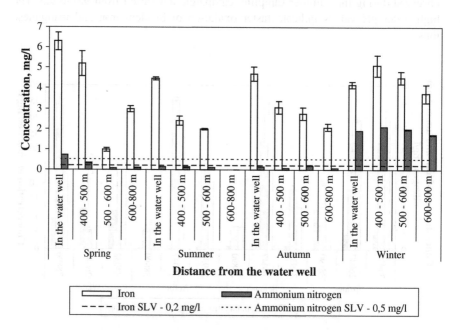

Fig. 3 Concentrations of total iron and ammonium nitrogen in the tap water, mg/L

water stagnation in the pipelines. The analysis revealed that decrease of iron and ammonium nitrogen concentrations receding from the water well is more rapid in warm period, while during cold period the concentrations decline less.

9 Estimation of Environmental Impacts and Costs

The total requirement of drinking water for the analyzed village is 16,790 m^3/year. In addition, 1.1 m^3 of water is used in filter backwashing process. The flowchart of input–output analysis for estimation of environmental impacts and costs of water preparation is presented in the Fig. 4. The input–output analysis of groundwater pretreatment included evaluation of incoming material and energy flows as well as assessment of energy use and waste generated during the drinking water preparation procedure.

It was assumed that reverse osmosis filters will be suitable option for water preparation at each individual household (see Fig. 4). If water would be treated individually at each household, it would require 8,150.0 kWh of electric energy per year. Regular filter regeneration is performed by backwashing, and used filter medium makes 5,477.6 kg of non-hazardous waste yearly. Wastewater (1,116.3 m^3) after backwash is discharged directly into the surface water body.

The purchase and installation cost of iron removal filter for an individual household is 1,280 Euro, respectively, and yearly operational cost makes up 54.84 Euro.

For central iron removal system, the non-reagent technology, which implies oxidation of contaminants and their containment in the filters, was analyzed.

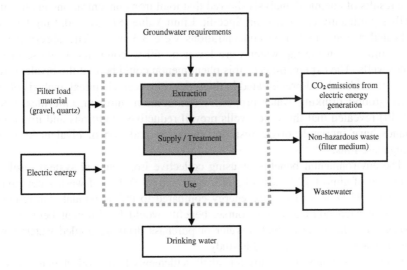

Fig. 4 Flowchart of input–output analysis

Table 1 Annual environmental impacts of alternative iron removal systems estimated for one household

	Input parameters		Output parameters		
	Amount of filter load material, m^3	Electric energy, kWh	Equivalent CO_2 emissions, kg[a]	Non-hazardous waste, kg	Wastewater, m^3
Individual household iron removal system	2.1	8150.0	4295.4	5477.6	1116.3
Collective iron removal system	0.6	263.4	138.8	1590.2	964.0

[a]www.defra.gov.uk/environment/climatechange/uk/individual/pdf/actonco2-calc-methodology.pdf

The total yearly electric energy requirement, estimated for one household, will be 263.4 kWh, respectively, and the CO_2 equivalent emissions will make 138.8 CO_2/kWh. Removal of contaminants by washing the filter will amount in 1,590.2 kg of non-hazardous waste (gravel, quartz) and 964.0 m^3 of wastewater (see Table 1).

The purchase and installation cost of collective ground drinking water quality conditioning system with automated iron removal filters and pumping station for the selected village is 29,000.00 Euro. Annual operational cost, including electric power consumption and water loses, makes 1,136.00 Euro.

10 Conclusion

The results of chemical analysis showed that total iron concentrations in all water wells significantly exceeded the Specific Limit Value ($SLV_{Fe} = 0.2$ mg/L). The indicated average total iron concentrations were 3.3 mg/L. The observed iron concentrations in the tap water were lower by 60 % compared to those in the water wells. The explanations of this phenomenon could be that the bivalent iron ions are oxidized and precipitated in the pipes. Physical–chemical analysis of the other ground drinking water properties (ammonium nitrogen, chlorides, ORP, pH, PI) revealed that the water wells prevail reductive conditions and iron compounds have inorganic origin resulting in faster oxidation of bivalent iron and manganese ions.

The environmental benefits of using collective iron removal system result in formation of almost 70 % less of solid waste and 13 % less of wastewater, and it consumes 97 % less of electric energy compared to the individual iron removal facility at each household. Of course, benefits would be different between the households which use untreated water or purchase drinking bottled water; however, it was not in the scope of this study.

Central iron removal systems for small settlements have evident benefits with reduction of costs. Estimated overall cost, including installation and operational costs, for collective iron removal system make up 1,335 Euro/year per household, and the respective cost for individual household iron removal facility is 390 Euro/year.

The analysis showed that central iron removal system is more beneficial compared to individual household iron removal system. The simplified approach used in this study provides expeditious assessment results and could serve as a model for sustainable groundwater use in small-/medium-scale villages facing high concentrations of iron.

Acknowledgments The study presented in this paper was performed in the frame of the EU Baltic Sea Region Programme 2007–2013 project "From theory and plans to eco-efficient and sustainable practices to improve the status of the Baltic Sea—WATERPRAXIS." The authors are grateful to project partners for insights and valuable comments when writing this paper.

References

Bong-Yeon C (2005) Iron removal using an aerated granular filter. Process Biochem 40:3314–3320

Council Directive 98/83/EC of 3 Nov 1998 on the quality of water for human consumption. The drinking water directive. Official J Eur Communities Ser L 330:32–54. Consolidated version of 07/08/2009

Das B, Hazarika P, Saikia G et al (2007) Removal of iron by groundwater by ash: a systematic study of a traditional method. J Hazard Mater 141:834–841

The Water Framework Directive (2000) Directive 2000/60/EC of the European Parliament and of the Council of 23 Oct 2000 establishing a framework for community action in the field of water policy. Official J Eur Communities Ser L, 327:1–73

Diliunas J, Jurevicius A, Zuzevicius A (2006) Formation of iron compounds in the quaternary groundwater of Lithuania. Hydrogeology 55:66–73

Ellis D, Bouchard C, Lantagne G (2000) Removal of iron and manganese from groundwater by oxidation and microfiltration. Desalination 130:255–264

Genius M, Hatzaki E, Kouromichelaki EM et al (2008) Evaluating consumers' willingness to pay for improved potable water quality and quantity. Water Resour Manag 22:1825–2834

HN 24:2003 (2003) Geriamojo vandens saugos ir kokybes reikalavimai (Lithuanian Hygiene Norm HN 24:2003 drinking water safety and quality requirements). V-455 Ministry of Health of the Republic of Lithuania, Vilnius

ISO 5667-5 2006 Water quality—Sampling. Part 5: guidance on sampling of drinking water from treatment works and piped distribution systems

ISO 5667-11 2009 Water quality—Sampling. Part 11: guidance on sampling of groundwaters

ISO 6332 1988 Water quality—Determination of iron—Spectrometric method using 1,10-phenanthroline

ISO 9297 1998 Water quality—Determination of chloride—Silver nitrate titration with chromate indicator (Mohr's method)

ISO 8467 1993 Water quality—Determination of permanganate index

ISO 7150-1 1984 Determination of ammonium—Part 1: manual spectrometric method

Katsoyiannis IA, Zouboulis AI (2004) Biological treatment of Mn(II) and Fe(II) containing groundwater: kinetic consideration and product characterization. Water Res 38:1922–1932

Katsoyiannis IA, Zikoudib A, Huga SJ (2008) Arsenic removal from groundwaters containing iron, ammonium, manganese and phosphate: a case study from a treatment unit in northern Greece. Desalination 224:330–339

Lindhe A, Rosen L, Norberg T et al (2011) Cost-effectiveness analysis of risk-reduction measures to reach water safety targets. Water Res 45:241–253

Munter R, Ojaste H, Sutt J (2005) Complexed iron removal from groundwater. J Environ Eng 131(7):1014–1020 ASCE

Pezzeta E, Lutman A, Martinuzzi I et al (2011) Iron concentrations in selected groundwater samples from the lower Friulian Plain, northeast Italy: importance of salinity. Environ Earth Sci 62:377–391

Serikov LV, Tropina EA, Shiyan LN et al (2009) Iron oxidation in different types of groundwater of Western Siberia. J Soils Sediments 9:103–110

Shah T (2005) Groundwater and human development: challenges and opportunities in livelihoods and environment. Water Sci Technol 51(8):27–37

Tekerlekopoulou AG, Vasiliadou IA, Vayenas DV (2006) Physico-chemical and biological iron removal from potable water. Biochem Eng J 31:74–83

Vaaramaa K, Lehto J (2003) Removal of metals and anions from drinking water by ion exchange. Desalination 155:157–170

van Halem D, Heijman SGJ, Johnston R et al (2010) Subsurface iron and arsenic removal: low-cost technology for community-based water supply in Bangladesh. Water Sci Technol 62(11):2702–2709

Vasudevan S, Jayaraj J, Lakshmi J, Sozhan G (2009) Removal of iron from drinking water by electrocoagulation: adsorption and kinetics studies. Korean J Chem Eng 26(4):1058–1064

Zekster IS, Everett LG (2004) Groundwater resources of the world and their use. UNESCO, Paris

Authors Biography

Linas Kliučininkas, Viktoras Račys, Inga Radžiūnienė and Dalia Jankūnaitė Authors of the paper represent the Department of Environmental Technology at Kaunas University of Technology, Lithuania. The research group has an extensive record of publications on different aspects of environmental science. A special interest is paid on water supply as well as wastewater treatment issues. The major part of publications is of an applied character and has an interdisciplinary nature. During last decade, the group has initiated and participated in a number of national and international programs and projects supporting environmental and sustainability studies and research in the Baltic Sea Region.

The Contribution of Education for Sustainable Development in Promoting Sustainable Water Use

Gerd Michelsen and Marco Rieckmann

Abstract Education for sustainable development (ESD) aims at enabling people to make a contribution to sustainable development. Its central educational goal is the development of sustainability key competencies. Water is one of the bases of life on earth and so an important topic for ESD. ESD can help to increase the awareness of water issues and to promote the careful use of water resources. The purpose of this article is to demonstrate the relevance of water as a topic for ESD and to show the approaches and methods that can be used in ESD to raise awareness of the importance of natural resources and develop competencies for the sustainable development of our global society. The International Decade "Water for Life" can be seen as a good context for these educational objectives.

Keywords Education · Sustainability key competencies · Sustainable development · Water

1 Introduction

Sustainable development is connected with comprehensive and far-reaching social transformation and fundamental changes in perspective (e.g., regarding humanity's relation with nature). These fundamental re-orientations and changes require a correspondingly far-reaching change in awareness on the part of individuals. This can only take place through learning and so this change of mindset should be systematically initiated and defined as a responsibility of the education system.

G. Michelsen
UNESCO Chair "Higher Education for Sustainable Development", Leuphana University of Lüneburg, Scharnhorststr. 1, 21335 Lüneburg, Germany

M. Rieckmann (✉)
Institute for Social Work, Education and Sport Sciences, University of Vechta, Driverstr. 22, 49377 Vechta, Germany
e-mail: marco.rieckmann@uni-vechta.de

© Springer International Publishing Switzerland 2015
W. Leal Filho and V. Sümer (eds.), *Sustainable Water Use and Management*, Green Energy and Technology, DOI 10.1007/978-3-319-12394-3_6

Education is an essential part of the sustainability processes; its contribution is explicitly called for in Chap. 36 of Agenda 21: "Education is critical for promoting sustainable development and improving the capacity of the people to address environment and development issues" (UN Department of Social and Economic Affairs 1992: 36.3). Education for sustainable development (ESD) is not possible without learning processes (cf. Vare and Scott 2007). Education should create awareness for problems related to sustainability, enable the acquisition of knowledge about these problems, and develop the necessary competencies to address them.

The goal of this article is to demonstrate the relevance of water as a topic for ESD and show the approaches and methods that can be used in ESD to raise awareness for the importance of natural resources and develop competencies for the sustainable development of our global society. Before dealing with these aspects, however, we will first explain in general terms the concept of ESD.

2 Education for Sustainable Development

If education is to meet these demands, then sustainable development must be seen in education as a crosscutting issue. This understanding formed the background in the 1990s as the concept of ESD was first put forward (cf. de Haan and Harenberg 1999).

Since then, a wide variety of efforts have been undertaken to integrate elements of ESD in all—formal, non-formal, and informal—educational sectors (cf. Michelsen 2006). At an international level, the United Nations proclaimed the UN Decade of Education for Sustainable Development (2005–2014) in 2005 (UNESCO 2005; cf. Combes 2009). The United Nations Economic Commission for Europe (UNECE) drew up a strategy for the implementation of ESD (UNECE 2005). In Germany, an important federal initiative involved integrating ESD in school education by means of the programs "Program 21" and "Transfer-21" (de Haan 2006, 2010). In the informal sector, there have also been many activities promoting ESD (cf. Rode et al. 2011).

The concept of ESD combines approaches found in educational programs focusing on the environment and development as well as peace, health, and politics. The contents and goals of each of these approaches are related to each other from the perspective of sustainable development. ESD thus attempts to make a contribution to an understanding of complex interrelationships that cannot be dealt with by environmental or development education alone.

The UNECE has articulated its understanding of ESD in its strategy as follows: "ESD develops and strengthens the capacity of individuals, groups, communities, organizations and countries to make judgments and choices in favor of sustainable development. It can promote a shift in people's mindsets and in so doing enable them to make our world safer, healthier and more prosperous, thereby improving the quality of life. ESD can provide critical reflection and greater awareness and empowerment so that new visions and concepts can be explored and new methods and tools developed" (UNECE 2005: 1). ESD is meant to empower individuals to become involved in sustainable development and critically reflect on their own actions in this effort. This requires individual competencies that learners should be able to acquire in ESD (cf. Barth et al. 2007; de Haan 2010; Rieckmann 2012; Wiek et al. 2011).

In addition to the development of sustainability-related competencies, Stoltenberg (2009) identifies further goals for ESD: alongside orientative knowledge and action-oriented knowledge, future-oriented knowledge plays just an important role as the critical reflection on values that are part of a vision of sustainable development (especially as related to the preservation of natural resources, human dignity, and justice). Furthermore, this involves gaining the experiences and knowledge that one can participate together with others in shaping one's own life and by taking action today can protect future generations.

3 Sustainability Key Competencies as Educational Goal

The increasing complexity, uncertainty, and dynamics of social change place high demands on the individual (cf. Rychen 2004), whether at the workplace or as an engaged volunteer or in everyday life. These changed conditions make it necessary to be able to take creative, autonomous action. Competencies describe what individuals require in order to take action in a variety of complex situations. They are individual dispositions that include cognitive, emotional, volitive, and motivational elements and are developed on the basis of reflecting on practical experiences (cf. for example, Weinert 2001). In contrast to domain-specific competencies, key competencies are seen as multifunctional and transferable competencies that are especially relevant for attaining important societal goals, are important for all individuals, and require a high degree of reflexivity (cf. Rychen 2003; Weinert 2001). Sustainable development can be seen as a normative framework for the selection of such key competencies.

In recent years, a number of different concepts have been developed which define and identify which key competencies should be acquired as an essential part of ESD (cf. for example, Rieckmann 2012; Wiek et al. 2011). In the German context, ESD centers on the concept of *Gestaltungskompetenz* (cf. de Haan 2006, 2010). "Gestaltungskompetenz means the specific capacity to act and solve problems. Those who possess this competence can help, through active participation, to modify and shape the future of society, and to guide its social, economic, technological and ecological changes along the lines of sustainable development" (de Haan 2010: 320). It includes a number of sub-competencies, which have been repeatedly modified and supplemented. There are now twelve sub-competencies (Fig. 1).

The concept of Gestaltungskompetenz is especially characterized by sub-competencies that enable an individual to shape sustainable development in a future-oriented and autonomous way. It emphasizes in particular the fact that sustainable development entails the necessity of fundamental societal changes.

At an international level, OECD in its DeSeCo project ("Definition and Selection of Competencies") has defined key competencies for living in an interdisciplinary and international knowledge society (Rychen and Salganik 2001, 2003). The project aimed at developing a conceptual framework and a theoretical

- To gather knowledge in a spirit of openness to the world, integrating new perspectives
- To think and act in a forward-looking manner
- To acquire knowledge and act in an interdisciplinary manner
- To deal with incomplete and overly complex information
- To co-operate in decision-making processes
- To cope with individual dilemmatic situation of decision-making
- To participate in collective decision-making processes
- To motivate oneself as well as others to become active
- To reflect upon one's own principles and those of others
- To refer to the idea of equity in decision-making and planning actions
- To plan and act autonomously
- To show empathy for and solidarity with the disadvantaged

Fig. 1 Sub-competencies of Gestaltungskompetenz. *Source:* de Haan (2010: 320)

Table 1 DeSeCo key competencies

Functioning in socially heterogeneous groups	Acting autonomously	Using tools interactively
Ability to relate well to others	Ability to act within the big picture	Use language, symbols, and text interactively
Ability to cooperate with others	Ability to form and conduct a life plan and personal projects	Use knowledge and information interactively
Ability to manage and resolve conflict	Ability to defend and assert one's rights, interests, responsibilities, limits and needs	Use technology interactively

Source: Rychen (2003: 85ff)

basis for the definition of key competencies which are crucial for the individual and social development of human beings in modern and complex societies. The DeSeCo framework defines three categories of such key competencies (Table 1).

The German discourse on Gestaltungskompetenz can be compared at an international level to the discussion on sustainability literacy (Parkin et al. 2004) and sustainability competencies (Wiek et al. 2011). The key competencies found in different national discourses are comparable; however, there is often a different ranking of their importance, as can be seen in a comparison between key competencies in Europe and Latin America (Rieckmann 2012).

4 The UN Decade of Education for Sustainable Development

Following the recommendation of the World Summit for Sustainable Development in Johannesburg (2002), the General Assembly of the United Nations proclaimed a World Decade of Education for Sustainable Development for the period 2005–2014 to be coordinated by the United Nations Educational, Scientific and Cultural

Organization (UNESCO) (UNESCO 2005; cf. Combes 2009). Its goal is to develop educational measures to contribute to the implementation of the Agenda 21, which was adopted at the 1992 Rio Summit and then reaffirmed at the 2002 Johannesburg Summit, and anchor the principles of sustainable development in national educational systems around the world. All member nations of the United Nations are called on to develop national and international educational activities that will show pathways to preserving and enhancing the living conditions and chances of survival for both existing and future generations.

The German Commission for UNESCO has taken on a coordinating function in Germany similar to the UNESCO role in the United Nations. In order to implement the UN Decade ESD, the Commission has convened a National Committee, chaired by Prof. Dr. Gerhard de Haan (Free University of Berlin) with 35 experts from universities, culture and media, representatives of the German parliament, the German government, and the Conference of Ministers of Education as well as other notable individuals to publicize and promote the idea of sustainability. The work of the Committee is to bundle the numerous existing initiatives, facilitate the transfer of good practice to the broader community, create a closer network of different actors, strengthen public awareness of ESD, and encourage international cooperation (UNESCO 2004a).

At the beginning of the Decade in 2005, the National Committee submitted a National Action Plan, the main goal of which is to anchor the idea of sustainable development in all educational sectors in Germany (UNESCO 2004b). In order to reach this far-reaching goal, the following four strategic objectives are being pursued (ibid.): developing and bundling activities as well as transferring good practice to the broader community, improving public awareness of ESD, and strengthening international cooperation. In 2008 and 2011s, third revised versions of the National Action Plan were published.

Projects and communities receive awards from the German Commission for outstanding engagement in the area of ESD. These awards have helped make the UN Decade of ESD better known throughout Germany, creating a "map" of project locations showing how ESD is being anchored across the country. At the same time, it provides local support for individual actors of ESD. Over 1,900 projects have already received the award Official Project of the UN Decade of ESD, making them members of the Alliance of Sustainability Learning—just as have the 21 Decade Municipalities. Further contributions to promoting the Decade in public have been the ESD Portal (http://www.bne-portal.de) and the nationwide Days of Action: ESD (http://www.bne-aktionstage.de), which have taken place annually in September since 2008.

5 Water—A Crucial Topic in Education for Sustainable Development

Although the acquisition of competencies is of central importance in ESD, the choice of subject matter for developing these competencies should be carefully made. The topic should be a current one, and it should also be crucial for sustainable, future-oriented development processes, or critical moments in those processes.

De Haan (2002: 16f.) proposes four general criteria for selecting topics for ESD:

1. *Crucial local and/or global topic for sustainable development processes*: The focus should be on the critical debate surrounding the impact, the causes, and possible approaches to solving the global problem. What is important here is that it is possible to create a link between the global problem and an individual's own lived experience. It is also important from an educational point of view that students are able to grasp the reciprocal relationships between local action and global change.
2. *Long-term importance*: As the focus is on the possibilities of shaping the future, ESD should make use of topics that involve persistent challenges. Current topics are also appropriate if they can be shown to have an importance over the long term.
3. *Complexity of knowledge*: Topics should be chosen that already have a complex knowledge base so that it can be emphasized that there is a plurality of approaches to sustainable development.
4. *Potential for action*: It is particularly important that topics have a potential for taking action and so offer specific possibilities for engagement and participation in development processes. The potential for taking action motivates students to take the topic seriously.

Water—as a fundamental condition of life itself—is such a topic. It is of central importance not only at a global level but also at a local level; it has long-term importance, a complex knowledge base, and contains the potential for action—as will be explained below. The problem of water is characterized in general by two aspects: water scarcity and water pollution.

Freshwater is a scarce good in nature. Of all the water on the planet, only 3 % is freshwater, and of that, only 0.3 % is directly available as surface water—most of which, almost 70 %, is contained in the ice cap and in glaciers (Strigel et al. 2010). Human water consumption has risen as a result of increases in population, industry, and agriculture. In particular, with 70 % of worldwide water consumption, agriculture is responsible for the scarcity of freshwater (UNESCO 2012). The availability of water around the world varies greatly. Alongside areas with high precipitation, such as in North America, there are other areas with severe water scarcity, in large parts of Africa and Asia, for example. Approximately 40 % of the world population now lives in the water-poor regions of the world (Simonis 2012). If there is a further increase in population in these parts of the world, then water scarcity will be exacerbated. Since water plays such a crucial role for human life, water scarcity can lead to social discontent and armed conflict (ibid., cf. Menzel 2010).

According to the most recent data of the World Health Organization (WHO) and UNICEF, 884 million people worldwide either have no access or have insufficient access to safe drinking water; and about 2.6 billion people do not have access to toilets or other basic sanitation facilities (WHO and UNICEF 2010).

Alongside scarcity and lack of access to clean water, pollution is a part of the global water problem. The pollution of surface and groundwater causes considerable problems. Although the pollution of bodies of water by industrial and urban wastewater in North America and Western Europe has been reduced considerably, fertilizer and pesticide emissions from agriculture remain a serious problem. "Agrochemicals in particular have had a detrimental impact on water resources throughout the region as nitrogen, phosphorus, and pesticides run into water courses" (UNESCO 2012: 9). Another reason is the contamination of water by pharmaceuticals (e.g., antibiotics) and pathogens from the healthcare system, which also have a feedback effect on the healthcare system (cf. Kümmerer 2009; Schuster et al. 2008; Vollmer 2010).

Contaminated water leads to serious health problems especially in developing countries. Approximately 3.5 million people die every year due to shortages of clean water or from diseases related to contaminated water. Every day, 5,000 children die—roughly 1.8 million every year—from diarrhea and other diseases caused by contaminated water and the lack of sanitation facilities (UNESCO 2012; cf. WHO and UNICEF 2010).

Water is part of many different complex global interrelationships, which are intensified by "global change" (cf. Kaden 2010).[1] For example, the progressive deforestation of the Amazon region is related to global impacts on the hydrological cycle (cf. Simonis 2012). Another example is the interaction between anthropogenic climate change and the local availability of water (cf. Maurer and Moser 2010; Menzel and Matovelle 2010; UNESCO 2012). According to Hoff (2010): 92f), "(...) climate change (will) increase the precipitation variability and further decrease the water supply, particularly in the critical arid regions". These changes will also have impacts on global food security (ibid.). Future patterns of precipitation are difficult to predict since future population levels are uncertain and global climate models only use very large grid spacing (Maurer and Moser 2010).

In addition, the high water use for the production of food products, industrial, and consumer goods for people in industrial and emerging economies contributes to water scarcity in other parts of the world (cf. August 2010). Hoekstra and Mekonnen (2012) showed that one-fifth of global water use in the period 1996–2005 was due not to household consumption but to exports.

In this context, the concepts of "virtual water" and "water footprint" (WF) are significant. "Virtual water is the amount of water needed for the production of food, industrial, and consumption goods" (August 2010: 88). The water footprint "is an indicator that accounts for both the direct and indirect water use of a consumer or producer and shows how much water and where water is used when a product or service is consumed" (ibid.: 88f.). Water footprint data show that water consumption is unequally distributed across countries. "The WF of the global

---

[1] To investigate the consequences of global change for the water cycle, the German Federal Ministry for Education and Research commissioned the research program "Global Change in the Water Cycle" (http://www.glowa.org).

average consumer was 1,385 m³/y. The average consumer in the United States has a WF of 2,842 m³/y, whereas the average citizens in China and India have WFs of 1,071 and 1,089 m³/y, respectively" (Hoekstra and Mekonnen 2012: 3232).

Against this background, the central challenges for sustainable development in the water sector are (Simonis 2012):

- Ensuring safe water and sanitary conditions for everyone
- Safeguarding a sufficient water supply for agriculture and industry
- Promoting effective water management with measures for water conservation and water resource protection
- Improving international cooperation and providing sufficient means for a preventative global water strategy

The relevance of water as a topic of ESD was underlined when the United Nations proclaimed in December 2003 the International Decade for Action "Water for Life" (2005–2015). The goal of the Decade is to promote awareness of water issues among decision makers and the general public around the world and to encourage implementation of commitments that have already been made. In this context, the seventh Millennium Development Goal is of extremely importance, i.e., to "halve, by 2015, the proportion of the population without sustainable access to safe drinking water and basic sanitation" as well as to end non-sustainable forms of water use. This goal was already reached in 2010. "Between 1990 and 2010, over two billion people gained access to improved drinking water sources, such as piped supplies and protected wells" (United Nations 2012: 4). On July 28, 2010, through Resolution 64/292, the General Assembly of the United Nations adopted the right to water in the Universal Declaration of Human Rights.[2]

6 Learning Sustainability with a Focus on Water

ESD contributes to raising awareness about water issues and promotes a preventive approach to water resources (cf. UNESCO 2009). As discussed above, aspects of water that can be addressed in ESD include, for instance, water as a basis of life, water scarcity, water pollution, equitable access to water, water distribution, water as a resource for agriculture and industry, and complex global interrelationships such as climate change, water exports (virtual water, water footprint), and water resource management.

Engaging in serious discussions on these topics not only enables students to learn about water and sustainable development but also to develop

---

[2] Introduced by Bolivia, the resolution received 122 votes in favor and zero votes against, while 41 countries abstained from voting, one of them the United States. The Resolution calls upon States and international organizations to provide financial resources, help capacity-building and technology transfer to help countries, in particular developing countries, to provide safe, clean, accessible and affordable drinking water and sanitation for all.

sustainability-related key competencies. If, for example, the complex interrelationships discussed in the section above become part of the focus of ESD, then students will be able to improve their competency for systems thinking. Discussions about the possible consequences of climate change for precipitation variability can also contribute to developing their competency for thinking and acting in a forward-looking manner and competency in dealing with incomplete and overly complex information. If, for example, the ecological, economic, and health aspects of water and their interrelationships are discussed, then this can contribute to interdisciplinary learning. When learners are confronted with the fact that 92 % of the global water footprint is due to the consumption of agricultural products—22 % of which involves the production of meat (Hoekstra and Mekonnen 2012)—then there is an obvious connection to be made to their own consumption habits, in particular to their nutritional behavior. And there are also possibilities to develop the competency in referring to the idea of equity in decision-making and planning actions. Table 2 shows an overview of examples of possible relationships between important issues concerning water and ESD.

As competencies cannot be taught but only developed by means of practical experiences and subsequent reflection upon them, it is essential to provide space for learners to be able to make their own experiences, try things out, organize things on their own, and face up to challenges. In this context, particularly appropriate educational approaches include independent study, project-oriented learning (in real-life situations), multi-perspective and interdisciplinary thinking and working, as well as developing the capabilities for participation, dialog, and self-reflection (cf. Stoltenberg 2009).

The following three projects received awards in the UN Decade program (cf. http://www.dekade.org/datenbank) and are specific examples illustrating how water can be dealt with as a topic in ESD.

Project "Virtual Water—Hidden in your Shopping Basket" [3]: In spring 2008, the "ideas competition" sponsored by the German Association of Water Protection invited young people above the age of 10 years old to discover and then reveal to others the amount of water hidden in our food and products we use every day. They were also able to do research on the relationship of our lifestyle with water scarcity in other countries and show how a more environmentally aware way of living could contribute to a more responsible use of water. Questions such as "How much water is in the products we use and consume every day, and where do they come from?" or "How much water do we really use if we include virtual water?" are among those discussed in the project. The project thus contributes to raising awareness about such aspects as high levels of water use and water export. Moreover, young people are encouraged to reflect on their own behavior. Some of the competencies that can be developed include competencies in acquiring knowledge and acting in an interdisciplinary manner, in dealing with incomplete and

---

[3] http://virtuelles-wasser.de (18.02.2014).

Table 2 Water issues and education for sustainable development

Water issues	ESD relevance	ESD key competencies
Water as a basis for life (e.g., water as a resource for agriculture and industry)	Retinity principle[a]	Competency in acquiring knowledge and acting in an interdisciplinary manner
Water scarcity/water pollution/water resource management	Managing natural resources, intergenerational justice, orientation toward the future	Competency in thinking and acting in a forward-looking manner
		Competency in referring to the idea of equity in decision-making and planning actions
Access to safe water/water distribution/conflicts over water use	Intergenerational justice	Competency in showing empathy for and solidarity with disadvantaged people
Cultural understanding of water and its use	Cultural perspectives	Competency in showing empathy for and solidarity with disadvantaged people
		Competency in reflecting upon one's own principles and those of others
Complex global interrelationships, e.g., related to climate change	Global networks, global change, complexity	Competency in acquiring knowledge and acting in an interdisciplinary manner
		Competency in dealing with incomplete and overly complex information
Climate change and precipitation variability	Uncertainty, orientation toward the future, intergenerational justice	Competency in thinking and acting in a forward-looking manner
		Competency in dealing with incomplete and overly complex information
Water export (virtual water, water footprint)	Consumption behavior, global networks, complexity, intergenerational justice, managing natural resources	Competency in referring to the idea of equity in decision-making and planning actions
		Competency in acquiring knowledge and acting in an interdisciplinary manner
		Competency in showing empathy for and solidarity with disadvantaged people
		Competency in dealing with incomplete and overly complex information
Interrelationship between, for example, ecological, economic, and health aspects of water	Complexity	Competency in acquiring knowledge and acting in an interdisciplinary manner

[a]In its 1994 Report on the Environment, the German Advisory Council on the Environment emphasized the importance of the key principle of "retinity", that is the holistic linking of all human activities and products with their basis in nature (SRU 1994)

overly complex information, in referring to the idea of equity in decision-making and planning actions, and in showing empathy for and solidarity with disadvantaged people.

Project "Water School"[4]: The goal of the water foundation is to empower people in water-poor regions to secure their own supplies of water. The focus of this mission is the "Water School" project, the first of which was in Eritrea. In a relatively small area around a school, experts, partners, and local people develop a concept for regional water use and supply. School students are the most important disseminators in this concept. Partnerships between project schools in Germany and in developing countries, but also within developing countries, serve to further the exchange of experiences. Some of the competencies that can be developed include competencies in cooperation, in participation, in gathering knowledge in a spirit of openness to the world and integrating new perspectives, and in planning and acting autonomously.

Project "Bread and Fish: Caring for the Baltic"[5]: This project of the Ecumenical Foundation for the Conservation of Creation and Sustainability focuses on environmental ethics and communication in the whole Baltic Sea region. The terms "bread" and "fish" stand for agriculture and fishing. The central instrument of the project is the "Bread + Fish Days", an innovative event that serves to further intercultural exchange and strengthen the relationships among the different regions around the Baltic Sea and the relevant social and political institutions in each country. The goal of the project is thus the creation of a common ethos toward sustainability in the countries of the Baltic Sea region, a deeper understanding of ecological and economic conflicts afflicting the agriculture and fishing industries, as well as providing stimulus for exemplary projects and types of networking. Some of the competencies that can be developed include competencies in gathering knowledge in a spirit of openness to the world, integrating new perspectives, in cooperation, and in reflecting upon one's own principles and those of others.

In order to emphasize the particular importance of water for ESD, the National Committee set up by the German UNESCO Commission for the Decade ESD selected water to be the issue highlighted in 2008. As a contribution to this year's topic, the Commission organized, as part of the nationwide Days of Action of the UN Decade of ESD, a conference on water in ESD in Hanover on September 22, 2008.

And finally, there is a variety of educational and teaching material for deepening the understanding of water as part of ESD. As part of a free service provided to teachers, the Federal Ministry of the Environment provides materials for lessons about "Water in the twenty first Century". The topics "A river is more than water" and "Lifestyle and water" show school students the importance of acting in a preventive and responsible manner with water as a crucial resource in the context of scientific, geographic, and social problems. In addition, six of the workshop

---

[4] http://www.wasserstiftung.de (18.02.2014).

[5] http://www.bread-and-fish.de (18.02.2014).

materials that are part of the federal-state "Program 21" are related to the topic of water: renaturalization of streams and rivers, city park ponds, water, swimming ponds, water project week, and stream sponsorships.

The material "Resources—Use and Waste"[6] introduces school students to the problem of using resources with a number of different sets of issues. There are a number of worksheets on the topic available in the section "Water Belongs to Everybody". In the educational offering "Lifestyle and the global water crisis",[7] the focus is on such topics as virtual water consumption, lifestyle, and consumption behavior—in a 4-h lesson unit that is meant to encourage school students not only to discuss the issue but also to act.

7 Conclusion

ESD aims at empowering people to make a contribution to sustainable development. The central educational objective is to develop sustainability key competencies, such as the competency for thinking and acting in a forward-looking manner and the competency for acquiring knowledge and act in an interdisciplinary manner or participation competency. Within the framework of the UN Decade of ESD (2005–2014), measures have been taken in all areas of the education system to promote sustainable development. In Germany, more than 1,900 projects have been already recognized as official projects of the UN Decade of ESD and are thus members of the Alliance Learning Sustainability.

Water is a fundamental condition for life on earth. Water scarcity and pollution are key challenges of sustainable development. The importance of water and the complex global interrelationships make it an ideal topic for ESD. ESD can help increase awareness of water issues and promote the careful use of water resources. Aspects of the water topic which can be addressed in ESD include, for instance, water as a basis for human life, water scarcity, water pollution, (equitable) access to water, complex global interrelationships with climate change, and water export (virtual water, water footprint). By dealing with these topics, the development of key competencies relevant to sustainability can be promoted, such as the competencies for interdisciplinary knowledge acquisition, in thinking and acting in a forward-looking manner, dealing with incomplete and complex information or referring to the idea of equity in decision-making and planning actions. As many projects, e.g., the Project "Bread and Fish: Caring for the Baltic", and materials indicate the topic of water is suitable to raising awareness of the importance of natural resources and to developing competencies for the sustainable development of our (global) society. The International Decade Water for Life can be seen as a good context for these educational objectives.

---

[6] http://www.institutfutur.de/transfer-21/daten/materialien/tamaki/t2_ressourcen.pdf.

[7] http://www.transfer-21.de/daten/themen/28_E.1.1_Virtuelles%20Wasser_sp.doc.

References

August D (2010) Virtuelles Wasser—Woher stammt das Wasser, das in unseren Lebensmitteln steckt? In: Strigel G, von Eschenbach ADE, Barjenbruch U (eds) Wasser—Grundlage des Lebens. Hydrologie für eine Welt im Wandel. Schweizerbart, Stuttgart, pp 88–90

Barth M, Godemann J, Rieckmann M, Stoltenberg U (2007) Developing key competencies for sustainable development in higher education. Int J Sustain High Educ 8(4):416–430

Combes BPY (2009) The United Nations decade of education for sustainable development (2005–2014): learning to live together sustainably. In: Chalkley B, Haigh MJ, Higgitt D (eds) Education for sustainable development. papers in honour of the United Nations decade of education for sustainable development (2005–2014). United Nations, London, pp 215–219

de Haan G (2002) Die Kernthemen der Bildung für nachhaltige Entwicklung. ZEP—Zeitschrift für internationale Bildungsforschung und Entwicklungspädagogik 25(1):13–20

de Haan G (2006) The BLK '21' programme in Germany: a 'Gestaltungskompetenz'-based model for education for sustainable development. Environ Educ Res 12(1):19–32

de Haan G (2010) The development of ESD-related competencies in supportive institutional frameworks. Int Rev Educ 56(2):315–328

de Haan G, Harenberg D (1999) Gutachten zum Programm Bildung für nachhaltige Entwicklung. Bund-Länder-Kommission für Bildungsplanung und Forschungsförderung, Bonn

Hoekstra A, Mekonnen MM (2012) The water footprint of humanity. PNAS 109(9):3232–3237

Hoff H (2010) Wasser und Nahrungsmittel. Gefährdet Wasserknappheit die Ernährungssicherheit? In: Strigel G, von Eschenbach ADE, Barjenbruch U (eds) Wasser—Grundlage des Lebens. Hydrologie für eine Welt im Wandel. Schweizerbart, Stuttgart, pp 91–96

Kaden S (2010) Hydrologie—vom sektoralen Denken zu komplexen Ansätzen. In: Strigel G, von Eschenbach ADE, Barjenbruch U (eds) Wasser—Grundlage des Lebens. Hydrologie für eine Welt im Wandel. Schweizerbart, Stuttgart, pp 43–49

Kümmerer K (2009) The presence of pharmaceuticals in the environment due to human use—present knowledge and future challenges. J Environ Manage 90:2354–2366

Maurer T, Moser H (2010) Klimawandel und Wasser. Auswirkungen der Erderwärmung auf den Wasserhaushalt. In: Strigel G, von Eschenbach ADE, Barjenbruch U (eds) Wasser—Grundlage des Lebens. Hydrologie für eine Welt im Wandel. Schweizerbart, Stuttgart, pp 104–111

Menzel L (2010) Globale Entwicklung—Wasser als limitierender Entwicklungsfaktor. In: Strigel G, von Eschenbach ADE, Barjenbruch U (eds) Wasser—Grundlage des Lebens. Hydrologie für eine Welt im Wandel. Schweizerbart, Stuttgart, pp 82–88

Menzel L, Matovelle A (2010) Current state and future development of blue water availability and blue water demand: a view at seven case studies. J Hydrol 384:245–263

Michelsen G (2006) Bildung für nachhaltige Entwicklung. Meilensteine auf einem langen Weg. In: Tiemeyer E, Wilbers K (eds) Berufliche Bildung für nachhaltiges Wirtschaften. Konzepte—Curricula—Methoden—Beispiele. Bertelsmann, Bielefeld, pp 17–32

Parkin S, Johnston A, Buckland H, Brookes F, White E (2004) Learning and skills for sustainable development. Developing a sustainability literate society. Guidance for Higher Education Institutions. London. https://www.upc.edu/sostenible2015/documents/la-formacio/learningandskills.pdf. Retrieved 10 Feb 2014

Rieckmann M (2012) Future-oriented higher education: which key competencies should be fostered through university teaching and learning? Futures 44(2):127–135

Rode H, Wendler M, Michelsen G (2011) Bildung für Nachhaltige Entwicklung (BNE) in außerschulischen Einrichtungen, Wesentliche Ergebnisse einer bundesweiten empirischen Studie. Leuphana Universität Lüneburg, Lüneburg

Rychen DS (2003) Key competencies: meeting important challenges in life. In: Rychen DS, Salganik LH (eds) Key competencies for a successful life and well-functioning society. Hogrefe, Cambridge, pp 63–107

Rychen DS (2004) Key competencies for all: an overarching conceptual frame of reference. In: Rychen DS, Tiana A (eds) Developing key competencies in education: some lessons from international and national experience. UNESCO, International Bureau of Education, Paris, pp 5–34

Rychen DS, Salganik LH (eds) (2001) Defining and selecting key competencies. Hogrefe, Seattle

Rychen DS, Salganik LH (eds) (2003) Key competencies for a successful life and well-functioning society. Hogrefe, Cambridge

Schuster A, Hädrich C, Kümmerer K (2008) Flows of active pharmaceutical ingredients originating from health care practices on a local, regional, and nationwide level in Germany—is hospital effluent treatment an effective approach for risk reduction? Water Air Soil Poll Focus 8:457–471

Simonis U (2012). Wasser. Lokal eine Freude—global ein Problem. http://www.deutscheumwelts tiftung.de/index.php?option=com_phocadownload&view=file&id=74:udo-e-simonis-wasse r&Itemid=229. Retrieved 10 Feb 2014

SRU—Rat von Sachverständigen für Umweltfragen (1994) Umweltgutachten 1994. Für eine dauerhaft-umweltgerechte Entwicklung. Metzler-Poeschel, Stuttgart

Stoltenberg U (2009) Mensch und Wald. Theorie und Praxis einer Bildung für nachhaltige Entwicklung am Beispiel des Themenfeldes Wald. ökom, München

Strigel G, von Eschenbach ADE, Barjenbruch U (2010) Hydrologische Tatsachen—was untersuchen Hydrologen? In: Strigel G, von Eschenbach ADE, Barjenbruch U (eds) Wasser—Grundlage des Lebens. Hydrologie für eine Welt im Wandel. Schweizerbart, Stuttgart, pp 7–11

UN Department of Social and Economic Affairs (1992) Agenda 21. https://www.un.org/esa/dsd/agenda21/index.shtml. Retrieved 10 Feb 2014

UNECE—United Nations Economic Commission for Europe (2005) UNECE strategy for education for sustainable development. http://www.unece.org/env/documents/2005/cep/ac.13/cep.a c.13.2005.3.rev.1.e.pdf. Retrieved 10 Feb 2014

UNESCO—United Nations Educational, Scientific and Cultural Organization (2004a) unesco aktuell: "Allianz Nachhaltigkeit Lernen" stellt sich am 2. November in Berlin vor, 51/04

UNESCO—United Nations Educational, Scientific and Cultural Organization (2004b) unesco heute online: DUK koordiniert VN-Dekade "Bildung für nachhaltige Entwicklung in Deutschland". November/December 2004

UNESCO—United Nations Educational, Scientific and Cultural Organization (2005) United Nations decade of education for sustainable development (2005–2014): international implementation scheme. UNESCO, Paris

UNESCO—United Nations Educational, Scientific and Cultural Organization (2009) Water education for sustainable development. http://unesdoc.unesco.org/images/0018/001853/185302e. pdf. Retrieved 10 Feb 2014

UNESCO—United Nations Educational, Scientific and Cultural Organization (2012) World water assessment programme. The 4th United Nations world water development report: managing water under uncertainty and risk, vol 1. UNESCO, Paris

UNITED Nations (2012) The millennium developments goals report 2012. http://mdgs. un.org/unsd/mdg/Resources/Static/Products/Progress2012/English2012.pdf. Retrieved 10 Feb 2014

Vare P, Scott W (2007) Learning for a change: exploring the relationship between education and sustainable development. J Educ Sustain Dev 1(2):191–198

Vollmer G (2010) Disposal of phamaceutical wastes in households—a European survey. In: Kümmerer K, Hempel M (eds) Green and sustainable pharmacy. Springer, Heidelberg, pp 165–174

Weinert FE (2001) Concept of competence: a conceptual clarification. In: Rychen DS, Salganik LH (eds) Defining and selecting key competencies. Hogrefe, Seattle, pp 45–65

WHO—World Health Organization, UNICEF—United Nations Children's Fund (2010) Progress on sanitation and drinking water: 2010 update. WHO, UNICEF, Geneva

Wiek A, Withycombe L, Redman CL (2011) Key competencies in sustainability: a reference framework for academic program development. Sustain Sci 6(2):203–218

Authors Biography

Prof. Dr. Gerd Michelsen is Holder of the UNESCO Chair of Higher Education for Sustainable Development at the Leuphana University Lüneburg, Germany. His major research and teaching interests are as follows: education and sustainability, higher education for sustainable development, and sustainability communication.

Prof. Dr. Marco Rieckmann is Assistant Professor of University Teaching and Learning at the Institute for Social Work, Education and Sport Sciences of the University of Vechta, Germany. His major research and teaching interests are as follows: university teaching and learning, competence development, and (higher) education for sustainable development.

Water Security Problems in Asia and Longer Term Implications for Australia

Gurudeo A. Tularam and Kadari K. Murali

Abstract This paper reports on water security issues in Asia that has long-term security implications for Australia. Asia's water problems are severe with one in five people not having access to safe drinking water. Water security is defined as the availability of an acceptable quantity and quality of water for health, livelihoods, ecosystems and production, coupled with an acceptable level of water-related risks to people, environments and economies. It is a function of access to adequate quantities and acceptable quality, for human and environmental users. This analysis shows many Asian countries will face greater challenges than present from population explosion, shifts of populations from rural to urban areas, pollution of water resources and over-abstraction of groundwater. These challenges will be compounded by the effects of climate change over the next 50 years. It is then necessary to mobilise technologies, techniques, skills and research to aid security issues in Asia now. Otherwise, population growth, rapid urbanisation and climate change issues will worsen placing strong demands on water resources, thus creating water refugees, and this will affect countries close to Asia such as Australia. Reducing water's destructive potential and increasing its productive potential is a central challenge and goal for the sake of future generations in Asia and Australia.

Keywords Water · Water security · Asian water security · Water threats · Governance of water · Climate change · Water pricing

G.A. Tularam (✉) · K.K. Murali
Environmental Futures Research Institute, Science Environment Engineering and Technology [ENV], Griffith University Brisbane, Brisbane, Australia
e-mail: a.tularam@griffith.edu.au

© Springer International Publishing Switzerland 2015
W. Leal Filho and V. Sümer (eds.), *Sustainable Water Use and Management*,
Green Energy and Technology, DOI 10.1007/978-3-319-12394-3_7

119

1 Introduction

Water is crucial to all life on earth, economical activities, and environmental and agricultural systems (Sadoff and Muller 2009; Tularam and Ilahee 2007; Tularam 2010). People require water to drink, produce food items and conduct other real-life activities (Beppu 2007; Tularam and Singh 2009; Tularam 2012; Tularam and Keeler 2006). It is one of the most fundamental requirements for survival of all life on earth (George et al. 2008) in that all earthly systems depend on the reliability and quality of water (Beppu 2007; Indraratna et al. 2001; Tularam and Ilahee 2007). The effects of global warming are drying out many regions making clean water supply precious (Jones 2009; Renner 2010); yet, interestingly, Biswas and Seetharam (2008) argued a fortunate position exits, in that a drier and more crowded world could still have enough water to meet the needs, if water supply is regulated nationally and internationally.

Water security can be defined as "the reliable availability of an acceptable quantity and quality of water for health, livelihoods and production" (Sadoff and Muller 2009, p. 11). History shows that water problems have wiped out civilisations such as Mohenjo in India and others in the past (Tularam 2012). Indeed, there have been some boarder conflicts around the world arising because of water (Beppu 2007; Tellis 2008; Tindall and Campbell 2010). A mild border conflict of water rights and pumping from rivers presently exist in Australia. The potential for future conflicts arising from water security issues is real, and as such, it is important to consider water security concerns in Asia and examine possible consequences and implications for Australia (Wouters 2010; Tularam and Properjohn 2011).

Growing populations and economies including some fundamental changes in the hydrologic cycle have together increased demands on global water supply threatening biodiversity, food production and other everyday needs (Barnett 2003; Renner 2010). Increasing demands of water has led to water shortages, with more than one billion people being without adequate drinking water even when water is essential for maintaining adequate food supply and productive local environments (Smith and Gross 1999). The global water stress situation particularly in many parts of Africa and Asia demands higher level planning and strategies if we are to overcome major future water crises (Guthrie 2010; Smith and Gross 1999). In recent times, water security is gaining attention but governments need to act quickly as water is a strategic resource that can help achieve sustainable growth and progress (Sadoff and Muller 2009; Tookey 2007).

This paper specifically addresses water issues in countries close to Australia such as those in Asia. The critical analysis is conducted to consider reasons why flow of persons (so-called water refugees) may occur from Asia towards greener pastures such as Australian shores, thus considering longer term implications for Australian boarders. This is an issue that has had very little attention in the past but requires important strategic visionary planning for sustaining harmonious relationships with our neighbouring countries in future.

1.1 Water Security

The concept of water security is framed in terms of seven interconnected concerns that are relevant to water allocation in order to evaluate the overview of water security framework (Briscoe 2009; De Loe et al. 2007):

- Ecosystem protection;
- Economic production;
- Equity and participation;
- Integration of resource;
- Water conservation;
- Climate variability and change; and
- Trans-boundary sensitivity.

As noted, water security is a multi-dimensional concept that recognises that sufficient good-quality water is needed for social, economic and cultural uses as a whole. It seems that an effective, efficient and equitable water allocation system is critical in achieving water security (De Loe et al. 2007; Pigram 2006; Renner 2010). Essentially, water security involves protection of vulnerable water systems, protection against water-related hazards, sustainable development of water resources and actively safeguarding access to water functions and services (Biswas and Seetharam 2008; De Loe et al. 2007; Raj 2010). Adequate water is required to sustain human activities and enhance ecosystem functions as well as maintain border sensitiveness and security where rivers, for example, pass through boarders.

Figure 1 shows the Asiatic region and the position of Australia of in it. The high level of population growth in Asia compared with the rest of the world is one of the main reasons that water security could become a great global challenge not only for Asia but also for Australia (Wouters 2010). For example, the population increase in India and China has contributed to doubling of irrigated areas and tripling of water withdrawals from groundwater in the region (Raj 2010; Wouters 2010).

Figure 1 shows populated water stress Asian countries close to our region such as India and Pakistan, Nepal, China and Indonesia, for example. The impacts of climate change such as rising sea levels, extreme flooding and droughts and decline in agricultural productivity are occurring in Asian countries (Nuttall 2005; Vorosmarty et al. 2010). There already exists some tension within Asia concerning access to water, public health and global economic growth (Tellis 2008; Tindall and Campbell 2010). Therefore, it is important to critically analyse water resources, accessibility, capacity, quality and use in Asia (Biswas and Seetharam 2008; De Loe et al. 2007) if only to identify gaps in water security, and to address them to aid water security of the region (Briscoe 2009; Tellis et al. 2008).

The main aim of this paper was to identify water security issues in Asian countries to consider possible longer term implications for the Australian shores by "water refugees". This is done by examining critically the water security issues of populated and water-stressed countries in Asia. Particular attention is paid to populated Asian countries that experiencing serious domestic water challenges and are

Fig. 1 Asian countries close to Australian borders (adapted from BugBog 2001) (http://www.bu
gbog.com/maps/asia/asia_map.html)

in proximity of Australia. Today, boat migrants are increasing daily but they are
not water refugees, yet the complexities in the region need a closer examination in
terms of water security problems to comprehend underlying issues that may cause
flow of water refugees. A background of water is presented next that is followed
by country-by-country analysis in terms of water stress. A critical analysis of the
issues related to water security follows factors such as causal factors and solutions
and their concerns, use of technology and so on. Some important questions are
then posed, and possible solutions are explained. This is followed by the conclusion including some implications for water security in Asia and Australia.

2 Background

More than 97.5 % of the world's water is salt water in the oceans and seas, leaving 2.5 % as freshwater and much of it is contained in glaciers, deep aquifers or
soil moisture (Fig. 2; UNEP 2010). Water is not uniformly distributed throughout
the world (Nettle and Lamb 2010; UNEP 2010; Tularam and Keeler 2006), and
as such, accessibility to water cannot be separated from notions of human rights
(Pimentel et al. 2004; Renner 2010; Smith and Gross 1999). The security of supply for the next 50 years appears to be critical in overcoming water security problems in the region.

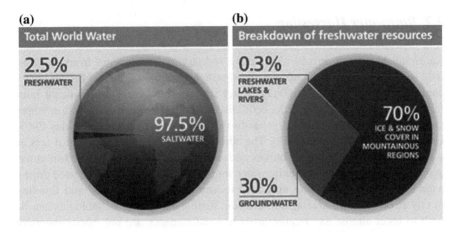

Fig. 2 a, b Total amount of world water (UNEP 2010)

Although dams constructed in water catchments, and rivers and streams are the main source of fresh potable water in most countries (Indraratna et al. 2001; Qadir et al. 2007), there has been a significant drop in water stored due to the lack of rainfall in many Asian countries over the past decades. Such countries then need to rely on alternative water resources to avoid restrictions and service interruptions to domestic water supply. Where the population density is high and water demand is increasing, desalination, rainwater harvesting, groundwater and surface water have all been considered (Aditi 2005; Guthrie 2010; Pigram 2006), but there are advantages and disadvantages in each case.

2.1 Desalination

Desalination is a process that converts sea water or highly brackish groundwater into good-quality freshwater and used where rain water, storm water supply or the supply of sea water remains abundant (Tularam and Ilahee 2007). In water-scarce environments, potable water can be obtained from desalination plants. There are some advantages such as low whole of life cycle cost and highly reliable water supply source (Pigram 2006; Raj 2010; Saliby et al. 2009). But equally, there are a number of disadvantages for it requires significant amount of energy primarily heat and electricity that can be expensive. There are negative environmental impacts such as air pollution mainly due to greenhouse gas emissions and sea water pollution caused by the production of highly concentrated brine solution (Pigram 2006; Raj 2010; Saliby et al. 2009; Tularam and Ilahee 2007).

2.2 Rainwater Harvesting

Rainwater harvesting involves collecting and storing of rainwater from rooftops, land surfaces or rock catchments using simple techniques (jars and pots) to highly engineered ones. It is an important water source where there is significant rainfall but the process usually lacks centralised supply system. It is an option where good-quality surface water or groundwater is lacking (Barron and Salas 2009; Qadir et al. 2007; Rezwan 2013). The advantages include the following: it acts as a supplement to other water sources and utility systems; it can be used in emergency or breakdown of the public water supply systems; and it can be used during natural disasters. Harvesting may help reduce storm drainage load and flooding in city streets, often reducing soil erosion; finally, harvesting technologies can be built to meet almost any requirements and maintenance are not labour intensive (Barron and Salas 2009; Qadir et al. 2007; Rezwan 2013). Harvesting has disadvantages such as supply contamination by bird/animal droppings on catchment surfaces and guttering structures; requires constant maintenance and cleaning/flushing before use; contamination of water by algal growth; and finally invasion by insects, lizards and rodents—thus becoming possible breeding grounds for disease vectors (Barron and Salas 2009; Qadir et al. 2007).

Water can also be stored in an aquifer for use during dry periods and for storage reservoir during wet periods. The use of surface supplies is encouraged when surface water availability is plentiful especially in winter months and during wet years (Sadoff and Muller 2009; Tularam and Krishna 2009). As such, pumping of aquifers should be comparatively less during these wet periods, allowing them to refill naturally or through replenishment efforts such as aquifer storage and recovery (ASR). When stream flows are less (during dry periods), groundwater can be tapped to meet irrigation or urban demands. Thus, surface water can be used for low-demand periods, while groundwater maybe pumped at other times (Aditi 2005; Tularam and Krishna 2009).

Such conjunctive water management plays an important role in addressing sustainable water resource management issues. However, it is important to recognise both the strengths and the limitations of the water management in terms of water security (Aditi 2005; Aswathanarayana 2007). The advantages include the following: improved security of water access; flexibility of switching between more than one water source according to relative availability; capture and conserving of surplus water supplies when available; and delivery of water management and environmental targets (MacKenzie 2009). The main disadvantages are as follows: evaporation, sedimentation, environmental impact of surface reservoirs; flooding of agricultural land; and distribution of water from the reservoir is expensive (Aditi 2005; Renner 2010).

Figures 3 and 4 show groundwater withdrawal and consumption highlighting Asia and India as the greatest user of groundwater. Although the groundwater availability is high, the largest population on earth also lives in this region, thus placing much pressure on groundwater levels.

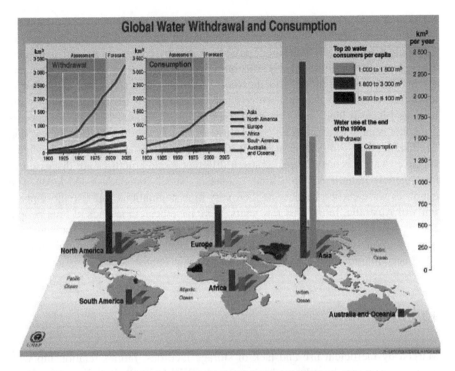

Fig. 3 Groundwater withdrawal and consumption (MacKenzie 2009)

Fig. 4 Freshwater resources
(Groundwater in km^3;
MacKenzie 2009)

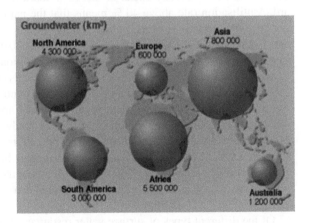

3 Water Stress in Asia: Country-by-Country Analysis

As noted earlier, water security implies affordable access to clean water for agricultural, industrial and household usage, and access for such use is an important part of human security (Grey and Sadoff 2007). Increasing populations, poverty, industrialisation, urbanisation and negative effects of climate change together with inefficient water use have all added to water security problems in Asia. Increase in settlements, grazing and deforestation in Asian mountains appears to have all adversely affected even the rainfall patterns in nearby countries (Aswathanarayana 2007; Jones 2009; Nuttal 2005). Unsupervised withdrawal of water from the rivers or lakes solves problems in the short term but when rivers run dry, ecosystems such as wildlife habitats are critically affected, thus degrading not only the habitats but also land use. Therefore, monitoring rivers and lake waters appears critical if Asia is to avoid problems with water security in the future (Biswas and Seetharam 2008; Pimentel et al. 2004; Xia et al. 2007). To develop a balanced perspective, water security issues of seven different Asian countries are discussed in turn, namely India, Pakistan, China, Indonesia, Nepal, Bangladesh and Japan. The countries were chosen because of their very high populations and also close proximity of Australia. In the past few years, there have been thousands of boat people arrivals and most of them considered refugees. The majority are from the countries selected of analysis excepting Japan. Japan is a developed country; it will provide a balanced picture for it also has a number of water security issues.

3.1 India

Marked by inefficient use and lack of storage facilities, India's relationship with its water resources has always been unsteady (Walsh 2009). India's water utilisation rate is 59 %, already ahead of the 40 % mark that has been set as the standard. A utilisation rate above 40 % means that the natural mechanisms in place do not have the capacity to recharge adequately. That is, water is being used at a rate that is unsustainable (Aswathanarayana 2007; Raj 2010). Water security for India implies effective responses to changing water conditions in terms of quality, quantity and uneven distribution (George et al. 2008; Tellis et al. 2008).

India's water resources are a combination of groundwater resources and surface water resources. However, surface water resources are present in the country in much greater volume when compared to the groundwater resources (Qadir et al. 2007). While rivers form the means of support of most of the cities, towns and villages across the country, groundwater is vital to India's people (Tularam and Krishna 2009). As majority of the rivers in the country are not perennial, groundwater sustains much of the population during lean months (Aditi 2005; Walsh 2009).

Of the different types of surface water resources, rivers constitute the most valuable and huge part. The majority of India's rivers are rain-fed with the exception of those originating in the Himalayas. The Himalayan rivers are perennial owing to the glacier melt that feeds India's fortunes throughout the year. While other rivers in the country are seasonal in nature, due to their dependence on rainfall, the Himalayan rivers flow all year round (Aditi 2005; Shrestha 2009). It seems that Ganges and the Brahmaputra are the most important rivers in the country (Biswas and Seetharam 2008; Jones 2009; Pigram 2006).

3.2 Pakistan

Pakistan problems are initially related to overgrazing of their lands along road corridors in dry regions that have resulted in erosion and landslides as well as dust storms. Such erosion leads to degradation of land resources having a severe impact on agricultural land. The erosion in turn affects the available water resources including groundwater, thus causing water security problems (Grey and Sadoff 2007; Pimentel et al. 2004).

There have been long-term problems in the irrigation sector in Pakistan, which has been further complicated by trends in the upper reaches of the Indus River Basin. The deterioration of infrastructure has led to much seepage from irrigation canals, suggesting only 36 % of the water drawn for irrigation reaches crops. This loss of water makes the system highly inefficient, requiring large quantities of water to be withdrawn in order to grow crops (De Loe and Bjornlund 2008). An estimated 40 % of irrigated land has been affected by poor maintenance and naturally poor drainage, which has led to waterlogging and increased salinity of irrigation water (Sadoff and Muller 2009).

Human population pressures together with unchecked development that facilitates logging, for example, have impacted upon the biodiversity and the ability of water sheds to handle monsoon floods (Rezwan 2013). Some predictions state around 80 % of the productive land may become severely impacted by developments that are deemed to occur by 2030 (Nuttall 2005). Also, the water levels of its rivers in Pakistan may decrease by 30–40 % in future (Funabashi and Shimbun 2010).

In the dry, mountainous region of the upper basin, population growth appears to have caused rapid depletion of groundwater and deforestation that are both linked to more sediment downstream diminishing the quality of water stored in downstream reservoirs and irrigation canals. Sediment build-up has also reduced Indus River Basin reservoir storage capacity by about 20 % (Asia Society 2009).

3.3 China

China's massive population is expected to grow from 1.33 billion in 2010 to 1.42 billion by 2050, causing significant water resource challenges. The country's diverse landscape and large landmass make water problems distinct from the arid north to the more water-rich south (Briscoe 2009). Northern China is characterised primarily by desert and grasslands, but it is experiencing population growth that is accelerating the exploitation of scarce water resources. Severe desertification is further eating up land that once was used for agricultural production, thus choking the more heavily relied upon rivers (Xia et al. 2005). The North China Plain is home to almost 40 % of China's cultivated land area, but 40 % of the population holds only 7.6 % of the country's water resources. Yet agricultural and industrial water demands are growing by more than 10 % a year and are expected to increase by 40 % by 2020 (Xia et al. 2007). Officials have failed to curtail industrial dumping and sewage discharge into the plain's three major river basins, namely the Huai, Hai and Huang (Yellow) Rivers, severely polluting these rivers.

Water shortage in North China is a major issue facing the country due to over-committed water resources. These include drying up of rivers, decline in groundwater levels and degradation of lakes and wetlands and water pollution (Xia et al. 2005). Its water scarcity is characterised by insufficient local water resources as well as reduced water quality due to increasing pollution both of which have caused serious impacts on society and the environment. Thus, the problem of water shortage in North China has become the significant limiting factor affecting sustainable development (Jianyun et al. 2009; Renner 2010; Xia et al. 2007).

Moreover, climate change effects have increased aridity in drought seasons and will further deteriorate the short water supply in the northern and north-western parts of China (Ahmed 2009). The catchment management authorities (CMA's) statistics show the temperature in China has risen 0.4–0.5 °C during the past century, slightly lower than the global average increase of 0.6° (Henebry and Lioubimtseva 2009).

In southern China, meltwater from the Himalayan glaciers bordering the Tibetan Plateau feeds some of Asia's greatest water sources including the Yangtze, Yellow, Ganges, Brahmaputra and Mekong rivers. However, the quality and quantity of water from these rivers are threatened by pollution and overwithdrawal including the effects of climate change. Climate change is projected to decrease China's glacial coverage by 27 % by 2050, seriously diminishing water availability for communities throughout south-eastern China (Henebry and Lioubimtseva 2009). South-eastern China is a major site of global manufacturing, as well as agricultural and industrial pollution of its water. The pollution restricts access to good-quality water and creates threats to human health and fisheries, which has in past resulted in bans on Chinese fish exports (Xia et al. 2005).

The dense coastal population of south-east China is particularly vulnerable to forecasted sea-level rise. In north-west China, water access is becoming more salient as domestic, agricultural, and industrial usage. However, the rising sea level has caused salinity changes in the freshwater rivers that feed the ocean, affecting water access and quality, which could cause drying of rivers during non-rainy seasons. If urban infrastructure proves incapable of handling the encroaching sea, the epicentre of China's manufacturing region could be destroyed (Jianyun et al. 2009). As a whole, China is facing increasing water shortages as well as experiencing water resource overexploitation. The low water quality in many parts of China has the potential for serious environmental and socio-economic impacts (Tellis et al. 2008).

Table 1 shows that water demand in China has been rapidly increased compared to 1995 and 2030. These figures indicate that water consumption is going to increase in residential, industrial and agricultural sectors (Tellis et al. 2008). Meteorologists estimate that the western regions of China will lack about 20 billion cubic metres of water from 2010 to 2030, and in 2050, the regions would still

Table 1 Projected water demand in China (1995–2030 in billion tons)

	1995	2030
User type		
Residential	31	134
Industrial	52	269
Agricultural	400	665
Total	483	1,068

need 10 billion more cubic metres of water (Asia Society 2009). China will face a tougher challenge in its water security as global warming will further increase the evaporation of its seven major valleys, of which the annual natural run-off has kept falling as a whole during recent years (Moore 2009). Thus, responsibility needs to be taken to implement major water security measures to protect China's water supply and quality.

3.4 Indonesia

Although Indonesia enjoys around 21 % of the total freshwater available in the Asia Pacific region, many of the country's water security issues are tied to its rapid development, poor urban infrastructure and stretched institutional capacity (Arnell 2004; Bates 2008). The economic growth has not been accompanied by a corresponding expansion of infrastructure and institutional capacity. As a result, nearly one out of two Indonesians lacks access to safe water and more than 70 % of the nation's 220 million people rely on potentially contaminated sources (Wouters 2010). Significant land-use changes and a high level of deforestation have left many areas more vulnerable to extreme events such as monsoon floods (Jones 2009).

Urbanisation and economic development has made Indonesia a pollution hot spot (Asia Society 2009). Waste streams are evident due to growing industrial, domestic and agriculture sectors. Extractive industries account for much of the development, and waste from industrial and commercial processes has made its way into both surface water and groundwater supplies (Illangasekare et al. 2009). Around 53 % of Indonesians obtain their water from sources that are contaminated by raw sewage for people living in urban slums lack wastewater treatment tools, and the basic sanitation infrastructure necessary to prevent human excrement from contaminating water supplies is virtually non-existent, thus greatly increasing human susceptibility to water-related diseases (Jones et al. 2007).

Large barren hillside areas and underlying soils, which are subjected to heavy precipitation, greatly increase the likelihood and severity of floods and landslides. When flooding occurs, urban infrastructure is quickly overwhelmed, leading to sewage spillover. The post-event clean up and repair costs can be immense, and thus, managing water scarcity is a critical challenge for Indonesia and for surrounding Southeast Asian Nations with similar climates (Barnett 2003; Renner 2010; Sadoff and Muller 2009).

Located along the equator, Indonesia is surrounded by warm waters that create relatively stable year-round temperatures and monsoons drive seasonal variations. Yet climate change threatens to disrupt the regular, alternating periods of rain and arid dryness. The dry season may become more arid, driving water demand, while the rainy season may condense higher precipitation levels into shorter periods, increasing the possibility of heavy flooding. This results in decreasing the ability to capture and store water (Benjamin et al. 2006).

Vulnerability to extreme events and water quality in the capital regions of Indonesia has deteriorated sharply because of sea water intrusion. Pollution and compromised sanitary conditions in much of the country may lead to epidemics and severe health problems, testing institutional capacities. The enormous challenge of environmental degradation directly feeds into many of Indonesia's water security problems (Guthrie 2010).

3.5 Nepal

Nepal lies in the middle of the Ganges and the Brahmaputra (South Asia's major river systems) and is one of the countries with the highest level of water resources. For many Nepalese who live in the hills, the water flowing in the large valleys below is out of reach (Asia Society 2009). Only half of all farmland is irrigated, and more than a third of the population has difficulties in obtaining water in this country (Renner 2010).

Shrestha (2009) noted that many of the rivers in the region have already been affected by deforestation and increased use of water for irrigation, which has been fuelled by existing infrastructure developments. Large areas of pristine wildlife habitats have been laid bare because the river has run dry as a result of demand for water for irrigation by growing settlements. These rivers are the lifeline for the people, but seasonal scarcity of water is an increasing problem as are floods, as a result of land-use changes such as deforestation and intensive agriculture (Henebry and Lioubimtseva 2009; Nuttall 2005).

The uncontrolled dumping of wastes into flowing streams has turned the Himalayan waters into giant sewers. It is said that 80 % of the country's illness is due to contaminated water. Every year, many children die from the waterborne diseases such as dysentery, hepatitis and even cholera which are very common throughout the country. In recent years, another danger has been added to the list of the water pollutants say arsenic (Prasai 2007).

Nepal faces acute shortage of water and remains one of the poorest countries in the world. Nepal has the poorest drinking water and sanitation coverage for its population in South Asia, and a large percentage of its drinking water contains faecal coli forms. Waterborne disease is transmitted through contaminated water, and several bacterial, protozoal and viral waterborne diseases have posed serious public health problem in Nepal. According to the Nepal Country Environmental Analysis, diarrhoea, intestinal worms, gastritis and jaundice are the top five waterborne diseases in Nepal (Prasai 2007).

3.6 Bangladesh

Bangladesh is recognised worldwide as one of the most vulnerable countries to the impacts of global warming and climate change. This is due to its unique geographic location, dominance of floodplains, low elevation from the sea, high population density, high levels of poverty and overwhelming dependence on nature, its resources and services (Arnell 2004; Nishat 2008). Water scarcity of drinking water in Bangladesh is mainly due to reduced precipitation, prolonged dry season and droughts. The available freshwater resources are contaminated with saline water in the coastal aquifer (Beppu 2007; Renner 2010; Sadoff and Muller 2009).

Already challenged by a confounding physical landscape, Bangladesh is faced with unique obstacles to growth and stability because of rising population, urbanisation and poverty (Roberts and Kanaley 2006). According to the United Nations Development Program, four out of five Bangladeshis live below the poverty line (less than US$2/day), and one out of three lives in extreme poverty (less than US$1/day). Despite improvements in health, mortality and poverty rates, significant portions of the population lack access to clean drinking water and sanitation. Clearly, the accelerating urbanisation of Bangladesh will perpetuate the water security challenges (Biswas and Seetharam 2008).

3.7 Japan

Although a water-rich country, Japan is a large importer of mineral water. Japanese culture and society were shaped in a diverse natural environment with a climate ranging from subtropical to subarctic, and Japanese cities are experiencing severe ground subsidence issues due to excessive groundwater abstraction during that period of rapid economic growth. Overabstraction results primarily from population growth and urbanisation (Funabashi and Shimbun 2010). The overabstraction of groundwater has become the major concern in water security (Tularam and Krishna 2009).

Japan depends on imports of many goods. The quantity of water that is necessary for the production of the food that Japan imports is said to be the equivalent of tens of billions of cubic metres of water per year. The processing of grains and meat imported by Japan requires vast quantities of water is as about the same amount as that is being used by 1.3 billion people in developing countries (Asia Society 2009).

Japan is also a high-risk nation in terms of water, vulnerable to drought and flooding, because it lies in the monsoon climate region (Barnett 2003). Japan failed to control water levels due to increased population, and also, water quality is deteriorated due to lack of treatment techniques. Failure to efficient water usage, reuse of sewage treated water has shown impact on use of water resources and thus leads to impact on ecosystems (Jones 2009).

Table 2 Population size and density of Asian countries selected

Country name	Population	Area (km^2)	Population density (person/km^2)
India	1,184,639,000	3,287,590.00	360.34
Pakistan	170,260,000	803,940.00	211.78
China	1,339,190,000	9,596,960.00	139.54
Indonesia	234,181,400	1,919,440.00	122.01
Nepal	29,853,000	140,800.00	212.02
Bangladesh	164,425,000	144,000.00	1,141.84
Japan	127,380,000	377,835.00	337.13

Table 3 Annual freshwater withdrawal for agriculture (%)

Country	Agriculture % of total freshwater withdrawal
India	86.46
Pakistan	96.02
China	67.72
Indonesia	91.33
Nepal	96.46
Bangladesh	96.16
Japan	62.46

3.8 Analysis: Population, Size, Density and Water Withdrawal in Asia

Essentially, water shortages are occurring in Asia due to the rapid population growth and economic development (Table 2). There are shortages of the water necessary to sustain daily life, leading to serious food shortages caused by negative effects on the ecosystems (Beppu 2007; Jones 2009). Water pollution caused by groundwater withdrawal (Table 3), lack of wastewater disposal facilities, increases in population in areas that are subject to dangerous flooding and climate-change-related coastal sea level rises causing pollution of coastal aquifers have altogether led to water security problems in the Asian countries studied (Nuttall 2005).

4 Factors that Influence Water Security: Between Countries Analysis

As noted, water security is emerging as an increasingly important and crucial issue for the Asia Pacific region. The simultaneous effect of agricultural growth, industrialisation and urbanisation is now beginning to show moderate-to-severe water shortages (Tookey 2007). Clearly, this will be compounded by the predicted effects of climate change that will produce even more erratic weather patterns (De Loe et al. 2007; Tindall and Campbell 2010). There are a number of factors that influence or are affected by water security, and these factors need more analysis such as agriculture, industrialisation, environmental factors, demographic factors, economy and livelihood, health security and conservation factors, including migration and conflict. In the following, these factors are briefly analysed.

4.1 Agriculture

One issue that is consistently emerged is the impact of growing food demand on global water supplies (Aswathanarayana 2007; Nettle and Lamb 2010). Renner (2010) stated that the need for irrigation water is likely to be greater than currently anticipated and also warns that the food production will likely be seriously constrained by freshwater shortages in the next century. In Asia, the growing demand for food is a significant factor that will determine the supply of available freshwater. About half of the water that is used for irrigation is lost due to seepage and evaporation. Hence, irrigation can be a powerful tool for expanding crop yields, but it can also be wasteful when mismanaged (Nelson 2009). For example, erosion, waterlogging and salinisation of the soil all make the soil less able to produce crops (Jianyun et al. 2009; Nuttall 2005; Sadoff and Muller 2009).

De Loe and Bjornlund (2008) argued that the irrigation is an important determinant of water security because of the volumes of water used. For example, in India, a change in water security has a direct and immediate impact on agriculture. A majority of India's population, almost 58 %, is employed either directly or indirectly by the agriculture sector. India's primary crops, rice, wheat and maize are all water-intensive crops, especially rice. Since most crops are directly dependent on the monsoon, often the weak and delayed monsoons have caused disorder to India's farming prospects, reducing yield significantly. Apart from the immediate impact of lack of water on crops, there is also the problem of growing desertification due to depleting groundwater resources (De Loe and Bjornlund 2008; Nelson 2009; Raj 2010; Tindall and Campbell 2010) (Fig. 5).

4.2 Industrialisation

Aside from agriculture, another factor that influences the status of water security is degree of industrialisation. Industrial activities in Asian countries require massive amounts of freshwater for such activities as boiling, cleaning, air conditioning, cooling, processing, transportation and energy production (George et al. 2008; Pigram 2006). Also, we see a migration from rural to urban areas and it is expected that by 2030 more than 85 % of the population of Asian Countries will

Fig. 5 Percentage of freshwater use for different activities

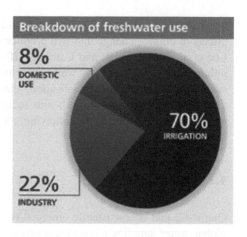

Breakdown of freshwater use

8%
DOMESTIC USE

70%
IRRIGATION

22%
INDUSTRY

live in urban centres. In addition, we see a rapid increase in the standard of living as well (more domestic water demand). Due to this, the consumption per capita as well as the need for urban and industrial water supply will increase over time (Asia Society 2009; Renner 2010).

While domestic and industrial water supply can be provided at a substantially higher price than agricultural water supply, we often see in cases where competition develops that water is shifted from the agricultural sector to the urban and industrial sector (Guthrie 2010; Tookey 2007).

4.3 Environmental Factors

Many Asian countries routinely dump human and industrial waste into their rivers and lakes, in which roughly 90–95 % of all domestic sewage and 75 % of all industrial waste are discharged into surface waters without any treatment (Tindall and Campbell 2010; Vorosmarty et al. 2010). The development of provisions for sanitation lags behind the development of urban and industrial water supply, resulting in substantial discharges of untreated wastewater with numerous impacts for the water quality and waterborne diseases (Pimentel et al. 2004). For example, in South Korea, more than 300 factories along the Naktong River illegally discharged toxic wastes directly into the river, whereas, in China, nearly three-fourths of the nation's rivers are so badly polluted in such a way that they no longer support fish life. Meanwhile, all of India's 14 major rivers are polluted, because they transport 50 million cubic metres of untreated sewage into India's coastal waters every year (Jones 2009; Wouters 2010; Xia et al. 2007). New Delhi alone is responsible for dumping more than 200 million litres of raw sewage and 20 million litres of industrial wastes into the Yamuna River as it passes through the city on its way to the Ganges (Biswas and Seetharam 2008).

Another potential environmental threat to water security in Asia is global warming and climate change (Ahmed 2009; Barnett 2003). Changing weather

Table 4 Possible impacts of climate change (adapted from Bates 2008)

Direct impacts of climate change	Implications of climate change
Temperature	Floods
Precipitation and seasonal shifts	Droughts
Variable stream flow and run-off	Water quality
	Loss in power generation
	Agriculture
	Rise in sea level

patterns could result in droughts in areas accustomed to plentiful rainfall and vice versa, but Southeast Asian countries may end up with little rainfall due to unusual weather patterns (Benjamin et al. 2006; Henebry and Lioubimtseva 2009). Many areas of china are likely to have much water when they do not need it (i.e. flooding during the rainy season) and too little when they do (the dry summer months) (Jianyun et al. 2009; Pimentel 2004). Thus, its impact will be felt in terms of reduced rainfall and run-off, leading to increased heat stress, drought and desertification (Renner 2010). Climate change and increasing demands from population growth will cause a worsening of water stress over the coming decades (Beppu 2007; Moore 2009) (Table 4).

For example, in India, climate change is expected to impact the Himalayan Rivers in two distinct ways. The rising temperatures will affect the glaciers at the mouth of rivers such as the Ganges and the Brahmaputra, accelerating the rate at which they melt. Global warming will impact monsoon patterns in such a way that rainfall is more intense and heavy, but concentrated on fewer rainy days. These two factors have already started to impact the two rivers that sustain themselves on rainfall and glacial melt (Ahmed 2009; Guthrie 2010; Moore 2009).

Rainfall is also expected to become more intense and concentrated on fewer days, which will lead to adverse situations such as flash floods. Due to the fewer days of rain, adequate amounts of water will not percolate down to the groundwater tables (Barnett 2003; Henebry and Lioubimtseva 2009). Increased temperatures will also increase the movement of water from the soil and vegetation into atmosphere through evaporation and transpiration reducing the actual amount of water that is available for human use (Aswathanarayana 2007; Nelson 2009).

Around 80 % of the annual rainfall of Nepal falls between June and September, and many people in the hills have to survive on less than five litres of water per capita per day. Months of the year are marked by long spells of drought (Sadoff and Muller 2009; Tindall and Campbell 2010). Nepal depends heavily on rainwater for irrigation, and only 35 % of its arable land has irrigation facilities. Too much water and too little water pattern is likely to continue in Nepal, owing to the unfavourable monsoon with changing weather patterns, erratic monsoons and rising temperatures (Tularam 2010; Shrestha 2009).

Climatic changes in Bangladesh are not only affecting rivers and the waters but also underground water is being affected (Ahmed 2009). Jones (2009) stated that dams and barrages constructed upstream in India are drastically reducing the availability of water in Bangladesh. As precipitation becomes uncertain and as rivers dry up, underground water is not being replenished. Minerals such as arsenic that are extremely harmful to all living beings including plants, animals and humans have been oxidised found in groundwater (Henebry and Lioubimtseva 2009). As sea levels rise and rivers dry up, salinity intrusion has also occurred, affecting land and groundwater (Briscoe 2009). During dry seasons, withdrawal of waters from already silted up rivers will make Bangladesh a desert, and during the monsoons, release of excess waters will flood the whole of Bangladesh since silted up rivers will be unable to carry waters to the sea (Nishat 2008; Nuttall 2005; Walsh 2009).

Land degradation is another variable that will influence the availability of water. For example, in India, land degradation has resulted in reduced aquifer recharge, even in areas that receive large amounts of annual rainfall. Village authorities in high rainfall regions have placed a petition to the central government for drought relief. Similar trends can be seen in China (Biswas and Seetharam 2008; Raj 2010). Xia et al. (2007) noted that more than 100 Chinese cities in Northern and coastal regions have experienced severe water shortages. Renner (2010) said that overpumping and inefficient irrigation techniques have led to sharply declining groundwater levels, loss of wetlands and salinisation of agricultural lands.

Deforestation is yet another challenge, and it is currently extensive in Asia. Deforestation is a major factor in water security because tropical forests protect weak soils from temperature and rainfall extremes, but if trees are removed, it can create a cycle of flooding and drought that results in extreme soil erosion and, in the most extreme cases, desertification (Pimentel et al. 2004; Xia et al. 2007).

4.4 Demographic Factors

At the beginning of the twentieth century, the world's population was roughly 1.6 billion people, but by 1990, it had increased to around 5.3 billion. Currently, the world's population is 6.77 billion and is expected to reach 8.5 billion by the year 2025. Roughly half of this population will live in Asia, but only 16 % of the world's total land surface is occupied by Asian countries. Population growth in Asia is seen as a major challenge for water security in the region and is a fundamental driver of natural resource stress. Asia's increasing population is straining the ecological systems that provide water for drinking, agriculture and other life-sustaining services, while causing a rapid increase in land degradation (Wouters 2010; Jones 2009). Massive urbanisation in Asia will present a new set of water management challenges in the coming decades (Beppu 2007; De Loe et al. 2007; Roberts and Kanaley 2006). Water is required not only for direct consumption and industrial use, but also for any kind of food production activity. Among other things, urbanisation is expected to shift water out of agriculture for growing

cities to supply drinking water (Grey and Sadoff 2007). China's growing industrialisation and urbanisation require increased amounts of water; this is the same
water that would have gone to agriculture (Tellis et al. 2008; Tookey 2007). Thus,
these countries should deploy water saving and water treatment technologies to
overcome water security issues in future since the demand for water is rising so
quickly.

4.5 Economy and Livelihood

The threat to food security due to water security issues will directly manifest itself
in India's economy. The rate of farmer suicides is high and is likely to increase,
placing an additional burden not only on the families of those farmers, but also
on the community and state (Jones et al. 2007). Apart from agriculture, there will
also be an impact on the fisheries and aquaculture sector in India (Nuttall 2005).
The lack of future food security will have an immediate and irreversible impact
on the economy of the nation. The livelihoods of hundreds of millions of workers,
and their families, who depend wholly on the agriculture and fisheries sectors for
their livelihood will be adversely affected (Aswathanarayana 2007; De Loe and
Bjornlund 2008; Raj 2010). It is clear that Pakistan, Bangladesh, China, Indonesia
and other Asian countries will also experience similar problems of an economic
nature described above.

4.6 Health Security

In India, for example, polluted water sources are also a leading cause of water-
related diseases. In the Ganges Basin, the poorest among the population often have
no choice but to drink and cook with seriously polluted water, causing numerous
diseases and stomach infections, such as diarrhoea and dysentery. Water shortages have devastating impact on human health, including malnutrition, pathogen
or chemical loading, infectious diseases from water contamination and uncontrolled water reuse. Due to lack of safe drinking water, people could end up in
using whatever water is available to them, including water tainted with sewage and
agricultural run-off or even, contaminated water and end up with diseases (Jones
2009; Raj 2010). In India, hundreds of millions of Hindus revere the Ganges. They
believe that bathing in the river purifies their souls. Unfortunately, the river is polluted with sewage. The concentration of pollutants is many times the permissible
level. Serious diseases occur, and recurring problems of diarrhoea among the wor-
shipers are common (Renner 2010). In a similar manner, other Asian countries
such as Indonesia, Bangladesh and Pakistan also have a number of health-related
issues due to lack of clean drinking water.

4.7 Conservation Factors

If the water was used more efficiently in agricultural, industrial and municipal settings, it could help assure water security. In many irrigation systems, as little as 37 % of the water used is actually absorbed by crops; the remainder is lost through evaporation, seepage or run-off. Water used for agricultural purposes could be saved if more efficient irrigation methods were utilised. Thus, the degree of wastage occurs is the key factor in water security (De Loe and Bjornlund 2008; Pigram 2006). The rapid industrialisation and urbanisation along with poor water resource management associated with a large population have severe impact on water security (Grey and Sadoff 2007). Water pollution in Asia resulting from factors such as population growth and greater demand from the agricultural and industrial sectors not only will contribute to increasing rates of food insecurity and land degradation, but also will have detrimental impacts on human health as noted earlier (Tindall and Campbell 2010).

4.8 Migration and Conflict

With parts of the country becoming increasingly water scarce, especially in North India, millions of people will be forced to move away from their homes in search of water supply. In the next two decades, many rural residents will be forced to abandon their hometowns due to the lack of water resources, and increased frequency of extreme weather events such as floods. Lack of job security in the agriculture sector due to water shortages will also force many farmers to leave their villages for a better life (Nettle and Lamb 2010; Raj 2010). With an increased number of people competing for scarce resources and jobs, an anti-outsider mentality will dominate creating a backlash against migrant workers. Inappropriate circumstances and negative external pushes can cause tension, and this could manifest into violence as has been noted in South Africa recently (Nettle and Lamb 2010; Raj 2010; Smith and Gross 1999; Walsh 2009).

The Himalayan River Basins (Ganges, Brahmaputra, Indus and Yangtze) in China, Nepal, India and Bangladesh are inhabited by around 1.3 billion people. These rivers were the lifelines of the ancient civilisations but presently the rivers are under threat. In the next two decades, the four countries in the Himalayan sub-region will face the depletion of almost 275 billion cubic metres of annual renewable water, more than the total amount of water available in Nepal today (Bates 2008). Water availability is estimated to decline by 2030, compared to present level by 13.5 % in case of China, 28 % in India, 22 % in case of Bangladesh and 35 % in case of Nepal. The contributing factors are as follows (Bates 2008; Jones 2009):

(a) about 10–20 % of the Himalayan Rivers (https://mail.bag-mail.de/owa/Henri k.Scheller@bertelsmann-stiftung.de/redir.aspx?C=9442d565e87c48b99e78 af220c0bdb8b&URL=http%3a%2f%2fblogs.ei.columbia.edu%2fwater%2f 2010%2f07%2f19%2fthe-glaciers-disappear-the-startling-photos-of-david-breashears%2f) are fed by Himalayan glaciers, and studies say 70 % of these

glaciers will be melted by the next century (https://mail.bag-mail.de/owa/He
nrik.Scheller@bertelsmann-stiftung.de/redir.aspx?C=9442d565e87c48b99e
78af220c0bdb8b&URL=http%3a%2f%2fblogs.ei.columbia.edu%2fwater%
2f2010%2f07%2f19%2fthe-glaciers-disappear-the-startling-photos-of-david-
breashears%2f) as a result of accelerating global climate change;

(b) glacial melting will eventually reduce river flow in the low season and
    increase in temperature in some areas leading to deforestation;
(c) disappearance of thousands of lakes;
(d) depletion of water resources due to pollution and natural reasons; and
(e) reduced river flows induce more deposit of silt in river bed lowering the depth
    of river, thus causing flooding.

Some implications of depletion of water resources are as follows: non-availability
of water leading to less productivity and massive reduction in the production of
rice, wheat, maize and availability of fish (Bates 2008; Nuttall 2005).

5 Ways to Overcome and Reduce Water Security Concerns

Water security is a critical factor in government planning in that the protection of
adequate water supplies for food, fibre, industrial and residential needs for expanding
populations is vital (Biswas and Seetharam 2008; De Loe et al. 2007). Maximising
water-use efficiency, developing new supplies and protecting water reserves in event
of scarcity are critical aspects of governance. A frequently proposed response to
water scarcity is water pricing. To ensure efficient agricultural use, the price of water
needs to be raised, but not to the point where it becomes too expensive for residents
or farmers to use (Briscoe 2009). The only way that we can cut down the enormous
amount of water wasted in farming is by applying more efficient agricultural practices such as drip irrigation and implementing regulations against the industrial pollution users (Walsh 2009). An effective water resource management can also help to
protect vulnerability especially when water scarcity may be more severe in the future
(Sadoff and Muller 2009; Walsh 2009). To improve drinking water quality, there are
two important factors. Firstly, there should be a community-level participation in the
management. Secondly, there should be appropriate protection of water sources and
waterways and maintenance and protection of infrastructure (Nuttall 2005; Qadir
et al. 2007). In the following, governance and regulation, cost of water and some
questions that need to be answered are considered in turn.

5.1 Governance and Regulation

Water security is not only about a sufficiency of water but also about recognising
the true value of water and managing it accordingly. There is a need for better governance and management at all levels, as well as at the catchment level in regard
to water security (Beppu 2007). Where rivers cross national boundaries or lakes are

shared between countries, a trans-boundary agreement for water allocation should be negotiated (Smith and Gross 1999). Importantly, when dealing with threats to global water security, negotiations should be sensitive to the individual country's political, social, economic, environmental, financial and cultural conditions (Raj 2010).

5.2 Cost of Water

Water has traditionally been regarded as a free resource. But any costs for it are usually associated with the cost of processing and delivering alone, rather than assigning any value to the resource. There is growing interest internationally in the use of water pricing to reduce demand as well as to generate revenue to cover the cost of providing water supplies and maintaining infrastructure (Pigram 2006; Roca and Tularam 2012; Sadoff and Muller 2009). The effectiveness of pricing in influencing demand depends on water users. For municipal water demand, pricing can be effective when combined with raising user awareness. In the case of water for irrigation, pricing is more complex because the amount of water consumed is difficult to measure and farmer behaviour may not be sensitive to price until the price of water is several times that of the cost of providing water. Large increases in the price of surface water may cause farmers to use groundwater instead, which is relatively unregulated by comparison (Sadoff and Muller 2009).

One of the central problems facing the construction and operation of water infrastructure is the price of water that is paid by the end user. In order to make water a universally accessible good, the National Development and Reform Commission (NDRC)'s Price Bureau has traditionally heavily subsidised and regulated the price of water for all users. However, low water tariffs often make it impossible for utilities and water companies to recover their capital expenditures through the collection of drinking and wastewater fees (Hu 2010). Within the sector, water tariffs differ based on use with industrial users typically subject to higher water prices than municipal or residential users, who pay still more than agricultural users. The first reason is economic in that the industrial wastewater can be more expensive to treat than municipal wastewater, while water used for irrigation is neither treated before being added back to the environment, nor requires as much treatment prior to use. The second reason is socio-economic, where the access to water is considered a basic right. Therefore, affordability remains central to residential prices and the problem is more acute in the agricultural sector, when small-scale subsistence farmers require large volumes of water for irrigation. Thus, water price reform can then have a direct and dramatic impact on livelihood and, in turn, on levels of social unrest (Asia Society 2009; Hu 2010).

Water is a social good, and so should be provided free of charge or at highly subsidised prices (Reza et al. 2013; Roca and Tularam 2012). In contrast, current studies indicate that without appropriate water pricing, the present vicious cycle of waste, inefficiency and lack of services to both the rich and the poor will continue. Lack of income of the utilities due to inadequate water pricing will ensure that the water systems are not properly maintained, and investment funds are not available for updating technology, improving management and technological capacities, expanding the networks and providing wastewater management. There is no question that the era when drinkable water could be provided to everyone free of charge or at highly subsidised rates on a long-term basis is now over (Hu 2010). The users pay for the services they want, the poor who cannot pay receive targeted subsidies, utilities provide water supply and wastewater management services efficiently and accountably, users cover the costs of the services, and public funds are used for public purposes. Of course, this does not mean that we now have all the answers on how water should be priced for different consumers and for different uses (Hu 2010; Tindall and Campbell 2010). There are a number of questions from the above analysis that are in line with those Biswas and Seetharam (2008) posed such as:

(a) How can it be ensured that the poor have adequate access to reliable water and sanitation services at affordable prices while the rich are not subsidised?
(b) How should water and sanitation services be managed in order to ensure the provision of reliable services, economic efficiency, universal access and maximisation of social welfare are met?
(c) What type of institutional frameworks and governance practices are needed to improve the delivery services?
(d) How quickly can all of the above be met allowing for regional social and political sensitivities?

The World Bank notes that many countries face a major challenge in developing and maintaining appropriate water systems infrastructure. Financial institutions are likely to play a key role in making up this shortfall. However, better information on likely costs and barriers to their implementation is needed. This may help to close the water supply–demand gap and help meet the Millennium Development Goals in the long run (Sadoff and Muller 2009).

6 New and Better Technologies, Techniques and Practices

Existing technologies need to be refined, developed and improved to address the challenges of research and development of water security issues. Also, an effort must be made to study the role that technology and the affect it may have on the ecosystem as well (Briscoe 2009; Guthrie 2010). In the following, managing variability, surface water storage, sustainable use of groundwater, water efficiency in agriculture and water efficiency in industry are discussed as possible ways of managing water security problems of Asia.

6.1 Managing Variability

High rainfall variability, high evapotranspiration rates, geographic separation of water resources and irrigation development together make the storage and delivery of water a major challenge to the Asian countries. However, water storage in rivers, lakes, reservoirs and aquifers provides a means of managing variability in water availability, allowing stored water to be used during dry periods. Historically, water resource development in Asian countries has taken a surface water focus, with the construction of large dams. There are opportunities in using aquifers as water banks in conjunction with surface water reservoirs to enable greater flexibility and efficiency in securing water supplies (De Loe et al. 2007; Nishat 2008). Enhancing recharge to aquifers during periods of above-average water availability provides a resource for access during droughts.

A conjunctive approach has benefits when compared with relying solely on large surface water storages. Using aquifers as storage is becoming a viable alternative considering the existing constraints of building new dams due to environmental concerns and general lack of suitable sites. Subsurface storage complements surface reservoirs, and while it may not replace large dams, it could be an alternative to expanding storage capacity using reservoirs alone (MacKenzie 2009; Tualram and Keeler 2006). The infrastructure costs are generally cheaper, and there is the potential that the aquifer material can filter and improve water quality. Although the concept appears simple, sustainable operations which protect groundwater quality require a sound understanding of the hydrological and biological processes involved, along with careful management. Thus, the storage in lakes and reservoirs can be managed to provide potential for the storage of excess flows during floods (Aditi 2005; Qadir et al. 2007).

6.2 Surface Water Storage and Sustainable Use of Groundwater

Surface water storage by means of dams can bring many benefits, but the benefits may come at social and environmental cost caused by the displacement of people or impacts on the ecosystem caused by changes in flow and continuity of rivers. There are many issues to be considered before the construction of a dam such as the need to gain public acceptance, address the impact of existing dams, sustain rivers and livelihood and share river resources for peace and development as well as security (Guthrie 2010). Storing and treating the river water in dams appears attractive. However, building dams in hills of Pakistan, Japan, Indian subcontinent or Indonesia, for example, may not be safe for the dam may sit in the seismically active zone. Moreover, steep gradients of the streams and siltation problems are challenging, and economic, social and environmental cost makes them unattractive (Shrestha 2009).

Another possibility is groundwater that is naturally replenished, or recharged, through rainfall and surface water. Excessive use of groundwater in Asia has been a major problem. Nonetheless, groundwater is an important supply of water for agricultural and domestic use. ASR is the process of storing excess water underground when it is available and recovering that water for use when supplies are short (Aditi 2005; Pigram 2006). This is a new technology but one that can be easily applied in most Asian countries.

6.3 Water Efficiency in Agriculture

It is our responsibility to manage the water use to meet the future needs. Measures such as mulching and conservation tillage will help in retaining soil moisture, especially to manage land cover if supported by soil conservation measures (Nettle and Lamb 2010). Small-scale rainwater harvesting helps to provide an additional source of water for crops. Improved surface irrigation methods such as level furrows, sprinkler and micro-irrigation methods, and the use of advanced techniques of irrigation scheduling and timing can help improve water management at farm level. By monitoring water intake and growth, farmers can achieve greater accuracy when necessary in water application and irrigating only. Remote-sensing schemes are beginning to allow farmers to detect their crops water taking into account meteorological data as well as soil moisture and biomass information (Aswathanarayana 2007; De Loe et al. 2007).

6.4 Water Efficiency in Industry

Water availability is becoming critical in the power industry for electricity generation. Water used for cooling by thermal and nuclear power plants is set to rise throughout the world as new power plants are commissioned. In some cases, it is not simply the availability of cooling water that is the issue, but that outflows from power stations can become warm enough to cause environmental damage on discharge. Water treatment and reuse on site can significantly reduce water abstraction (Tellis et al. 2008).

Achieving water security at the global, national, regional and local levels is a challenge the problem close to Australia must be recognised and met. The previous analysis of Asian countries allowed the identification of factors that influence water security. Factors that influence water security were studied, and new technologies and management issues were examined. Some questions posed regarding the issues that need addressing to achieve water security were posed; clearly, solving these will help build and manage the livelihoods of those who live in the region. A comprehensive and strategic plan to combat Asia's growing water scarcity and water quality problems must include commitment to the development of water infrastructure along with efficient water management guidelines, capacity and expertise. The analysis shows the key to mitigating the adverse impacts of climate change such as water scarcity and water-related disasters is increased understanding of the dimensions of water security infrastructure and management.

7 Conclusions and Recommendations

The analyses conducted in this paper show that population growth accompanied by increased water use will not only severely reduce water availability per person but also create stress on biodiversity in the entire global ecosystem. A fair allocation of water is a key factor while managing the water resources, considering the total agricultural, societal and environmental system, thus maintaining livelihoods of millions in Asia. It is clear that substantial withdrawal of water from lakes, rivers, groundwater and reservoirs to meet the needs of individuals, cities and industries has led to water stress in many parts of the Asia. The amount of water withdrawn from these resources (groundwater and stored water) both for use and for consumption in diverse human activities should be limited to be sustainable over time. Pumping of ground water in order to fulfil human requirements will lead to pollution of water resources and thus not only pose a threat to public and environmental health but also contribute to the high costs of water treatment. The rapid increase in freshwater withdrawals for agricultural irrigation and for other uses that have accompanied population growth has stimulated serious conflicts over water resources both within and between countries, and this has consequences such as creating water refugees. The best approach to conserve the world's water is to find the ways to facilitate the percolation of rainfall into the soil instead of allowing it to run off into streams and rivers. Such an approach will also help reduce flooding more generally during high rainfall periods.

Priorities to use water wisely in Asia could be as follows: (i) farmers should be the primary target for incentives to conserve water since agriculture consumes 70 % of the world's fresh water; (ii) farmers should implement water conserving irrigation techniques (such as drip irrigation to reduce waterwaste) and soil conservation practices (such as cover crops and crop rotations, to minimise rapid runoff); (iii) government and private industry should implement World Bank policies for the fair pricing of freshwater and should reduce or eliminate water subsides; and (iv) countries should develop better management guidelines, actions and policies to control water pollution and protect public health, agriculture and the environment.

The analysis shows that water security strategies depend upon pertinently developed and implemented water management plans and practices. The practices include plans for potable water sustainability, proper wastewater and waste disposal methods, distribution, water-use priorities and water resource development, to overcome water security issues in future. Water resource management must become economically more efficient, ecologically sustainable and also socially justifiable, especially in the regions of the world suffering from water crisis such as Asia. Clearly, the reliance on our rivers for water supplies should and the only way it may be achieved is by using storm water, recycling and desalination.

As noted in the paper, water security is the protection of adequate water supplies for food, industrial and residential needs for expanding populations. This requires maximising water-use efficiency, developing new supplies and protecting water reserves in event of scarcity due to natural, man-made or technological hazards. Thus, a major challenge for water-stressed Asian developing countries is the manner in which they coordinate all the concerned resource policies, legal and regulatory frameworks and institutions responsible for formulating and implementing these policies. For this to occur, there is a need to understand the dimensions of water security, namely household water, industrial and agricultural water, city water, healthy river, water disaster management and water governance.

Water security in Australia has become a major concern over the course of the late twentieth and early twenty-first century as a result of population growth, severe drought, fears of the effects of global warming on Australia, environmental degradation from reduced environmental flows, competition between competing interests such as grazing, irrigation and urban water supplies and competition between upstream and downstream users. Australia also has a major role to play in Asia in terms of aid, building infrastructure, providing capacity and technology transfer. It is in the interest of all Australians that the developing countries in close proximity to Australia have little or no water security issues soon, but for this to occur, a number of steps need to be taken. These steps will not only help here in Australia but also provide leadership in the region.

Australian should continue to recognise the water resource issues in Asia and should take steps to promote sustainable water use by implementing set of policy goals and prioritising the steps undertaken in managing the water resources in the region:

- Australia should establish policy objectives for water resource management, including strengthening river basin management, protecting drinking water sources, combating trans-boundary water pollution, enhancing water saving in agriculture and increasing the treatment rate of urban sewage by 2050;
- provide leadership in the promotion of efficient management of existing available water supplies for agricultural, urban and industrial purposes as well as utilising wastewater and recycling technology; and
- establish the need for public engagement, education and awareness in raising the subject of water security in the region.

Finally, there are a number of more specific water security considerations and concerns that need to be included in the framework such as ecosystem protection—monitoring and enforcement for protection; economic production—stable allocation rules, economically sound decisions, reallocate water between users, sectors and regions; equity and participation—issue of access and equity, stakeholder and public, development of rules to address conflicts, etc; integration of resource—integration of surface and groundwater, quality and quantity, land use and water allocation; water conservation—promotion of more efficient and less consumptive use, inclusion of conservation practices; climate variability and change—investments to understand effects of climate change, development and application of adaptive strategies; and finally trans-boundary sensitivity— coordination of water allocation systems across political and country boundaries, respecting state sovereignty and being sensitive to indigenous customs, etc.

Appendix

See Table 5.

Table 5 Water security concerns and impacts in Asia

Country	Element	Leads to	Impacts
INDIA	Desertification Dumping of unwanted waste into the rivers Excessive Population growth Climate change Land degradation	Depletion of water resources Badly polluted Water Stress Reduced rainfall, run-off and droughts Reduced aquifer recharge	Lack of water on crops Diseases Lack of access to clean water Severe water shortages
PAKISTAN	Erosion Landslides Duststorms Population growth	Agricultural Land and available water resources Rapid depletion of groundwater	Increased salinity issues and water logging problems
CHINA	Population growth Desertification Climate change	Scarce water resources Agricultural production Dimishing water availablity Sealevelrise Pollution Global Warming	Drying up of rivers Decline in groundwater levels Degradation of lakes and wetlands Salinity Issues Threats to human health and fisheries Floods and droughts
INDONESIA	Increase in Population Land use changes and deforstration Rapid urbanisation + Economic development Increased rainfall + flood conditions	Economic growth Floods Pollution Lack of access to clean water and sanitation	Lack of access to safe water Industries have polluted Water related diseases Spread of diseases
NEPAL	Unsustainable land use practices Climate change	Widespread landslides Flooding Uncontrolled dumping of waste into streams	Human lives and infrastrucure got damaged Water borne diseases
BANGLADESH	Climate change	Global warming Sea-level Rise Reduced precipitation and prolonged dry season	Impact on availiable resouces and services Indundation of coastal plains Increase in river and coastal erosion Increase in vector borne diseases Water scarcity
JAPAN	Rapid economic growth+ population growth+ urbanisation Heavy water demand	Excessive groundwater pumping	Groundwater subsidence Sewater Intrusion and sinking lands

References

Aditi M (2005) Groundwater intensive use. Taylor & Francis, Spain

Ahmed AU (2009) Implications of climate change in relation to water related disasters in South Asian countries. Earth Environ Sci 6(29):292020. doi:10.1088/1755

Arnell N (2004) Climate change and global water resources: SRES emissions and socioeconomic scenarios. Glob Environ Change 14:31–52

Asia Society (2009) WATER: Asia's next challenge. Asia Society, New York

Aswathanarayana U (2007) Food and water security. Routledge, India

Barnett J (2003) Security and climate change. Glob Environ Change 13(1):7–17

Barron J, Salas JC (2009) Rainwater harvesting: a lifeline for human well being. UNEP/Earthprint, UK

Bates B (2008) IPCC technical paper on climate change and water. Cambridge University Press, Cambridge

Benjamin LP, Ramasamy S, Ian M, Janice B (2006) Climate change in the Asia/Pacific region. CSIRO, Victoria

Beppu Declaration (2007) Asia's next challenge: securing the region's water future. Asia Society, Singapore

Biswas AK, Seetharam KE (2008) Achieving water security for Asia. Int J Water Res Dev 24(1):145–176

Briscoe J (2009) Water security: why it matters and what to do about it. Innovations 4(3):3–28

BugBog (2001) Asia map. Retrieved Dec 2012. http://www.bugbog.com/maps/asia/asia_map.html

De Loe R, Bjornlund H (2008) Irrigation and water security: the role of economic instruments and governance. Man Tech Pol II:35–42

De Loe R, Varghese J, Ferreyra C et al (2007) Water allocation and water security in Canada: initiating a policy dialogue for the 21st century. Report prepared for the Walter and Duncan Gordon Foundation. Guelph Water Management Group, University of Guelph

Funabashi Y, Shimbun A (2010) Global water security: Japan should play key role. Asahi Shimbun: Japan. http://www.eastasiaforum.org/2010/09/15/global-water-security-japan-should-play-key-role/. Retrieved 25 Jan 2013

George BA, Malano HM, Raza Khan A et al (2008) Urban water supply strategies for Hyderabad, India-future scenarios. Environ Model Assess 14(6):691–704

Grey D, Sadoff CW (2007) Sink or swim? Water security for growth and development. Water Policy 9:545–571

Guthrie P (2010) Global water security-an engineering perspective. The Royal Academy of Engineering, London

Henebry GM, Lioubimtseva E (2009) Climate and environmental change in arid Central Asia: impacts, vulnerability, and adaptations. Arid Environ 73(11):963–977

Hu Y (2010) Foreign investment in China's water infrastructure. China's Strat Dev 6(1):39–48

Illangasekare TH, Mahutova K, Barich JJ (2009) Decision support for natural disasters and intentional threats to water security. Springer, Netherlands

Indraratna B, Tularam GA, Blunden B (2001) Reducing the impact of acid sulphate soils at a site in Shoalhaven Floodplain of New South Wales, Australia. Quart J of Eng Geo and Hydrogeo 34:333–346

Jianyun Z, Guoqing W, Yang Y et al (2009) Impact of climate change on water security in China. Adv Climate Change Res 5:1673–1719

Jones JA (2009) Threats to global water security: population growth, terrorism, climate change, or commercialisation? In: Jones JA, Vardanian TG, Hakopian C (eds) Threats to global water security. Springer, Netherlands, pp 3–13

Jones JA, Vardanian TG, Hakopian C (2007) Threats to global water security. Springer, Netherlands

MacKenzie S (2009) Project groundswell. Retrieved from Dec 2012. http://projectgroundswell.com/2009/11/19/linking-water-security-and-climate-change/

Moore S (2009) Climate change, water and China's national interest. China Secur 3(3):25–39

Nelson GC (2009) Climate change: impact on agriculture and costs of adaptation. International Food Policy Research Institute, USA

Nettle R, Lamb G (2010) Water security: how can extension work with farming worldviews? Extension Farming Syst 6(1):11–32

Nishat A (2008) Climate change and water management in Bangladesh.In: International conference on global climate change and its effects. International Union for Conservation of Nature (IUCN), pp 1–38

Nuttall N (2005) Asia's water security under threat. Nairobi/Bangkok, UNEP, UK

Pigram JJ (2006) Australia's water resources from use to management. CSIRO, Collingwood

Pimentel D, Berger B, Filiberto D, Newton M et al (2004) Water resources: agricultural and environmental issues. Bioscience 54(10):909–918

Prasai T (2007) Nepal: water borne disease a major health problem in Nepal. IRC, Nepal

Qadir M, Sharma BR, Bruggeman A et al (2007) Non-conventional water resources and opportunities for water augmentation to achieve food security in water scarce countries. Agric Water Manage 87:2–22

Raj A (2010) Water security in India: the coming challenge. Future Directions International Pvt Ltd., Australia

Renner M (2010) Troubled waters. World Watch, 14–20: May/June

Reza SR, Roca E, Tularam GA (2013) Fundamental signals of investment profitability in the global water industry. Available at SSRN: http://ssrn.com/abstract=2371115 or http://dx.doi.org/10.2139/ssrn.2371115. Accessed 22 Dec 2013

Rezwan (2013) http://xoomer.virgilio.it/dinajpur/b10c/banglanews432en.htm#asia. Retrieved 12 Jan 2013

Roberts B, Kanaley T (2006) Urbanisation and sustainability in Asia. ADB, Manilla

Roca E, Tularam GA (2012) Which way does water flow? An econometric analysis of the global price integration of water stocks. Appl Econ 44(23):2935–2944

Sadoff C, Muller M (2009) Water management, water security and climate change adaptation: early impacts and essential responses. Global water partnership. Elanders, Sweden. http://www.gwp.org/Global/GWP-CACENA_Files/en/pdf/tec14.pdf

Saliby IE, Okour Y, Shon HK et al (2009) Desalination plants in Australia, review and facts. Desalination 47:1–14

Shrestha SD (2009) Water crisis in the Nepal Himalayas. Tribhuvan University, Kathmandu

Smith PJ, Gross CH (1999) Water and conflict in Asia?. Asia-Pacific Center for Security Studies, Hawaii

Tellis AJ, Kuo M, Marble A (2008) Asia's water security crisis: China, India, and United States. The National Bureau of Asian Research, Washington

Tindall JA, Campbell AA (2010) Water security-national and global issues. USGS-Science for a changing world. http://pubs.usgs.gov/fs/2010/3106/. Retrieved Dec 2012

Tookey DL (2007) The environment, security and regional cooperation in Central Asia. Communist Post-Communist Stud 40(2):191–208

Tularam GA (2010) Relationship between El nino southern oscillation index and rainfall. Int J Sustain Dev Plann 5(4):378–391

Tularam GA (2012) Water security Issues of Asia and their implications for Australia. Fifth International Groundwater Conference, Artworks, India

Tularam GA, Illahee M (2010) Time series analysis of rainfall and temperature interactions in coastal catchments. J Math Stat 6(3):372–380

Tularam GA, Ilahee M (2007) Environmental concerns of desalinating seawater using reverse osmosis. J Environ Monit 9(8):805–813

Tularam GA, Keeler HP (2006) The study of coastal groundwater depth and salinity variation using time-series analysis. Environ Impact Assess Rev 26(7):633–642, Elsevier, United States

Tularam GA, Krishna M (2009) Long term consequences of groundwater pumping in Australia: a review of impacts around the globe. Appl Sci Environ Sanitation 4(2):151–166

Tularam GA, Properjohn M (2011) An investigation into water distribution network security: risk and implications. J Secur 4:1057–1066

Tularam GA, Singh R (2009) Estuary, river and surrounding groundwater quality deterioration associated with tidal intrusion. Appl Sci Environ Sanitation 4(2):151–166

UNEP (2010) UN water. http://www.unwater.org/statistics_res.html. Retrieved 25th Jan 2012

Vorosmarty CJ, Mclntyre PB, Gessner MO (2010) Global threats to human water security and river biodiversity. Nature 467(7315):555–561

Walsh B (2009) Water fight. Time international, Asia edn, vol 36. http://www.time.com/time/magazine/article/0,9171,1891640,00.html. Retrieved Jan 2012

Wouters P (2010) Water security: global, regional and local challenges. Institute for Public Policy Research (IPPR), 1–17

Xia J, Liu MY, JIA S-F (2005) Water security problems in North China: research and perspective. Web Sci Pedosphere 15:563–575

Xia J, Zhang L, Liu C et al (2007) Towards better water security in North China. Wat Res Manage 12(233):247

Authors Biography

Gurudeo A Tularam is a Senior Lecturer in Mathematics and Statistics teaching both in Griffith Sciences and in particular the Griffith School of Environment. He is a senior research member of the Environmental Futures Research Institute, Griffith University. Dr Tularam is a pure and applied mathematics researcher working mainly with problems involving partial differential equations. He has worked a number of areas concerning groundwater flow and transport. Anand's work has been mainly in the area of environmental applications including transport of pollutants and flow in media. In more recent times he has been working with time series and stochastic calculus methods and applying them to areas such as rainfall, water flow and flood. Anand has also worked in advanced modelling in financial applications such as investment, portfolio analysis and risk assessment. He has completed the supervision of more than 10 Phd, MPhil, MSc, and Honours students.

Kadari K Murali was a Masters student in Environmental Sciences studying ground water pumping and pollution. He is presently teaching Mathematics and continues his interest in research in applied mathematics.

Social Networks in Water Governance and Climate Adaptation in Kenya

Grace W. Ngaruiya, Jürgen Scheffran and Liang Lang

Abstract In many sub-Saharan countries, studies indicate that water scarcity is caused by institutional and political factors. However, despite implementation of a decentralized and integrated approach in water governance, additional water stress from climate-related impacts now threaten to fuel water insecurity. Borrowing from social network theory, this chapter seeks to investigate how synergy among water governance actors influences adaptation status in rural Kenya. Network data from Loitokitok district in southern Kenya is collected using the saturation sampling method and analyzed for density, structural holes, and suitable brokers. Results indicate that rural water security is augmented mainly by individual rain water harvest, effective irrigation techniques, community-based water-point protection, and intermittent capacity building. However, the integrated governance strategy fails to aid interconnective and coordinative actions among actors thus hindering spread of adaptation strategies to the wider community and results in independent implementation of water conservation measures. Consequently, we call for deliberate linkage among local stakeholders to upscale adaptation measures and enhance water security in Kenya.

Keywords Climate adaptation · Decentralization · Integrated water management · Social networks · Structural holes · Kenya

G.W. Ngaruiya (✉) · J. Scheffran · L. Lang
Institute of Geography, University of Hamburg, Research Group "Climate Change and Security", Grindelberg 7, 20144 Hamburg, Germany
e-mail: grace.ngaruiya@zmaw.de

© Springer International Publishing Switzerland 2015
W. Leal Filho and V. Sümer (eds.), *Sustainable Water Use and Management*,
Green Energy and Technology, DOI 10.1007/978-3-319-12394-3_8

1 Introduction

Water is a natural resource whose supply wholly depends on the hydrological cycle and proper management of natural environments. It is often multifunctional and heterogeneous in nature as provision of sufficient water of adequate quality caters for human well-being and sustains biodiversity (Millennium Ecosystem Assessment 2005). A country that manages to maintain such a self-sufficient situation is deemed to have attained water security. However, Kenya is far from being water secure and is considered as a water-scarce country with less than 647 m^3 of water available per capita compared with the international benchmark of 1,000 m^3 per capita (Government of Kenya 2009b). Many reasons are forwarded to explain this scarcity including: First, arid and semiarid lands (ASAL) constitute about 80 % of the total Kenyan land area (Mutunga 2001) causing a natural shortage of water. Second, climate change impacts evidenced by recurrent drought episodes (Altmann et al. 2002) have dried up many water bodies and lowered the water table. Third, widespread deforestation of most watersheds (Wamalwa 2009) and overgrazing have interfered with microclimates and precipitation patterns. Finally, poor land management and lack of political-will to fund water infrastructure expansion have resulted in haphazard subdivision of land with subsequent negative impacts on water availability. Such diverse issues confirm findings from water security studies that indicate water scarcity in Africa is not only caused by a physical shortage of water but also mainly by institutional and political factors (UNEP and WRC 2008; Ward and Michelsen 2002). Interestingly, the cause for alarm is not only about poor water availability but also about social cohesion since researchers predict increased risk of conflicts due to water competition between social groups (farmers and nomadic herders), economic sectors, and administrative units in arid and semiarid regions of Africa (Carius 2009; Schilling et al. 2012).

To combat water scarcity, many efforts have been directed at diverse water management activities that are collectively termed under soil and water conservation (SWC) in Kenya. These activities are categorized into water conservation, harvesting, and management techniques (Mutunga 2001). For this analysis, we will use water conservation to cover all activities that promote efficient or economical management of available surface and subsurface water resources. Indigenous communities had their own small-scale methods of water conservation but these methods are not currently viable due to population increase and land subdivision that have increased the number of water users, water pollution, and obstructed access to water bodies. Therefore, donors and other water stakeholders advocate for communal projects such as water pans, rock catchments, subsurface (sand) dams, and spring protection so that more users are able to access safe water at the community level. Implementing such a project requires a governance structure that will facilitate multi-level interaction across various organizations (public, private, and civic) and actors (informal and formal) to formulate policies and legislations to access funds and give ownership of the constructed water source to the community. However, identifying diverse stakeholders to constitute a rural water governance committee is not such a straightforward process (Prell et al. 2010). This is because actors seeking to be enjoined to local resource governance networks to implement their adaptation and mitigation projects under the climate agenda have increased. Consequently, rural water governance structures must formulate ways to integrate these additional actors for effective knowledge transfer that will build resilience and cohesion in the community.

Few published works using case studies exist on role of actors on adaptation performance in the water sector in Kenya. For example, Wamalwa (2009) studied the Mara watershed and recommended the need for a more effective coordination arrangement. Ngigi (2009) looked at climate adaptation options in water management and called for increased investment in complementary strategies that enhance adaptation by vulnerable smallholder farmers. We continue with this discourse by analyzing rural actor water networks further to answer the question "is there synergy among actors implementing the rural water conservation and adaptation agenda?" Studies based on social relational theory are increasing due to the discovery of the role of social relationships in shaping environmental outcomes, and this study is no exception as we employ social network analysis to evaluate actor linkage and their water conservation and adaptation activities in the rural community of Loitoktok.

2 Key Issues

2.1 Climate Change Impacts and Water Supply

Climate change influences virtually every element of the global hydrological cycle through changes in precipitation, evaporation, and snowmelt, to threaten global and regional water security (WBGU 2008). Schewe et al. (2013) state "Depending on the rates of both population change and global warming, the level of water scarcity may be amplified by up to 100 % owing to climate change in some regions. This means 5–20 % of the global population is likely exposed to absolute water scarcity at 2 °C of global warming". Already many African countries experience physical water scarcity defined as a state of having less than 1,000 m^3 per capita per annum or suffer water stress defined as a state with less than 1,700 m^3 per capita per annum (Adger et al. 2007; Ngigi 2009; WBGU 2008). Subsequently, climate change impacts like; drought, heat waves, accelerated glacier retreat and hurricane intensity initiate climatic trends not previously experienced will increase future water stress (Adger et al. 2007).

In Kenya, long-term climatic analysis (1976–2000) of the ASAL region of Loitoktok in Kenya revealed a dramatic increase in the mean daily maximum temperature than did daily minimum (0.2775 °C vs. 0.071 °C per annum) especially during months with higher average maximum temperatures such as February and March but with no long-term trend (Altmann et al. 2002). This is majorly attributed to the continual landscape changes and transformation of the landscape surrounding Kilimanjaro into agricultural land that contribute regional forcing to the climatic conditions (Thompson et al. 2009). Such large-scale changes in landscape patterns have effects on regulating ecosystem services of carbon storage (global warming) and erosion control (percolation of rainfall) to significantly affect water security through temperature and rainfall behavior (Millennium Ecosystem Assessment 2005). Thus following consecutive years of major droughts, i.e., 1984, 1992, 1997, 2001/2002, 2006, 2009, and 2011, the climate change discourse in Kenya has adopted drought as the priority problem. This is because the recurrent drought episodes (Altmann et al. 2002) have dried up many water bodies and become the driving force behind migratory movements of people especially in the arid and semi-arid regions of Kenya. Impacts of natural disasters such as floods, landslides and wind storms destroy the water infrastructure and disrupt water supply in urban centres (Government of Kenya 2009b). Besides, the water crisis in Kenya is further exacerbated by water-related health problems (Government of Kenya 2009b), particularly diarrhoea and cholera in heavily populated informal settlements.

3 Adaptation Strategies in the Water Sector

There are no quick fixes for water scarcity except negotiated, proactive strategies that are economically feasible and sustainable, collectively termed as adaptation (Ngigi 2009). These strategies involve making adjustments in social and environmental processes in response to or in anticipation of climate change, to reduce potential damages or to realize new opportunities (Adger et al. 2007). According to UNECE (2009), a successful water conservation and adaptation strategy should be based on five pillars that address all stages of climate-based water crisis. These pillars are discussed below using appropriate activities implemented in Kenya.

- *Prevention* measures are long-term actions taken to avert negative effects of climate variability on water resources. For example, afforestation of the Mau Forest complex to re-establish previous flow of the Mara and Sondu rivers, timely relocation warning for Budalangi community members to avoid destruction of property and loss of lives during the annual floods season, and creation of Lake Naivasha Management Committee to promote sustainable use of the wetland.
- Measures to *improve resilience* aim to reduce negative effects of climate change on water resources. For example, livelihood diversification is clearly seen in pastoralists who have embraced crop farming using drought resistant crops to improve their incomes and living standards.
- *Preparation* measures decrease negative effects of extreme events on water resources. These include water storage, which is a prevalent activity in many Kenyan households. The government is still trying to institute an effective early warning system.

- *Response* measures alleviate direct effects of extreme events. These include evacuation, safe drinking water, and sanitation facilities inside or outside affected areas during extreme events. In Kenya, these actions are covered by non-governmental organizations such as Red Cross in coordination with government agencies during droughts or floods.
- *Recovery* measures seek to restore (but not necessarily back to the original state) economic, societal, and natural water systems after an extreme event. These include reconstruction of infrastructure especially water pipes following floods or landslides and introduction of insurance packages by Equity Bank Ltd. to act as a risk transfer mechanism.

If the five adaptation measures are incorporated into a single water governance plan, then adapting to climate variability will make economic sense because development priorities such as infrastructure quality and settlement plans will be included. However, identification and subsequent collaboration of diverse individuals with conservation knowledge and technical skills become the biggest hurdle to achieving water security in the community.

4 Integrated Water Resource Management

Traditional water management schemes concentrated on increasing water supply to growing populations (Adger et al. 2007) but poor translation of policy into strategic interventions to enhance water supply to the poor, marginalized, and rural communities created the need for new effective governance schemes. This is because studies on water governance indicate that poor management of water resources largely contributes to water scarcity in Africa. There are two such scarcity types:

- Economic water scarcity in hydrologically rich regions caused by
  - Unaffordability of water by the community when water prices are too high in relation to their incomes.
  - Nonexistent political-will causing lack of money, manpower, and other resources required for developing, allocating, transporting, and purifying water in a region (Ward and Michelsen 2002).
- Ecological water scarcity occurs when withdrawal of water resources for human use is so great that it threatens the integrity of ecosystems, and people who depend on the services of these ecosystems suffer damage (Smakhtin et al. 2004).

The need to reform water management has culminated in the formulation and implementation of the integrated water resource management (IWRM) approach in many African countries. In practice, the central government through the Ministry of Water (MoW) devolves power to regional and local administrative actors or institutions at lower levels (Ribot 2002). This strategy is founded on the decentralization principle that assumes sub-national governments are more apt at identifying water needs at rural levels and are well placed to respond to them and can be held accountable by the resident population (Smoke 2003; Wily and Dewees 2001). There are varied forms of decentralization such as political, administrative, fiscal, and institutional. Of concern to this study is institutional decentralization that is endorsed by the Kenyan IWRM. This form of decentralization denotes creation of mechanisms that devolve power from formal government bodies to other local and intergovernmental actors—traditional local authorities, non-governmental organizations, private sector partners, etc—in promoting development (Smoke 2003).

Born and Sonzogni (1995) divide the term "integration" into four aspects, namely comprehensive, interconnective, strategic, and coordinative, for successful water governance. In summary, the *comprehensive* feature involves consideration of critical biological, chemical, and human aspects for a detailed understanding and resolution of the problem. The *strategic/operational* feature identifies key goals to direct attention for ease of planning and achievement. The *interactive/coordinative* component entails informed negotiation and bargaining among parties in an interorganizational dimension, while the *interconnective* feature lays emphasis on the interrelationships among multiple resource users within the watershed. This integrative approach is seen as a solution, due to failure by traditional approaches to handle complex water resources challenges (Wamalwa 2009).

5 Social Network Analysis

Growing literature on governance focus on understanding power structures and institutions—how they develop, how they adapt to meet new challenges, and how they impact decision making at different levels of society (UNEP 2005). Thus, studying social networks of water governance can reveal deficiencies in existing water management that can be used to align water management explicitly to future challenges. Social network analysis is an approach that analyses relationships among various social actors as real interactions with local potentials and liabilities that influence success of any decision-making process (Lourenço et al. 2004). Guided by network theory, we have selected two measures of a social network to quantify patterns of interactions and indicate level of synergy among actors involved in water conservation and adaptation implementation.

5.1 Structural Holes

Structural holes are "empty spaces in social structure" that exist between two actors when either party is unaware of value available if they were to coordinate on some point (Burt 2011). There are various ways of measuring structural holes including, bridge counts, constraint values, hierarchy, and ego betweenness. Since we are analyzing information flow and actors with highest influence in the community, we use ego betweenness that examines the extent to which an actor is between other actors in the network (Everett and Borgatti 2005). The equation for calculating ego betweenness (C_B) is given below (Burt 2008; Prell 2012).

$$C_B(k) = \sum_{i \neq j \neq k} \frac{\partial_{ikj}}{\partial_{ij}}$$

where

∂_{ikj} is the number of paths linking actors i and j that pass through actor k
∂_{ij} is the number of paths linking actor i and j.

Having a high ego betweenness value is highly correlated with having many structural holes that subsequently gives potential information control to principal actors. Simply put, structural holes give competitive advantage to actors whose relationships span the holes as they gain the ability to "broker" information to other actors and thereby influence the level of collective knowledge in the community. Thus, few structural holes indicate a well-connected network with high information flow that could be beneficial for increasing adaptive capacity and community resilience. However, an ideal network not only has fewer structural holes but also must have actors with links across sectors and power levels. This is because structural holes measurement also infers to the quality of information circulating in a network as connections among similar actors assume high circulation of redundant information. Conversely, diversity of actors connected in a network suggests high quality of information that facilitates spread of new ideas and behaviors in the community.

5.2 Density

Network density is the average strength of connection between contacts (Burt 2008). It is an indicator of how actors are linked together (Prell 2012). The density (d_i) formula below calculates the proportion of ties present in a network and helps to understand the community behavior, attitudes, and performance.

$$d_i = \frac{L}{n(n-1)/2}$$

where

n is the number of actors connected to actor i
L is the number of lines between the actors.

Density scores indicate levels of cohesion among actors because the higher the density, the higher number of ties between actors based on the assumption of close communication in the community. For example, poor adaptation project implementation may be due to fragmentation in the community that can be solved by applying social network analysis to identify where to deliberately create the missing links.

6 Analytical Framework

Presence of diverse actors in a single network implementing water conservation activities increases adaptive capacity of the entire community. Therefore, we posit that synergy will enhance coordination among actors towards achieving strategic goals for comprehensive water security in the community. This study uses the social relational approach centered on quantitative social network analysis to investigate how patterns of social relations among actors enable and constrain actors and processes in Kenya. Figure 1 illustrates the interaction between the identified key issues in water governance: climate change impacts, adaptation measures, governance aspects, and social network characteristics. Climatic change is a major challenge to water security together with demographic, economic, environmental, social, and technological forces in Africa. Secondly, the dynamism of water resources requires diverse actors to cope with anticipated changes and enhance governance. For example, unpredictable rainfall patterns in a poorly governed rural area with people having low adaptive capacity will not only experience physical shortage of water but also will practice activities that pollute surface water and in the long run reduce their ground water levels. Consequently, analysis of the water network will diagnose the problem in such a community linkage that could either be poor *representation* (structural holes) by diverse actors and *low participation* (network density) of stakeholders in decision making.

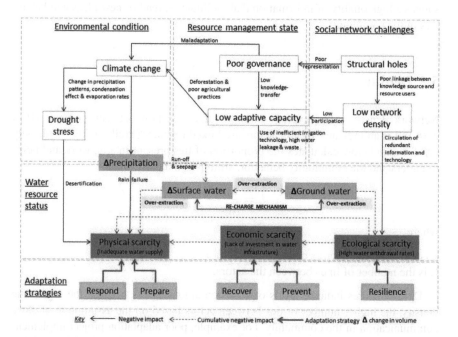

Fig. 1 Analytical framework that combines climate change, water governance, social network concepts, and five pillars of adaptation for enhanced water security

The solution to these two network structural problems is to identify *brokers* who are actors with the highest ability of bringing together a range of actors from different sectors (formal and informal) and power levels (Ernstson et al. 2010). By bridging two unconnected alters, an actor becomes capable of filtering and acquiring diverse information so that they transfer accurate and timely information through the network. Such actors seal structural holes and enhance ability of the community to respond effectively to water security issues. Thus, identification of brokers with accurate knowledge about the five pillars of water adaptation strategies and high actor linkage will equip the entire community with information to address the three types of water scarcity and promote capacity building for long-term water security.

7 Institutional Framework for Water Governance in Kenya

Kenya's water reforms initiated a paradigm shift from a narrow and sectorial approach to collaborative and multi-institutional approach in watershed management. This process structure was enshrined in the Water Act of 2002, which provides the legal framework for establishing new institutions at the national and community level, described below (Fig. 2).

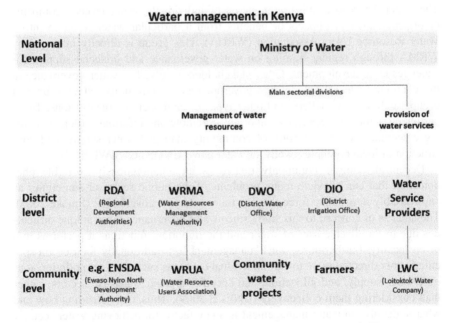

Fig. 2 Hierarchical arrangement of institutions involved in water governance in Kenya

7.1 National and Regional Level

The main institution in charge of water issues in Kenya is the MoW. Decentralization introduced other institutions such as Water Services Trust Fund (WSTF), the Water Resources Management Authority (WRMA) and departments like Water Services Board (WSB), and Water Resource Management, Irrigation and Drainage and Land Reclamation that have separate mandates within the water sector (UNDP and SIWI 2007). The government had previously established several regional development authorities to ensure equitable development through implementation of integrated programs and projects. Six institutions are charged with implementation of sustainable development within the ecosystems of major rivers and the coastline of Kenya, namely: Tana-Athi Regional Development Authority (TARDA), Kerio Valley Development Authority (KVDA), Lake Victoria Basin Development Authority (LBDA), Ewaso Nyiro North Development Authority (ENNDA), Ewaso Nyiro South Development Authority (ENSDA), and Coastal Development Authority (CDA). Their main objective is to complement the MoW projects and programs in water resource management. This chapter will focus on ENSDA established in 1989 by the Act of Parliament, Chapter 447 of the laws of Kenya and started operations in 1991.

7.2 Community Level

The Water Act also enables aggregation of individuals, water project, company, or organization that impacts or benefits from a particular water resource into a Water Resource Users Association (WRUA). This group is directly managed by WRMA through regular training on water governance and financial support for water resource development. Other stakeholders involved in water governance at the community level include, non-governmental organizations (NGOs), interest groups such as water-sellers, and other types of civil society organizations. These peripheral actors are especially important in remote and informal settlements during emergency relief, provision of community managed water supply, and construction of local boreholes, wells, or water pans (UNDP and SIWI 2007).

Decentralization is commonly treated as an unambiguously desirable phenomenon that can alleviate many problems of the public sector or sometimes as an invariably destructive force that frustrates effective government (Smoke 2003). The latter is of concern to this study since water governance faces unique difficulties due to the diverse uses of water and the important functions it performs in a given locality. Furthermore, water management has traditionally focused on specific factors directed more toward individual concerns such as water pollution control, water supply, and allocation, and specific targeted water-use sectors, rather than considering them collectively (UNECE 2009). Thus, understanding how and what to devolve in water management is a key factor for achieving water security especially under unreliable climate conditions.

8 Case Evidence on Water Adaptation Practices Under the IWRM Strategy

This case study is based on Lake Amboseli Internal Drainage Basin located in Loitoktok district in southern Kenya (Fig. 3). Loitoktok has two key rainfall seasons in the area, i.e., heavy rains in October to December and light rains in March to May. In addition, the area hydrologically benefits from the presence of Mount Kilimanjaro in two ways. First, evapotranspiration and condensing capacity of montane forests on the mountain add up to 10–25 % of the total annual rainfall (Grossmann 2008). This contribution results in uneven rainfall distribution in Loitoktok whereby the lowest elevation receives about 500 mm while the mountain slopes record an average of 1,250 mm (Government of Kenya 2009a). Similarly, temperature varies with altitude from as low as 10 °C on the eastern slopes of Mt. Kilimanjaro to a mean maximum of about 30 °C around Lake Amboseli. Secondly, the Amboseli basin also receives both surface run-off and groundwater (recharged at the forest zone between 1,500 and 3,000 m above sea level) from Mount Kilimanjaro (Grossmann 2008). Since precipitation is not enough to support the growing agricultural sector and emerging economic development in the case area, stakeholders have constructed 9 major irrigation schemes, 20 small-scale irrigation projects, 5 water system projects, 3 community water pans, 25 boreholes, 5 urban piped water schemes, and 300 shallow wells to increase local water supply (Government of Kenya 2009a). Loitoktok also supplies water through the 100-km-long old railway pipeline that transmits 17 L/s and the 262-km-long Noolturesh pipeline that transmits 200 L/s to other nearby towns

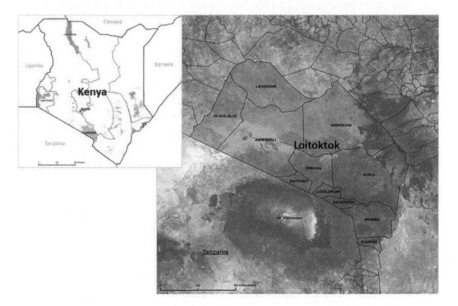

Fig. 3 Geographical location and administrative units of the study area in Kenya

such as Kajiado, Machakos, and Athi River (Grossmann 2008). Demand for water in Loitoktok district is growing rapidly due to accelerated land subdivision that is also complicating water allocation strategy between human consumption, livestock, wildlife, irrigation, and infrastructure construction.

8.1 Loitoktok Water Governance Actors

Figure 4 gives a sociogram of how actors are connected to each other while implementing different activities within the water sector in Loitoktok. A link between two actors constituted of their interaction in three activities—financial support, research and technology development, and/or project implementation in water governance at Loitoktok. To make the network more understandable, the actors were categorized according to their type and the number of structural holes for each actor was used to determine the size of the actors.

- Public actors
  Five government agencies manage the water resources in Loitoktok district, namely
  - District Water Office (DWO)—deals with domestic water data and is concerned with management of Rombo and Galeren rivers.
  - District Irrigation Office (DIO)—deals with all irrigation issues in the district together with the District Agricultural Office (DAO).
  - Water Resource Management Authority (WRMA)—responsible for protection of water sources, i.e., the Noolturesh River; administers water resource regulation, e.g., permits for water extraction and discharge and training of WRUAs.
  - Noolturesh Loitoktok Water Company (LWC)—provides piped water to urban centers in Loitoktok, i.e., Kimana, Loitoktok, and Isara. It is also commercially developing Gama springs and Kikelewa Springs
  - Ewaso Nyiro South Development Authority (ENSDA)—deals with protection and rehabilitation of Itila natural springs in the area.
  Other government agencies involved in water management are Kenya Wildlife Service (DKWS) for resolving wildlife conflicts and Kenya Forest Service (DKFS) for managing the water catchment forest zone at the slopes of Mt. Kilimanjaro.
- Private actors
  These are foreign government or international organization actors that usually collaborate with the Kenyan government. They are as follows:
  - Red Cross is responsible for establishing a canal lining project to reduce loss of irrigation water,
  - AMREF (African Medical and Research Foundation) carries out excavation of shallow wells for rain storage and conducts training on clean water and sanitation,

- SNV (Netherlands Development Organization) is involved in water provision,
- United Nations Children's Fund (UNICEF) conducts the water, sanitation, and hygiene (WASH) programs in the community.
- Non-governmental actors (NGOs)
  These actors work to increase water provision, distribution, and building capacity in the district. The non-governmental organizations that are well established in the community are Lewet-Kenya, Noomayianat, and Nia.
- Civic actors
  The civic actors comprise of 66 water project groups that are registered by the District Social Development Office (DSdO) and several business groups that are licensed by Local Government (DLG) to sell water to the community. Loitoktok has two groups of WRUA that are funded and regularly trained by WRMA in water management and resolution of water conflicts.

8.2 Loitoktok Social Network Analysis Results

The social network data was analyzed for density, structural holes, and suitable brokers. The network has a density of 0.13 confirming low linkage between stakeholders involved in water governance. The actor with the highest linkage "power" in the network is DWO who is 55 % linked to the rest of the stakeholders (Fig. 4). In terms of structural holes, DWO has an ego betweenness value of 109, DLG

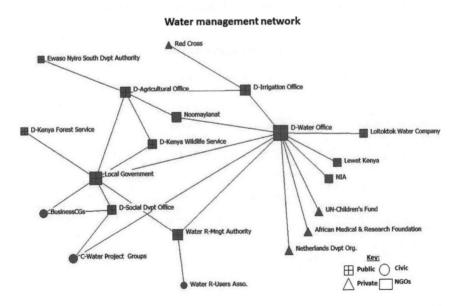

Fig. 4 Sociogram of water governance actors at Loitoktok

has 37, DAO has 18, WRMA 4 while water project groups and Noomayianat each have 2, and the rest of the actors have zero values. This means that an actor like DKWS with an ego betweenness value of 0 and an ego density of 1 is fully connected to his neighbors and thus assumedly could be implementing and sharing redundant governance information. Secondly, incomplete linkage among the Loitoktok actors has created many structural holes that hinder new information introduction and contribute to poor community representation in the network. Consequently, potential knowledge brokers in the network are DWO, DLG, DAO, WRMA, and Noomayianat that can be able to bridge these "information holes" for enhanced resource governance in the community.

8.3 Water Conservation and Adaptation Activities

Figure 5 gives the measures that are implemented by stakeholders to enhance water security in Loitoktok. Dominant activities include rain water harvest at individual homesteads and construction of water pans in communal areas for livestock and wildlife. The private actors promoted improved sanitation and hygiene practices as a way of reducing waterborne diseases. UNICEF, SNV and Red Cross, and smaller community-based organizations also initiate leadership interventions for increased collective actions in water infrastructure. Most of the technical knowledge on water harvesting and efficient irrigation came from the government agencies. This good performance in water conservation clearly indicates that the Loitoktok community is equipped with the technical know-how and knowledge in water governance but the question remains are they integrated actions?

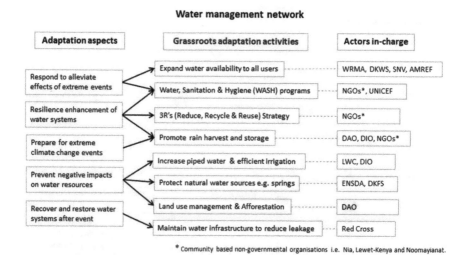

Fig. 5 Implemented adaptation activities and the respective actors in Loitoktok

The policy framework of IWRM facilitates diversity of actors to work in a single location to enhance water security for effective resource management. We use Born and Sonzogni (1995) definition to critique the level of integration in water management in Loitoktok district. First, the IWRM has a *comprehensive* aspect because the implemented measures encompass the five adaptation pillars for continuous water security. Secondly, the scheme is *strategic* in its operation because the implemented activities target-specific water issues (source, resource and user) that contribute to local water security. However, the integrated strategy fails to aid interconnective and coordinative actions among the actors resulting in independent implementation of water conservation and adaptation measures. This setback becomes a hindrance to building adaptive capacity and resilience in the community.

The analysis answers the research question by revealing lack of synergy among rural actors involved in IWRM. Though water conservation and adaptation activities are being implemented in the district, they are done in an independent manner that reduces coverage and access in the entire community.

9 Conclusion

Including concepts of social network theory in investigating rural water governance allows this chapter to contribute to the adaptation discourse through revealing weakness in decentralization that affect uptake of adaptation measures in rural communities. It is apparent from the study that IWRM in Kenya displays significant decentralization in institutional arrangements. However, effectiveness of the decentralized structures in pushing the climate adaptation agenda leaves much room for improvement. The case study of Loitoktok confirms that diverse actors have implemented solutions to the three factors instigating water security in the district namely, insufficient water supply, long distance to water points, and encroachment of water catchment areas. These solutions revolve around water storage, effective irrigation, water-point protection, and building capacity on improved sanitation and hygiene practices. However, this implementation is done independently due to lack of synergy in the Loitoktok IWRM strategy. This has resulted in small-scale gains in securing water for the region.

At the national level, there is need for alignment between national development objectives and rural strategy plans because as population increases and more urban centers are constructed then competition for water for the domestic, agriculture, and industrial sectors will intensify. Early planning and participatory technology development to develop and implement simple cost-effective water conservation measures can safeguard against future water crisis. Finally, we call for an IWRM plan that expedites linkage among local stakeholders to upscale existing adaptation measures. Such actions will strengthen rural water security, reduce resource conflicts, foster cooperative solutions, and even open opportunities for livelihood diversification.

Acknowledgments Research for this article was funded in parts by the German Science Foundation (DFG) through the Cluster of Excellence CliSAP (EXC177), Deutscher Akademischer Austauschdienst (DAAD), National Council for Science and Technology-Kenya (NCST) and Centre for a Sustainable University-Hamburg.

References

Adger WN, Agrawala S, Mirza MMQ, Conde C, O'Brien K, Pulhin J, Takahashi K (2007) Assessment of adaptation practices, options, constraints and capacity. In: Parry ML, Canziani OF, Palutikof JP, van der Linden PJ, Hanson CE (eds) Climate change 2007: impacts, adaptation and vulnerability. Contribution of working group II to the fourth assessment report of the inter-governmental panel on climate change. Cambridge University Press, Cambridge, pp 717–743

Altmann J, Alberts SC, Altmann SA, Roy SB (2002) Dramatic change in local climate patterns in the Amboseli basin. Kenya Afr J Ecol 40(3):248–251

Born SM, Sonzogni WC (1995) Integrated environmental management: strengthening the conceptualization. Environ Manage 19(2):167–181

Burt R (2008) Measuring access to structural holes. University of Chicago Graduate School of Business, Chicago

Burt R (2011) Brokerage and closure: an introduction to social capital, 2nd edn. Oxford University Press, Oxford

Carius A (2009) Climate change and security in Africa challenges and international policy context. United Nations, Berlin, p 15

Ernstson H, Barthel S, Andersson E, Borgström ST (2010) Scale-crossing brokers and network governance of urban ecosystem services: the case of Stockholm. Ecol Soc 15(4):28

Everett M, Borgatti SP (2005) Ego network betweenness. Soc Netw 27(1):31–38

Government of Kenya (2009a) Loitoktok district development plan. Office of the Prime Minister, Ministry of Planning National Development and Vision 2030, Nairobi

Government of Kenya (2009b) Ministry of water and sanitation: ministerial strategic plan 2009–2012. Government printers, Nairobi Kenya

Grossmann M (2008) The Kilimanjaro aquifer (working paper). German Development Institute, Germany, pp 91–124

Lourenço N, Rodrigues L, Machado C (2004) Social networks and water management decision-making: a methodological approach to local case studies (no. EVK1—2000—00082). Universidade Atlântica, Italy, p 55

Millennium Ecosystem Assessment (2005) Ecosystems and human well-being, vol 5. Island Press, Washington

Mutunga K (2001) Water conservation, harvesting and management (WCHM)—Kenyan experience. In: Stott DE, Mohtar RH, Steinhardt GC (eds) Sustaining the global farm. 10th International Soil Conservation Organization Meeting. Purdue University and the USDA-ARS National Soil Erosion Research Laboratory, pp 1139–1143

Ngigi SN (2009) Climate change adaptation strategies: water resources management options for smallholder farming systems in sub-Saharan Africa. Columbia University, MDG Centre, New York

Prell C (2012) Social network analysis: history, theory and methodology, 1st edn. Sage, London

Prell C, Reed M, Racin L, Hubacek K (2010) Competing structure, competing views: the role of formal and informal social structures in shaping stakeholder perceptions. Ecol Soc 15(4):34

Ribot JC (2002) African decentralization: local actors, powers and accountability. United Nations Research Institute for Social Development, Geneva

Schewe J, Heinke J, Gerten D, Haddeland I, Arnell NW, Clark DB, Kabat P (2013) Multimodel assessment of water scarcity under climate change. In: Proceedings of the National Academy of Sciences (special issue), p 6. doi:10.1073/pnas.1222460110

Schilling J, Akuno M, Scheffran J, Weinzierl T (2012) On raids and relations: climate change, pastoral conflict and adaptation in Northwestern Kenya. In: Bronkhorst S, Bob U (eds) Climate change and conflict: where to for conflict sensitive climate adaptation in Africa? Human Sciences Research Council, Durban

Smakhtin V, Revenga C, Döll P (2004) A pilot global assessment of environmental water requirements and scarcity. Water Int 29(3):307–317

Smoke P (2003) Decentralization in Africa: goals, dimensions, myths and challenges. Public Adm Dev 23(1):7–16

Thompson LG, Brecher HH, Mosley-Thompson E, Hardy DR, Mark BG (2009) Glacier loss on Kilimanjaro continues unabated. Proc Natl Acad Sci 106(47):19770–19775

UNDP and SIWI (2007) Improving water governance in Kenya through the human rights-based approach: a mapping and baseline report (mapping and baseline report). Governance in Development International on behalf of UNDP, Nairobi, p 186

UNECE (2009) Guidance on water and adaptation to climate change. United Nations Publications, New York

UNEP (2005) Hydropolitical vulnerability and resilience along international waters. United Nations Environment Program, Nairobi

UNEP and WRC (2008) Assessment of transboundary freshwater vulnerability. Division of Early Warning and Assessment, North America

Wamalwa IW (2009) Prospects and limitations of integrated watershed management in Kenya: a case study of Mara watershed (masters' program in environmental studies and sustainability science). Lund University, Sweden

Wangai P, Muriithi J, Koenig A (2013) Drought related impacts on local people's socioeconomic life and biodiversity conservation at Kuku group Ranch, Southern Kenya. Int J Ecosyst 3(1):1–6. doi:10.5923/j.ije.20130301.01

Ward FA, Michelsen A (2002) The economic value of water in agriculture: concepts and policy applications. Water Policy 4(5):423–446

WBGU (2008) Climate change as a security risk. Earthscan, London

Wily L, Dewees PA (2001) From users to custodians: changing relations between people and the state in forest management in Tanzania, vol 2569. World Bank Publications, Washington

Authors Biography

Grace W. Ngaruiya did her Ph.D. on quantitative social network analysis of rural actors involved in climate governance and natural resource management in the ASALs of Kenya and is now based at Kenyatta University as a lecturer in Conservation Biology.

Prof. Jürgen Scheffran is head of the KlimaCampus Research Group Climate Change and Security (CLISEC) at the Institute of Geography of the University of Hamburg, with a focus on climate vulnerability and adaptation in the water–food–energy nexus in Northern Africa and Southern Asia.

Liang Lang did his Ph.D. on climate change impacts and social adaptation, focusing especially on water risk assessment and urban responses simulation in coastal areas of China.

Eco-feedback Technology's Influence on Water Conservation Attitudes and Intentions of University Students in the USA: An Experimental Design

Janna Parker and Doreen Sams

Abstract Water conservation is a universal issue. One segment of the population with a significant future impact on water conservation is college-aged students. College students in the United States of America (USA) are typically not held responsible for their individual utility bills when living in dorms and there are little to no incentives to conserve resources. At one small public liberal arts university in the Southeastern (USA), water has been used in alarming amounts over the last few years. A sample of college students ($n = 208$) from the university participated in an experiment to determine their attitudes, behaviors, and intentions. This paper discusses relevant literature and explains the research methodology (2×2 between subjects randomized across treatments experiment) that examined attitudes and intentions as to purchasing eco-feedback technology and the role of marketing in consumers' choices. The paper identifies hypothesized relationships to be measured. It presents the findings as to the influence of novelty of eco-feedback technology, personal value (economic and emotional), attitude toward environmentalism (substantive and external), price, and knowledge of green living products influence on intentions to purchase. Further, it reports conclusions, limitations, and practical implications.

Keywords Eco-feedback technology · Water conservation · Marketing · Environmentalism

1 Introduction

It is an undeniable fact that clean water is a limited resource and that conservation is vital to long-term survival of life on Earth (Kappel and Grechenig 2009). For example, from 1987 to 1992, California, USA was in a severe drought and in some

J. Parker (✉) · D. Sams
Georgia College & State University, Milledgeville, GA, USA
e-mail: janna.parker@gcsu.edu

© Springer International Publishing Switzerland 2015
W. Leal Filho and V. Sümer (eds.), *Sustainable Water Use and Management*,
Green Energy and Technology, DOI 10.1007/978-3-319-12394-3_9

places water was rationed. The current drought in California reinforces the idea that the responsibility for water conservation lies with everyone. The governor of California issued a statewide request for all individuals to cut water usage by 20 %. In an unprecedented move, he also declared that water would not be sent from the State reservoirs to local communities. This policy change affected the drinking water supply of 25 million residents and the irrigation of over one million acres of farmland (Williams and Dearen 2014). California is the number one dairy State in the USA and is also responsible for producing 50–99 % of many fruits, vegetables, and nuts in the USA. The long-term effects of the drought cannot be fully predicted; however, higher food prices across the nation are expected (CDFA 2013). This is but one example of how vital water conservation is to life on Earth. Examples of current water conservation programs in San Francisco, California focus not only on education, but also the providing of free equipment such as low-flow showerheads, free books on water reduction for gardening, and rebates for water efficiency certified appliances for both residential and business customers (SFPUC 2014). While the results are encouraging from these water conservation initiatives in California, it is important to note that San Francisco's water usage reduction is based mainly on financial incentives and free equipment. Many communities or universities throughout the USA do not have the financial resources to fund programs such as those offered by the San Francisco Public Utilities Commission.

Just as State and local communities must look for strategies that will increase the awareness for the need for water conservation, universities are focusing on water conservation as well. Reduced funding for higher education in many States has created a need to find savings in budgets. The administrations of many universities have identified water and energy as areas for reducing costs in the overall budget. Therefore, this research study examined the perceptions of college students (future business leaders, industry leaders, influencers, and residential consumers) of two types of eco-feedback water saving technologies marketed to change water saving behavior to determine the students' perceptions of, and intentions to purchase the devices, as well as the economic and psychological drivers of their consumption choices. Understanding what marketing draws millennials (18–30 year olds) in the USA to purchase water saving devices provides valuable insights into how to market to the millennial generation of consumers. Although many studies have investigated the effects of environmental learning requirements (ELRs) on college students (millennial generation) (Kagawa 2007; Moody and Hartel 2007), they have shown mixed results. This indicates that education alone may not be the answer to water conservation.

The university participating in this study reported water usage that over an extended period of time has been higher per capita than many other universities in the USA. The university's operations department and university housing office personnel conducted studies to determine where the majority of the water was being used. The source of excessive water usage was identified as occurring in the dorms. In an effort to resolve the water usage problem, management at the subject university investigated changing the existing showerheads to low-flow showerheads and/or installing other water saving devices. However, previous research revealed that low-flow showerheads are not extremely efficient in reducing water usage in the home; therefore, other options were considered by the subject university (Heaney et al. 1999). Budget constraints, previous research on low-flow showerheads, and the university's inability to retrofit buildings with individual room water meters which would enable the university to charge students for their individual water consumption led to the current study.

This study sought solutions to water conservation on a university campus where students have no direct responsibility for paying for water usage. Saving water is strictly voluntary. Although most students have the knowledge that they are *wasting a scarce resource,* encouraging water conservation must take a *carrot approach* (eco-feedback technology), as the *stick approach* (individual fiscal responsibility) is impossible.

To date, many water conservations strategies (e.g., advertising campaigns and educational events) at the university in this study have not met campus expectations in water usage reduction. The subject university's environmental education efforts of: (1) maintaining an active sustainability council (approximately 7 years), (2) teaching sustainability at many levels across disciplines, (3) bringing in outside speakers, (4) mandating sustainability measures across campus, and (5) sponsoring events related to sustainability have not been shown to have a significant impact on students' behavior to date. This begs the question, what factors will influence students to conserve water? The current study is part of a larger study to determine ways to influence behavior in order to reduce unnecessary water usage and combat soaring utility costs. The university that participated in this study is in the process of pilot testing eco-feedback water technology for the dorms in an attempt to reduce water usage, thus reducing costs to the university. Before purchasing and installing the technology, the researchers were asked to conduct a study to determine whether the university would benefit from this purchase and if students were open to using the technology.

2 Millennials' Conservation Attitudes

Kagawa (2007) found "dissonance" between students' understanding and agreement with environmental actions, and actual behavior is based on financial and personal convenience and/or comfort. Kagawa found that students generally believe in collective sustainability efforts, but as to personal behavior changes, their proposed individual lifestyle changes do not align with their principled beliefs.

A recent study by Telefonica (2013) in conjunction with the *Financial Times* surveyed 12,171 millennials (age 18–30) in 27 developed and developing countries. The study asked millennials to rank several important problems/issues facing the world. The rankings revealed that the environment did not make the top two in any of the countries surveyed. In the USA, the economy was number one (46 %) for the respondents. This suggests that in general, millennials may not place great concern on environmental issues (Telefonica 2013). Moody and Hartel (2007) investigated the impact of an ELR at a major university in the Southeastern USA. They found that implementing a university ELR increased student knowledge, but more importantly 26 % of students self-reported that they had made environmentally related behavioral changes due to increased knowledge of environmental issues after taking courses.

2.1 Consumers' Choice Behavior

Many companies recognize that there is a growing group of consumers that care as to whether the products they buy are sustainable (Kagawa 2007). A widely known economic theory of the consumer is one of a rational maximizing models describing how consumers should choose. However, there are situations in which consumers act in a manner that appear inconsistent with economic theory creating systematic errors in behavior predictions. In order to facilitate the economic value (e.g., it saves money or is a high-quality product based on its price), it is important to remember that price is not the sole determinant and that the functional quality offered by the product may be equally as important (Thaler 1980). Another dimension of personal value is emotional value. Emotions and functional quality of the product/service have a significant impact on purchase decisions. Emotions, defined as a state of physiological arousal, include a cognitive aspect that is context specific (Consoli 2009). Shoppers do not always choose products that just meet a need, but also choose based on emotional satisfaction. Emotional needs and functional needs thus align with psychological value of product ownership (e.g., I am saving the planet) (Consoli 2009). Consumers may not necessarily intend to purchase these products based solely on their environmental benefits. However, drivers of environmentally sustainable product purchases encompass more than intentions of saving the planet. For example, other considerations influencing purchase intention behaviors have been identified as personal values (i.e., emotions and economics) (Kaplan 2000). The assumption that good motives lead to good behavior is a dangerous assumption taken in isolation. The message of giving up something as a sacrifice for the betterment of the future of all (altruistic behavior) is often perceived as personal unhappiness and that materialism and waste are more fun (Kaplan 2000). As a pure altruistic mindset is less than realistic to incite change (e.g., sustainability), other behavioral factors must be examined.

3 Interventions to Sustainable Consumer Behavior

From a behavioral science perspective, studies have shown two sets of interventions (i.e., (1) goal setting, information, commitment, and (2) modeling and consequences) in decision making both of which are noteworthy. However, consequences of student goal setting cannot be considered in this study, as change cannot be forced and thus it is not feasible or realistic (Kappel and Grechenig 2009). The subject university has clearly made the goal of water reduction known to the student body, but it is a university goal and due to construction constraints cannot be suggested as a measurable goal for individual students. The subject university has modeled sustainable behavior for the students by monitoring and reporting water leaks and breakage of sprinklers and faucets in a timely fashion.

4 Experimentation Study Focus

The current study focused on students' individual goals by examining whether students hold intentions to change their sustainable behavior by purchasing water conservation eco-feedback technologies after reading about and seeing eco-feedback technology. One of the technologies is relatively inexpensive and not novel, whereas the other is relatively expensive and novel. For the study, attitude toward environmentalism was measured across two dimensions: external and substantive. This was measured to determine whether significant differences in responses of those with high scores on attitude toward environmentalism as opposed to those with lower scores existed. External involves the individual's perception of the severity of environmental problems, whereas substantive is the weight of the knowledge of green living products on the individual personally (Banerjee and McKeage 1994). Personal value was also examined as a multidimensional construct consisting of: (1) economic—cost verses benefit and (2) emotional benefits—how does it make me feel. Personal value is relevant to this study, as research has shown that customer-perceived value (i.e., cost vs. benefit) (Zeithaml 1988) is weighed as to both monetary and non-monetary price factors such as risk of poor performance (Liljander and Strandvik 1993), whereas research has shown that consumers may not necessarily intend to purchase products based solely on economic value (Kaplan 2000).

5 Experimental Factors

One goal of this research study was to show whether students who have been exposed to multiple educational and marketing strategies about conservation of water usage would respond positively to either of two types of eco-feedback technology. Two types of water conservation eco-feedback technologies (i.e., high novelty and no novelty, and low and high price) were examined. This is an exploratory study that sought to determine whether knowledge, attitudes, consumer type, price, and/or personal values are drivers of intentions for behavioral change. A randomized between subjects (2×2) experimental design that examined two levels of eco-feedback technologies (High Novelty—light emitting-diode (LED) showerhead; No Novelty—manual shower timer) at two price levels (low/high) for each type of feedback technology was conducted. Informational advertisements were presented for each treatment across four scenarios to 208 undergraduate business students at a small liberal arts university in the Southeastern USA. The current study was conducted to explore whether various forms of eco-feedback technology might encourage a mindset of water conservation among college students. In order to address this question, students' perceptions as to two types of personal value (i.e., economic and emotional) of two types of eco-feedback technologies were examined.

One factor measured in this study was a personal value construct (i.e., economic and emotional dimensions that can be examined together or in isolation as either of these dimensions of the construct may drive intentions to change). One motivator that has been lacking for students is direct immediate feedback. The two technologies in this study provide direct and immediate feedback. Eco-feedback technology is not new. For example, ambient displays creating energy awareness such as the dimming of the computer screen when not in use for a specific period of time have been available for several years; however, this technology does not have a "novelty" factor. The novelty experimental condition in this study used ambient light display showerheads (water pressure changes the color of the light over a set period of time) signaling in three light stages (green, yellow, and ending in red) when the shower should end. This color progression is familiar to USA college students as it is the same progression in traffic lights. This is a novel technology that may elicit emotions ranging from joy to annoyance depending on personal circumstances. Previous research found that novel eco-feedback technology had a positive impact on family water reduction as long as it did not appear to reward extensive usage over long periods of time, which then reduces its effectiveness (Froelich et al. 2012). The technology that is not novel used in this experiment requires observing the progress of what is equivalent to an egg timer in the shower noting when it runs out. This is not novel, thus the range of emotions may be less than for a "light show in the shower." This study examined whether novelty seekers intentions to purchase are influenced by personal value or the technology.

From literature reviewed for this study, the following were hypothesized.

H1:   College students' knowledge of green living products or product price does not influence their intention to purchase water conservation eco-feedback technology, whereas personal value has a significant influence on intentions to purchase water conservation eco-feedback technology.

H2:   College students' intentions to purchase water conservation eco-feedback technology are influenced by personal value and not by price or attitude toward environmentalism.

H3:   Novelty seeking college students' likelihood to purchase water conservation eco-feedback technology is influenced by personal value and not the novelty of the technology.

6 Study Methodology

Juniors and seniors (*n*, 208) from the business college took part in this laboratory experiment. These students were previously exposed to various forms of education on environmental sustainability through lectures, multiple events, and advertisements on campus over their 4 years at the university. Further, this sample was chosen because all of the students in the study have taken at least one and typically two economic courses prior to engaging in this study and thus are exposed to Milton Friedman's rational choice theory (i.e., balancing cost against benefits to maximize advantage in which consumption motivations is not considered) (Friedman 1953). These students are future paying consumers of utilities such as water. However, at this time many do not directly (i.e., do not pay for water usage outside of tuition and housing fees) hold any responsibility for the payment of water used let alone excessive use of water. This sample was also selected because as college graduates with business degrees, their influence based on income earning potential in the USA currently ranges from a starting average salary of around $41,400 to mid-career salary potential of approximately $70,000 (PayScale Inc. 2013). Therefore, as future consumers educated in environmental sustainability practices and economic theory earning business degrees affords them the potential for substantial earning power, they are an ideal sample for this study.

7 Research Design

A between subjects 2 × 2 experimental design is employed to test the hypothesized relationships (See Fig. 1).

8 Study Measures

The scales measuring the constructs in this study are established scales that have previously demonstrated validity and reliability. This exploratory study fills a gap in the literature by determining whether knowledge of green living products (i.e., eco-feedback technology), personal values, attitude toward environmentalism, price, or consumer type (i.e., novelty seekers) are drivers of intentions for purchasing water conservation eco-feedback technology.

According to most philosophers in knowledge theory, there are three kinds of knowledge (i.e., propositional [knowledge by facts], personal [knowledge by

Fig. 1 Eco-feedback novelty seeking technology

Price	Novelty	
	Novel/High Price	Not Novel/High Price
	Novel/Low Price	Not Novel/Low Price

acquaintance], and procedural [knowing how to do something]). Only two dimensions are relevant for this study (propositional and personal). Procedural was not part of the study as the usage procedure for each product was part of the product description in the experiment. The Mukherjee and Hoyer (2001) two-dimensional knowledge scale (i.e., propositional knowledge and personal knowledge) consists of two items measured on seven scale points (1 = not at all knowledgeable and 7 = very knowledgeable) and one item measured on seven scale points (1 = very little experience and 7 = a lot of experience) was adapted for this study of green living products. The scale reliability for the Mukherjee and Hoyer (2001) study was 0.81. The scale is amenable to either one product category or multiple product categories and thus is appropriate for this study. The personal value construct for this study was measured with a two-dimensional scale [i.e., (1) emotional—five items (e.g., *this eco-feedback technology offers value for the money*) and (2) economic—four items (e.g., *the eco-feedback technology makes me happy*)] (Sweeney and Soutar 2001). This scale was developed to measure consumers' perception of products prior to actual purchases or immediately after a purchase making it relevant to this study. This is a nine item scale with endpoints of 1 = strongly disagree to 6 = strongly agree. Scale reliability was high ranging between 0.80 and 0.94. These two-scale dimensions were correlated at 0.74 (CI) with a standard error of 0.03 demonstrating discriminant validity. An adapted attitude toward environmentalism scale [i.e., two of three dimensions: (1) substantive environmentalism and (2) external environmentalism)] was modified from the Banerjee and McKeage (1994) scale. Scale reliability for the substantive environmentalism dimension was reported as 0.79 and for the external environmentalism was 0.87. Both demonstrate high reliability. The substantive environmentalism dimension examines, "individual perceptions of the severity of environmental problems." The external environmentalism dimension examines, "convenience, economic trade offs, and external perception of environmental problems (e.g., media attention)" (Banerjee and McKeage 1994, p. 149). The excluded dimension of the scale measured internal environmentalism was believed by the researchers to foster social responding bias. Price was measured with the actual and doubled prices of the product. For the non-novel product, the prices were ($5.99–$11.99) and for the novelty product, the prices were ($29.95–$59.95). Novelty seeking is broadly defined to include social coolness (i.e., good, hip, or fashionable) and technical coolness (i.e., technologically interesting or advanced) (Bodine and Gemperle 2003). This three-item novelty seeking seven-point (1 = strongly disagree to 7 = strongly agree) scale was used to measure consumer type in the current study (Oliver and Bearden 1985). The Oliver and Bearden scale (reliability 0.72) was originally used to measure the effect of heavily promoted time-released diet suppressants. The Burton et al. (1999) scale measuring the intention to purchase in this study is an established scale that has demonstrated validity and reliability. The scale was appropriate for this study as it is measures information read about the product to measure respondents' intentions to purchase (e.g., the eco-feedback technology). This three-item semantic differential scale has endpoints of more likely/less likely; very probably/not probable; and very likely/very unlikely. The Burton et al. (1999) study revealed a high reliability of 0.89. However, analysis of scale validity was not described in the original study.

9 Experimentation Study Materials

Students consisting of juniors and seniors across six classes in the college of business of the subject university were given the opportunity to participate in the experiment. Incentives were provided in the form of extra credit points. Those wishing not to participate were given alternative opportunities to gain points.

The procedures used by the researchers were consistent across courses and sections. Students were given folders that were randomized as to treatment order and students were instructed to not open them until otherwise instructed. The product photos and descriptions (1) novel product and (2) non-novel product were randomized within the folders. Each folder contained two separate scenarios for a water conservation eco-feedback product at different prices. Packet #1 [scenario #1—high price/novel; scenario #2 low price/not novel, etc.] Treatments were randomized to address order bias. Participants signed an agreement of confidentiality and two consent forms prior to beginning the experiment. Throughout the experiment, the students saw one treatment at a time. After each treatment, students filled out a questionnaire measuring the relevant variables discussed above.

10 Scale and Model Purification

The three-item novelty scale demonstrated reliability of 0.678, which was not acceptable and thus was not included in the Confirmatory Factor Analysis (CFA). As demonstrated in the Bergkvist and Rossiter (2007) study, a single-item predictor can have better predictive validity than a multiple item measure for concrete concepts. Thus, if scale items #1 and #3 were removed, the reliability would have been 0.797; therefore, it was determined that scale item #2 would be used as the measure (i.e., *I am usually among the first to try new products*). This item was the only scale item specifically related to the purchase of a novel product. The other two were generic in nature. A CFA was conducted using AMOS2 for these four constructs: Personal value [i.e., economic value (EconVal) and perceived emotional value (EmotVal)], knowledge (product category of green living products—know), attitude toward environmentalism (i.e., substantive environmentalism and external environmentalism), and purchase intentions (i.e., likelihood to purchase water conservation eco-feedback technology—likely). After running the full model, it was determined that some items needed to be dropped in order to achieve a "good fit," acceptable reliability, and average variance extracted for one of the constructs. Two items were dropped from the external environmentalism dimension and three items were dropped from substantive environmentalism dimension of the attitude toward environmentalism scale leaving a minimum of three scale items per dimension as needed for the constructs to be identified. Items were selected for deletion due to low standardized regression weights (Hair et al. 2010). Table 1 contains the results of the CFA for both the "full" model and the "respecified" model.

All indices for the respecified model fell within accepted ranges (Hair et al. 2010). After determining that the measurement model had good fit, the reliability and validity were assessed. The estimated loadings (E.L.), standard errors (S.E.), critical ratios (C.R.), standardized regression weights (S.R.W.), reliability (α), and average variance extracted (AVE) are found in (Table 2).

Construct reliability was established using coefficient alpha and all constructs met the minimum level of 0.7. Convergent validity was established by determining the AVE for each construct. The desired level is 0.50 (Hair et al. 2010). All constructs met this requirement with the exception of the substantive environmentalism dimension of the attitude toward environmentalism construct (Substantive Env) at 49.10 %. Since only one dimension of one construct missed this by less than a percentage point, the measurement model should be considered to have convergent validity. Additionally, the critical values for each variable were significant ($p < 0.05$). Discriminant validity was assessed by comparing the AVE for each construct with the squared inter-construct correlation estimates (SIC).

Table 1 Results of confirmatory factor analysis $n = 208$

Measurement of fit	"Full" model	"Respecified" model
Chi square	566.112	345.051
Degrees of freedom	309	194
Probability	0.000	0.000
CMIN/DF	1.832	1.779
Comparative Fit Index (CFI)	0.906	0.937
Root mean square error of approximation (RMSEA)	0.063	0.061
Confidence interval for RMSEA	(0.055; 0.072)	(0.051; 0.072)

Table 2 Indicators of reliability and validity

Construct	Variable	E.L.	S.E.	C.R.	S.R.W.	Reliability	AVE (%)
PriceHi						**0.88**	**71.84**
	Price1R	1.000[a]			0.919		
	Price2	0.823	0.055	15.028	0.811		
	Price3R	0.907	0.061	14.895	0.808		
External env.						**0.79**	**49.10**
	Prob2R	1.000[a]			0.796		
	Prob3R	1.114	0.126	8.838	0.671		
	Prob4R	1.139	0.139	8.176	0.619		
	Prob5R	1.214	0.132	9.229	0.705		
Substantive env.						**0.87**	**53.73**
	Action1	0.532	0.075	7.136	0.533		
	Action3	1.366	0.137	9.976	0.835		
	Action4	1.000[a]			0.794		

(continued)

Table 2 (continued)

Construct	Variable	E.L.	S.E.	C.R.	S.R.W.	Reliability	AVE (%)
Likely						**0.87**	**69.38**
	Likely1R	1.000[a]			0.851		
	Likely2	1.054	0.073	14.468	0.859		
	Likely3R	0.961	0.074	12.956	0.787		
Know						**0.91**	**76.75**
	Know1	0.997	0.069	14.393	0.874		
	Know2	1.057	0.070	15.090	0.958		
	Know3	1.000[a]			0.789		
EconVal						**0.89**	**61.62 %**
	EconVal1	1.000[a]			0.692		
	EconVal2	1.087	0.100	10.878	0.820		
	EconVal3	1.301	0.109	11.926	0.927		
	EconVal4	0.953	0.105	9.058	0.674		
						0.86	55.43 %
	EmotVal1	1.444	0.130	11.123	0.887		
	EmotVal2	1.581	0.137	11.529	0.944		
	EmotVal3	1.084	0.131	8.267	0.626		
	EmotVal4	0.792	0.116	6.852	0.510		
	EmotVal5	1.000[a]			0.664		

[a]Loading set to 1.0. Not estimated

Table 3 Squared interconstruct correlation estimates

	Substantive	External	Likely	Knowledge	EconVal	EmotVal
Substantive env.	1.00					
External env.	0.31	1.00				
Likely	0.09	0.11	1.00			
Knowledge	0.00	0.00	0.00	1.00		
EconVal	0.03	0.08	0.41	0.00	1.00	
EmotVal	0.05	0.14	0.41	0.00	0.08	1.00

The AVE for constructs was higher than the SIC for each construct pair indicating discriminant validity. The SICs are found in Table 3. Nomological validity was assessed by evaluating the covariances between the constructs. Table 4 below contains the significant covariances between constructs.

Likely to purchase had a significant positive covariance with every construct except for *knowledge of green living products*. It would not be expected that all constructs would be significantly related to each other. It was important to find significant covariances between likely to purchase and the independent variables in the study. Additionally, since many green living products are new, many people are still seeking information about these product categories; therefore, the lack of a positive covariance is not surprising.

Table 4 Construct
covariances

Constructs	Estates	S.E.	C.R.	$p < 0.001$
Likely ↔ Substantive	0.324	0.093	3.483	***
Likely ↔ External	0.470	0.125	3.752	***
Likely ↔ EconVal	0.638	0.114	5.614	***
EmotVal ↔ Likely	0.641	0.104	6.141	***
Substantive ↔ External	0.462	0.086	5.381	***
EmotVal ↔ External	0.289	0.070	4.111	***
EmotVal ↔ EconVal	0.180	0.053	3.414	***

***Indicates significance at <0.001

11 Findings

11.1 Manipulation Check

Upon closing the experiment treatment folders, students were instructed to "not look back into the folder" when answering the manipulation check question (what was the price of the eco-feedback technology you saw last?). All respondents answered the manipulation check correctly.

11.2 Hypotheses Testing

Based on the knowledge that students at the subject university were exposed across their 4 years of education to sustainability education and initiatives, it was important to determine their level of concern for the environment. The findings (Table 5) demonstrate that all of the constructs directly related to environmentalism suggest that most of the students in the sample do have some level of concern for the environment. All constructs were measured with seven-point scales.

The CFA findings (Table 4) revealed that all covariances were significant except for knowledge of green living products. Results from an ANCOVA supported Hypothesis #1 [f 43.041, p 0.000—knowledge of green living products p 0.098 and personal value (i.e., emotional p 0.000 and economic 0.000), r^2 0.461, alpha 0.05]. Personal value was shown to have a significant influence on intentions, whereas knowledge of green living products or product price did not have a significant influence on their intentions.

Hypothesis #2 was supported based on the results from an ANCOVA. Findings revealed [f 24.538, p 0.000—personal value (i.e., emotional p 0.000 and economic 0.000), attitude toward environmentalism (i.e., substantive p 0.535 and external p 0.072), price 0.491, r^2 463, alpha 0.05]. Thus, it was concluded that the sample's intention to purchase water conservation eco-feedback technology is influenced by personal value not price or attitude toward environmentalism.

Hypothesis #3 was not supported based on the results from the ANCOVA. Findings revealed [f 41.929, p 0.000—novelty seeking p 0.469, type of technology

Table 5 Descriptive statistics

	N	Minimum	Maximum	Mean	Std. deviation
Substantive Env.	207	1.33	7.00	5.6554	1.13936
External Env.	208	1.75	7.00	5.6274	1.00361
Knowledge	208	1.00	7.00	3.4810	1.28370
Goal	208	1.00	7.00	5.0756	1.03920
EconVal	208	1.00	6.00	3.7100	0.99978
EmotVal	208	1.00	6.00	4.2952	0.93373
Likely	208	1.00	7.00	3.1907	1.45406

p 0.291, personal values (i.e., emotional p 0.000 and economic p 0.000), r^2 0.452, alpha 0.04]. This finding revealed that novelty seeking college students' likelihood to purchase water conservation eco-feedback technology is not influence by personal value or the novelty of the technology. However, this finding should be taken with caution as the degree of novelty seeking was only 3.59 on a six-point scale (3 = somewhat disagree and 4 = somewhat agree), indicating a measurable portion of sample that may not have self-identified as being novelty seeking individuals.

12 Conclusions

Through this experimental design study, the researchers were able to establish a generalizable cause–effect relationship that is a true representation of actual behavior of college students (millennials) in the USA.

The 2010 Pew Research Center report was a comprehensive study of millennials. Millennials consist of 77+ million individuals. This makes their consumption potential greater than even that of the Babyboomers (i.e., *ages* 49–67 in 2013) (Pew Research Center 2010). Based on the sheer size of the population, understanding their "personality" as a group is important to the successful marketing of water conservation products. Water conservation products that meet the personal value standards along with water conservation education for this large demographic segment of the population (millennials) are expected to significantly reduce water management needs as the behaviors of the individual within a reference group (e.g., colleagues at work) is expected to influence other members of the group (Asch 1952).

The majority of the students (millennials) in the current study have been exposed repeatedly over four years of college to environmental sustainability issues; yet, this study found environmental knowledge among the respondents was not a significant driver of intention to purchase eco-feedback technology. This supports the findings of the Telefonica's (2013) study in conjunction with the *Financial Times* that millennials in the USA are not as concerned with societal issues, but rather focus on personal values. Therefore, one important takeaway from this study is that despite an ongoing educational program at the university, students' intentions to purchase water conservation eco-feedback technology regardless of price or novelty were influenced solely by personal values with no indication that educational experiences had a direct influence on their choice. It further supported the Kagawa (2007) study, in that college students believe in collective sustainability efforts, but not personal responsibility. One possible explanation for this is that although their beliefs do not consciously influence personal behavior, it may occur subconsciously and should be considered in future research.

Another interesting conclusion from the study is that students who self-reported as, "novelty seekers" were no more influenced by the "egg timer-type eco-feedback technology" than the "lightshow eco-feedback technology" novelty when it came to their likelihood to purchase water eco-feedback technology. The findings showed that it is all about the "personal value." Whether marketing a novel product or a non-novel product, findings of this study show that the millennial consumer seeks out personal value. As long as personal value is apparent, then the purchase decision between these two types will fall on the one with the most personal value to the consumer. From this, it is possible to draw the conclusion that the management of water resources on the college campus must address the personal value of using the technology properly (stopping when the red light goes out) coupled with continuing education.

Although previous research on advertising appeals (financial vs. green) for *green* products found that purchase intentions differ across levels of environmental involvement (Schuwerk and Lefkoff-Hagius 1995), the current study did not. From the current study findings, the research team found that advertising appeals that focus on personal value appeal to college of business students regardless of the extent of their individual environmental education. The findings of this study informed the research team that regardless of strength of environmental attitudes, perceived personal value influenced likelihood to purchase among this sample of college students. From this knowledge, the research team was able to develop advertisements (focusing on personal value rather than societal benefits) and select an eco-feedback technology, from the two in this study, to be pilot tested in an upcoming longitudinal experiment to be conducted on the subject college campus dorms.

This study was a first step in the investigation of a very complex water management issue in a highly bureaucratic government institution and was never intended to be all inclusive of such a massive issue. Although findings revealed that personal value is the most significant factor as to water conservation among the target population, the strong emphasis on environmental sustainability education must be considered as a contributing (although unconscious) factor. However, other variables should be included in future studies such as goal seeking and usage behavior overtime. This study informs research as to factors that influence whether eco-feedback technology has its place in water conservation among millennials. It also demonstrates the worthiness for testing eco-feedback water saving technology in a real-world application. This study was limited by the sample from only the college of business. Students in a college of business and those studying environmentalism and the sciences may report differing beliefs and purchase intentions. Therefore, an interdisciplinary study may be more revealing of the true environmental nature across millennials.

References

Asch S (1952) Social psychology, 2nd edn. Holt, New York

Banerjee B, McKeage K (1994) How green is my value: exploring the relationship between environmentalism and materialism. Adv Consum Res 21:147–152

Bergkvist L, Rossiter J (2007) The predictive validity of multiple-item versus single-item measures of the same constructs. J Mark 44(2):175–184

Bodine K, Gemperle F (2003) Effects of functionality on perceived comfort of wearables. In: IEEE international symposium on wearable computers (ISWC'03) proceedings of the seventh IEEE international symposium on source DBLP. Carnegie Mellon University, Pittsburgh, pp 57–60

Burton S, Garretson J, Velliquette A (1999) Implications of usage of nutrition facts panel information for food product evaluations and purchase intentions. JAMS 27(4):470–480

Consoli D (2009) Emotions that influence purchase decisions and their electronic processing. Annales Universitatis Apulensis Series Oeconomica 11(2):996–1008

Friedman M (1953) The methodology of positive economics. In Essays in positive economics. University Press, Chicago

Froelich J, Findlater L, Ostergren M, Ramanathan S, Peterson J, Wragg I, Larson E, Fu Fu, Bai M, Patell S, Landay J (2012) The design and evaluation of prototype eco-feedback display for fixture-level water usage data. In: Proceedings of the SIGCHI conference on human factors in computing systems, pp 2367–2376

Hair J Jr, Black W, Babin B, Anderson R (2010) Multivariate data analysis. Prentice Hall, New Jersey

Heaney JP, DeOreo W, Mayer P, Lander P, Harping J, Stadjuhr L, Courtney B, Buhlig L (1999) Nature of residential water use and effectiveness of conservation programs. Retrieved Mar 2013 from BCN: http://bcn.boulder.co.us/basin/local/heaney.html

Kagawa F (2007) Dissonance in students' perceptions of sustainable development and sustainability: implications for curriculum change. Int J Sustain High Educ 8(3):317–338

Kaplan S (2000) Human nature and environmentally responsible behavior. J Soc Issues 56(3):491–508

Kappel K, Grechenig T (2009) Show-Me: water consumption at a glance to promote water conservation in the shower. In: Conference proceedings of Persuasive'09, California, 26–29 April 2009

Liljander V, Strandvik T (1993) Estimating zone of tolerance in perceived service quality and perceived service value. Int J Serv Ind Manag 4(2):6–28

Moody G, Hartel P (2007) Evaluating an environmental literacy requirement chosen as a method to produce environmentally literate university students. Int J Sustain High Educ 8(3):355–370

Mukherjee A, Hoyer W (2001) The effect of novel attributes on product evaluation. J CONSUM RES 28:462–472

Oliver R, Bearden W (1985) Crossover effects in the theory of reasoned action: a moderating influence attempt. J Consum Res 12:324–340

PayScale Inc (2013). Majors that pay you back. Retrieved 18 Mar 2013 from PayScale Inc.: http://www.payscale.com/college-salary-report-2013/majors-that-pay-you-back

Pew Research (2010) A Portrait of the next generation. Retrieved Dec 2013 from Pew Research: http://www.pewsocialtrends.org/files/2010/10/millennials-confident-connected-open-to-change.pdf

Schuwerk M, Lefkoff-Hagius R (1995) Green or non-green? Does type of appeal matter when advertising a green product? J Advertising 24(2):45–54

SFPUC (2014) San Francisco water power sewer. Retrieved 3 Feb 2014 from San Francisco Public Utilities Commission: http://www.sfwater.org/index.aspx?page=129

Sweeney J, Soutar G (2001) Consumer perceived value: the development of a multiple item scale. J Retail 77(2):203–220

Telefonica (2013) Telefonica global millennial survey. Retrieved 7 Jul 2013 from Telefonica: http://survey.telefonica.com/globalreports/

Thaler R (1980) Toward a Positive Theory of Consumer Choice. J Econ Behav Org 1(1):39–60

CDFA (2013) California agricultural production statistics. Retrieved from California Department of Food and Agriculture: http://www.cdfa.ca.gov/statistics/

Williams J, Dearen J (2014) California drought: Governor Jerry Brown urges citizens to conserve water. Retrieved 3 Feb 2014 from Huff Post Green: http://www.huffingtonpost.com/2014/01/31/california-drought-jerry-brown_n_4699957.html

Zeithaml V (1988) Consumer perception of price, quality, and value: a means-end model and synthesis of evidence. J Marketing 3:2–22

Authors Biography

Janna Parker DBA (2013), Assistant Professor, Marketing (Georgia College & State University). Janna has published at conferences and in academic journals on sustainability. Prior to joining the Georgia College & State University faculty, she taught as an instructor at Louisiana Tech University and Cameron University. She has worked as a political consultant, an independent marketing consultant with local businesses and nonprofit organizations as well as the United States Air Force in Europe.

Doreen (Dee) Sams Ph.D. (2005), Associate Professor, Marketing (Georgia College & State University).Founder of GC Shades of Green, served on GC Sustainability Council, Sustainability Council for the State of GA Boys & Girls Club, developed a sustainability session for SMA, participated in Sustainability Across the Curriculum Train the Trainer Workshop, received 2011 Hometown Heroes "Go Green" award for sustainability, developed a session & served as track chair for 2012 MMA Fall Educators' Conference on "Teaching Sustainability Across the Business Curriculum." Board of Directors Heart of Georgia Energy Coalition & publishes articles at academic conferences and in academic journals on sustainability.

Part II
Case Studies in Sustainable Water Use and Management

Farm Management in Crop Production Under Limited Water Conditions in Balkh, Afghanistan

Paulo Roberto Borges de Brito

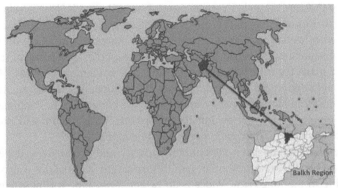

Source: http://www.ugo.cn/admin1/en/10117.htmhttp://www.ugo.cn/admin1/en/10117.htm

Abstract Water is the main limiting factor in crop production in the hot and dry summer period within arid and semiarid regions such as Balkh region, Afghanistan. Due to lack of irrigation water infrastructure and information about how much water should be applied at the field level given different levels of water availability, the individual farmer lacks information about the sustainability of the water supply, crop production, and how much net income he would get from this production. Thus, the main research question is how to find ways to optimize the use of water given different shortages in water availability at the farm level in Balkh, Afghanistan. The research uses a linear programming (LP) model to assess the economic impact of net returns given four different water availability scenarios at the farm level. The main result among others is that water value increases as the

P.R.B. de Brito (✉)
College of Business, Colorado State University, Fort Collins, CO 80521, USA
e-mail: paulo.brito1001@gmail.com

© Springer International Publishing Switzerland 2015
W. Leal Filho and V. Sümer (eds.), *Sustainable Water Use and Management*,
Green Energy and Technology, DOI 10.1007/978-3-319-12394-3_10

187

water availability decreases as expected. Water is binding in all scenarios, and harvesting labor is the most used input as water, and other resources/inputs become scarcer in the farmer's mixed crop production.

Keywords Water resources management · Sustainability · Crop production · Scarcity · Economic impact · Afghanistan

1 Introduction

Afghanistan faces a number of water and food security challenges related to irrigated agriculture. A short list of these challenges include a scarce capacity to implement rules and regulations relating to the water sector, a lack of acceptable data for formulating strategic plans for water resources development, and a lack of master plans for river basins. There is also a considerable shortage of staff and institutional support for establishing river basin councils. Moreover, the country also suffers from lack of capacity to implement a comprehensive national plan for drought mitigation as well as damaged irrigation infrastructure and poor overall performance of existing irrigation systems (Mahmoodi 2008).

Increasing Afghanistan's agricultural capacity and productivity depends upon addressing limited water resources and infrastructure and making water resources sustainable in the long term. Water is an essential resource for improving Afghan rural livelihoods. As of 2009, only 3.2 million hectare of a total estimated potential area of 7.7 million hectare or less than a half of land that can be used for crops is cultivated. The main limiting resource is inadequate access to water and irrigation (Ward and Torell 2009).

An important goal of the Ministry of Agriculture, Irrigation, and Livestock (MAIL) in addition to the Ministry of Energy and Water (MEW) in the country is to help alleviate poverty and to promote economic development (Mahmoodi 2008). Its goal is to support development of integrated water resources management (IWRM) at the river basin scale so as to encourage decentralized water management. However, implementation of IWRM poses numerous challenges. These challenges occur because of complex interactions between Afghanistan's hydrology, economics, agronomy, water-use sustainability, food security, and institutions.

Given the challenges above to improve the future of the country, the US Agency for International Development (USAID) has developed a research project from 2008 to 2011 called Afghanistan Water, Agricultural, and Technology Transfer (AWATT). The project was a partnership among four American Universities with New Mexico State University (NMSU) as the leader institution. The total project budget was 20 million dollars. Besides NMSU, Colorado State University (CSU), University of Illinois at Urbana Champaign (UIUC), Southern Illinois University-Carbondale (SIUC), and four Afghanistan Universities joined the collaborative effort (Jha and Pritchett 2008). The project was as a way to research and develop new tools to improve water-use efficiency, increase agricultural production for

Afghan farmers, and help inform Ministry personnel in better water management decisions. It uses information on available water, crop water requirements, crop prices, and production costs to estimate the most profitable use of water in irrigated agriculture. This article is part of the AWATT project conducted by CSU with a subcontract of 5.5 million dollars in the Balkh Province area in Afghanistan to determine the most profitable use of water in irrigated agriculture in Balkh. For this purpose, a survey was conducted with local farmers in Balkh and crop water requirements of several crops were estimated. It used a linear programming (LP) optimization model using several constraints to assess the economic impact of different shortages in water availability in Siagard Canal, Balkh, Afghanistan. Finally, the results also contribute to the estimation of net returns of the six crop activities chosen given the different four water shortage scenarios.

2 Water Scarcity in Balkh, Afghanistan

Water is the main limiting factor in crop production in the hot and dry summer period within arid and semiarid regions. When water resources are a limiting factor in crop production, irrigation programs need to be applied to enable maximum production per unit of irrigation water. Deficit irrigation is one way of maximizing water-use efficiency through higher yields per unit of irrigation water applied (Kiziloglu et al. 2009).

The irrigation time and water amounts are important determinants of maximizing crop yields and also future water sustainability. Predicting yield response to water use of crops is important in developing strategies and decision-making for farmers' use, their advisors, and researchers for irrigation management under limited water conditions. In arid and semiarid climatic regions, however, little attempt has been made to assess the water-yield relationships and optimum water management programs for different crops.

Also, there is a need to study the interaction of the water–yield relationship for optimum water amounts with the most profitable use of water in crop production at the farm level. Moreover, lack of irrigation infrastructure leads to lack of information of water supply at the farm level in Balkh, Afghanistan. This lack of information in water supply also leads to lack of knowledge in crop production and net income at the farm level. So the problem being researched is how to find ways to optimize the use of water given different shortages in water availability at the farm level in Balkh, Afghanistan.

The purpose of this research was to find ways to optimize the use of water in scarce water settings at the farm level such as the one above. More specifically, it has focused on the economic impact assessment of different shortages in water availability in Siagard Canal, Balkh, at the farm level and net returns estimation of each crop activity in all the different water availability scenarios.

Scarcity of water in Afghanistan is a government concern, especially in the agricultural setting. Economic analysis with different scenarios of water availability is an important contribution of this study to effectively help the government agencies in the country alleviate poverty and promote economic development for small farmers. More effective outreach depends on developing this information.

Most of the current research from the AWATT project is focused on the sectorial level of the economy in Afghanistan, using data from different sectors such as industrial, agricultural, and residential sectors (Ward and Torell 2009). Additional studies have been conducted focusing more on the agronomic side such as crop water requirements and crop productivity of major crops in Pakistan reported by Ahmad et al. (2005) and in Afghanistan by National Development Strategy (2005). However, to the knowledge of this researcher, no significant study to date has studied the economic impacts of different shortage levels in water availability at the farm level for Afghanistan. This is likely due to lack of data mainly and also due to greater attention to crop–water–yield relationships rather than the crop–water–economic impact relationships.

3 Constrained Profit Maximization and Use of Water

Economists frequently use standard economic theory based on rational choice and explain behavior. Choice theory fundamentally assumes that the decision an individual makes best enables them to meet their objectives. Farmers, in this case, are assumed to have as their primary objective the maximization of profits choosing different inputs in order to achieve this goal (Griffin 2006). Given the main research question of this study be how to find ways to optimize the use of water given different water shortages levels in Balkh region, the economic optimization model is very suitable for it.

A general standard simple model is assumed to be static, deterministic, no risk, and full information from farmer's side in order to simplify the model. However, the existing literature has a variety of other models including stochastic terms, dynamics, and spatial elements.

The neoclassical economic theory for a standard farmer's profit maximization model usually assumes markets are competitive with full information for outputs and inputs used. The output and input prices are vectors of strictly positive prices. Land is considered a fixed input, and the availability of water on farm is also fixed for consumptive use, and we assume the system has no loss.

In the case of simpler agricultural models, inputs are allocated to specific crop production activities; production is technically non-joint so that the allocation of inputs uniquely determines crop-specific output levels (Moore and Negri 1992).

The existing literature shows usually farmers choose and control input amounts in order to maximize their profits. The literature has shown that water and crop activities are the main resources to control (Griffin 2006). Land and labor are also choice variables found in previous studies.

The simplest profit maximization model is an unconstrained model where farmers usually choose the optimal amount of any inputs regardless of the limits they face. However, in real-world situations, farmers face several constraints (Nicholson 2005). Some of the main constraints found in the literature are land and labor (Manocchi and Mecarelli 1994; Moore and Negri 1992). A full description of assumptions can be found in Brito (2010).

The mathematical model described below has a theoretical basis that is fully described in Brito (2010). The farmer's profit maximization model is represented using a mathematical objective function; in this case, the maximization of profit or net return (π) choosing optimal levels of land area (l), for example. This objective is represented mathematically as follow:

$$\max_{l} \pi = \sum_{j=1}^{m} P_j Y_{aj} - \sum_{j=1}^{m} p_j^n n_j - \sum_{j=1}^{m} p_j^f f_j \tag{1}$$

where

π total profit or net return per year;
m total number to crop activities;
P_j crop price under jth crop activity;
Y_{aj} actual yield from the jth crop activity;
p_j^n price of labor input under the jth crop activity;
n_j labor hours required for the jth crop activity;
p_j^f fertilizer cost for the jth crop activity;
f_j quantity of fertilizer applied for the jth crop activity;

The standard profit maximization is usually unconstrained, but in real-world situations, this problem (Eq. 1) is subject to several constraints, for example:

1. Land constraints:

$$\overline{L} \geq \sum_{j=1}^{m} l_j \tag{1.1}$$

2. Water requirement constraints:

$$\overline{W} \geq \sum_{j=1}^{m} w_j \tag{1.2}$$

3. Labor cost constraints:

$$\overline{N} \geq \sum_{j=1}^{m} p_j^n n_j \tag{1.3}$$

4. Fertilizer cost constraints:

$$\overline{N} \geq \sum_{j=1}^{m} p_j^f f_j \tag{1.4}$$

And as the literature suggests, also there are other constraints such as cash and crop rotation. Then, a Lagrangian function, denoted $\mathcal{L}$, states the constrained maximization problem as:

$$
\mathcal{L}_{l,w,n,} = \sum_{j=1}^{m} P_j Y_{aj} - \sum_{j=1}^{m} p_j^n n_j - \sum_{j=1}^{m} p_j^f f_j + \lambda_1 \left(\overline{L} - \sum_{j=1}^{m} l_j \right) + \lambda_2 \left(\overline{W} \geq \sum_{j=1}^{m} w_j \right)
$$
$$
+ \lambda_3 \left(\overline{N} - \sum_{j=1}^{m} p_j^n n_j \right) + \lambda_4 \left(\overline{N} - \sum_{j=1}^{m} p_j^f f_j \right)
\tag{2}
$$

However, this constrained maximization problem involves also inequality constraints and a complete set of nonnegative constraints such as the constraint Eqs. (1.1–1.4) above. When we face this situation, we need to solve this constrained optimization problem with the so called Kuhn-Tucker formulation using the complementary slackness conditions (c.s.)*. This formulation is effective because if the constraints are binding, then $\overline{L} = \sum_{j=1}^{m} l_j$, $l_j > 0$, and if the constraint is not binding, then $\lambda_1 = \lambda_2 = \lambda_3 = \lambda_4 = 0$. Thus, this formulation allows the person, in this case, the farmer, not spending all his/her resources/inputs.

Then, first-order conditions are taken with respect to the input/or resource being optimized and set it equal to zero taking into account the c.s. conditions so that we get optimal levels of the inputs or resources chosen such as the following:

$$
\frac{\partial \pounds}{\partial l} = 0 \quad \text{c.s. } l \geq 0
\tag{2.1}
$$

$$
\frac{\partial \pounds}{\partial w} = 0 \quad \text{c.s } w \geq 0
\tag{2.2}
$$

$$
\frac{\partial \pounds}{\partial n} = 0 \quad \text{c.s } n \geq 0
\tag{2.3}
$$

$$
\frac{\partial \pounds}{\partial \lambda_1} = 0 \quad \text{c.s } \lambda_1 \geq 0
\tag{2.4}
$$

$$
\frac{\partial \pounds}{\partial \lambda_2} = 0 \quad \text{c.s } \lambda_2 \geq 0
\tag{2.5}
$$

$$
\frac{\partial \pounds}{\partial \lambda_3} = 0 \quad \text{c.s } \lambda_3 \geq 0
\tag{2.6}
$$

$$
\frac{\partial \pounds}{\partial \lambda_4} = 0 \quad \text{c.s } \lambda_4 \geq 0
\tag{2.7}
$$

The λ_i is the Lagrangian multiplier, and it measures the rate of change of the optimal value of the objective function with respect to the parameter given. It measures the effect of a unit increase in the variable being changed on the objective function, or in economic term, it measures the marginal value of an extra unit of input or resource to be used (Nicholson 2005).

When the optimization problem is unconstrained, $\lambda = 0$, and we say that the constraint in the use of inputs and/or resources is not binding. But, when λ must be ≥ 0, it means that the constraint is binding (Simon and Blume 1994).

4 Data and Methodology

Siagard Canal is part of the Balkh River Basin that is located in northwest Afghanistan. The Balkh River and the canals are fed by its water lie in the Jowzjan, Balkh, Sar-e-Pol, Samangan, and Bamian provinces.

The Balkh River supplies irrigation water to 14 canals, and each of these canals supplies water to the individual irrigators within the Balkh River Basin. The Balkh River is a (mostly) blind river, only reaching the Amu Darya Basin during very high-water periods (Ahmad et al. 2005).

The areas are dominated by rangeland, and irrigated crops are along the rivers. Weak economic frameworks for connecting land, water, and agriculture have left the Balkh River Basin region at considerable risk of food insecurity and have contributed to the inability of irrigated agriculture to provide acceptable wages to the large number of Afghans who make their living in production agriculture.

The data collected for the economic model in this study is partly from the on-site surveys from AWATT project and secondary data gathered in the crop budgets reported in AWATT Crop Budget Group (2009). Data regarding to output and input prices and quantities used per one-unit hectare of land are all from the AWATT Crop Budget Group (2009). We used yield and water requirements data for a one-unit hectare of land area for cropping from the estimations of crop water requirements developed by Brito (2010).

After gathering the data from the mentioned sources, we have chosen six different crops from Brito (2010) which we calculated total water requirements and actual yields for each. The chosen crops are winter wheat, alfalfa, cotton, melon, maize, and tomato.

The actual yields were calculated using functions of just water irrigation (mm) using 100, 75, 50, and 25 % water irrigation levels. The actual yield functions for the six crops are as follows:

$$\text{Winter wheat: } Ya = 9.0294 * \text{water irrigation} + 1.0331 \tag{3.1}$$

$$\text{Alfalfa: } Ya = 0.3431 * \text{water irrigation} + 61.864 \tag{3.2}$$

$$\text{Cotton: } Ya = 2.613 * \text{water irrigation} + 451.27 \tag{3.3}$$

$$\text{Melon: } Ya = 2.183 * \text{water irrigation} - 202.83 \tag{3.4}$$

$$\text{Maize: Ya} \ = \ 1.9017 \ * \text{water irrigation} \ - \ 231.29 \tag{3.5}$$

$$\text{Tomato: Ya} \ = \ 1.067 \ * \text{water irrigation} \ - 51.15 \tag{3.6}$$

The actual yields calculated were assumed to be completely demanded so the quantity sold of each crop is exactly the same quantity produced so that no storage assumption is in the model.

For simplification, just two inputs were selected from the enterprise budgets to build the empirical model. The first input is harvesting labor hour for each crop and the corresponding cost (per hour). The second input chosen was fertilizer. There were several categories of fertilizer on the report; however, it was chosen just manure due to be the fertilizer used in most of the crops chosen for this study.

The model also assumes the planting area corresponds to the total area available which is equal to 6 ha. No fallow area is assumed, and there is no crop rotation. The exogenous variables are as follows: output and inputs prices, actual yield for the different scenarios of water availability, quantity of inputs bought, and quantity of output sold. The choice variable for the model is crop activity land area (in hectares) indirectly choosing water availability level.

5 Linear Programming

After collecting, choosing, and gathering the data in Excel spreadsheets, the following step was to specify the empirical model, which was described in Eqs. (3.1–3.6).

As mentioned before, optimization of farm decision-making is usually undertaken by applying mathematical programming techniques. The technique used in this study was the one called LP in Excel Solver and based in Rae (1994).

According to McKinney et al. (1999), the LP approach has the advantage that it can be implemented with a minimum of data for those problems in which the fixed input assumption and linear constraints are reasonable approximations of reality.

Although spatial, dynamic, and stochastic models have been the new direction to future models, our model is partial equilibrium, deterministic, and static due to small data set and one of the fewer studies of this kind in Siagard Canal, Balkh region, Afghanistan, so that the model can be developed in more complex pieces in the future.

Given the theoretical model in the Eq. (1) and the assumptions in the previous section, the matrix formulated to solve the problem in LP is presented in Table 1. This table shows the initial values plugged into the cells for the 100 % water availability scenario. The other three scenarios had similar tables not shown in this document in order to be concise.

The empirical maximization model is then formulated in LP matrix with the initial values for a scenario of 100 % water availability in Table 1 as follows:

$$\max_{\text{HA}j} \pi = \sum_{j=1}^{m} \text{TR}_j - \sum_{j=1}^{m} \text{HVTC}_j - \sum_{j=1}^{m} \text{FertTCTR}_j \tag{4}$$

Table 1 LP matrix formulated for farmer net return maximization

	WH	AL	CO	ME	MA	TO				HV Labor cost	Fert cost			USED	RHS
π	10416.1	-19.92	-714.34	-2070.3	6167.11	3883.1				3	0.02			17661.8	
Output price	6.21	2.33	5.65	3.25	6.5	5.01									
Crop activities	WH	AL	CO	ME	MA	TO	SE_{WH}	SE_{AL}	SE_{CO}	SE_{ME}	SE_{MA}	SE_{TO}			
Land (Ha)	1	1	1	1	1	1							<=	6	6
Water (Eta per mm)	189.61	1587.08	975.3	1006.17	636.65	982.47							<=	5377.2	5377.2
HA Labor (hours)	72	460.8	240	691.2	57.6	345.6							<=	1867.2	
Ya (kg)	-1713.1	-601.46	-2996.8	-1990.6	-978.34	-990.26	1713.09	601.46	2996.8	1990.62	978.62	990.26	<=		0
Fert (kg)	1000	2000	0	0	1000	2400							<=	6604.5	
HV labor cost constraint	216	1382.4	720	2073.6	172.8	1036.8							<=	5601.6	5601.6
Fert cost constraint	20	40	0	0	20	48							<=	128	128

Target cells	
Changing cells	

π — net returns for cropping all the activities during the growing period choosing optimal levels of cropland areas;

TR_j — total revenue of crop j;

$HVTC_j$ — harvesting labor total cost of crop j;

$FertTC_j$ — fertilizer total cost of crop j.

Subject to

1. Land constraint:

$$HA_{WH} + HA_{AL} + HA_{CO} + HA_{ME} + HA_{MA} + HA_{TO} \le 6 \qquad (4.1)$$

2. Water requirement constraint:

$$ETa_{WH} * HA_{WH} + Eta_{AL} * HA_{AL} + Eta_{CO} * HA_{CO}$$
$$+ Eta_{ME} * HA_{ME} + Eta_{MA} * HA_{MA} + Eta_{TO} * HA_{TO} \le 5377.28 \qquad (4.2)$$

3. Harvesting labor cost constraint:

$$HVlabor_{WH} * HVcost_{WH} + HVlabor_{AL} * HVcost_{AL}$$
$$+ HVlabor_{CO} * HVcost_{CO} + HVlabor_{ME} * HVcost_{ME}$$
$$+ HVlabor_{MA} * HVcost_{MA} + HVlabor_{TO} * HVcost_{TO} \le 2346.8 \qquad (4.3)$$

4 Fertilizer cost constraint:

$$Fert_{WH} * fertcost_{WH} + Fert_{AL} * fertcost_{AL} + Fert_{CO} * fertcost_{CO}$$
$$+ Fert_{ME} * fertcost_{ME} + Fert_{MA} * fertcost_{MA}$$
$$+ Fert_{TO} * fertcost_{TO} \le 6604.5 \qquad (4.4)$$

where

WH	winter wheat
AL	alfalfa
CO	cotton
ME	melon
MA	maize
TO	tomato
HAi	land area per hectare for crop j
Yaj	actual yield of crop j
HVlabor$j$	harvesting labor of crop j
HVcost	harvesting labor cost of crop j
Fert	fertilizer quantity of crop j
Fert cost	fertilizer cost of crop j
SEj	sold quantity in kg of crop j
ETaj	actual Eta of crop j (mm)

where π, HA, SE, HVlabor, Fert, output and input prices are ≥ 0.

The same model was run for 100, 75, 50, and 25 % water availability in Solver, Excel Software. The Solver solution reported the answer report, sensitivity report with the values of shadow prices for constraints, and limits report for each scenario.

After running the empirical LP maximization model for the four scenarios, the next step of the methodology was to develop the analysis framework.

As said before, in neoclassical economic theory, a farmer tends to maximize profit. Given the limited land, water, harvesting labor, and fertilizer costs constraints for cultivation of crop during a growing season, a farmer needs to make decisions in order to best allocate their resources.

This farmer makes his decision based on two pieces of information:

- The marginal cost associated with adding one more extra unit of the input being used;
- The marginal benefit associated with adding one more extra unit of this input added.

Then, if the marginal benefit of the added input is equal or greater than the marginal cost of adding this extra unit of input/resource, a farmer interested in increasing his/her profits will add more of the input for their cropland area.

For non-priced inputs, the marginal value is analyzed looking at the Lagrangian multiplier value.

Thus, the analysis conducted in this study is the marginal analysis based on the principle of marginal value product (Griffin 2006). The results of the model estimation in Excel Solver were analyzed according to the marginal analysis approach just described.

6 Results of the Maximization Problem

The results indicated the objective function value and the marginal values of the cropland area used on the 6 ha farm. It was found that the farmer can obtain a maximum net return of US$54,414.35 per growing season period using 100 % of water available by cultivating winter wheat (WH), maize (MA), and tomato (TO). It also showed that the farmer would crop 5.1852 ha of winter wheat, 0.6131 ha of maize, and 0.2016 ha of tomato (see Table 2). Alfalfa (AL), cotton (CO), and melon (ME) have a reduced cost of −2,131.30. It means that if we add one more unit of land area to crop those activities, then net return would decrease by US$2,131.30. Alfalfa, cotton, and melon are not selected to be cultivated probably due to their higher water requirements. Thus, we can see that the farmer would probably almost mono-crop winter wheat with 86, 42 % of the 6 ha total area with 100 % of water available.

A net return of US$45,842.85 per growing season period using 75 % of water available was found by cultivating the same crops as in the 100 % water scenario plus an almost insignificant area of alfalfa (AL). The Table 5.1 shows that this addition of alfalfa area is 0.0085 ha corresponding to 0.14 % of a total of 6 ha. There is a decrease in wheat area for 4.8961 ha and a small increase of maize (0.7733 ha) and tomato (0.2972 ha) areas (Table 2).

A maximum net return of US$34,914.45 per growing season period using 50 % of water available was found by cultivating the same crops as with the 75 % water scenario. However, winter wheat decreases from 4.8961 to 4.4829 ha and tomato from 0.7733 to 0.651 ha. Alfalfa increases to 0.0727 ha and tomato increases to 0.3887 ha.

Lastly, a maximum net return of US$4,864.53 per growing season period using 25 % of water available was found by cultivating all the six crops with a decrease in wheat (2.6165 ha) but still being the main crop cultivated (Table 2).

We can also see from Table 2 that cotton and melon are never chosen in all water scenarios, except for the 25 % water case. These two crops would make net returns decrease if they were cropped by the amount of the reduced costs in each scenario.

Table 2 Level of activities and reduced cost

	100 % water available		75 % water available		50 % water available		25 % water available	
Cropland area	Value	Reduced cost	Value	Reduced cost	Value	Reduced cost	Value	Reduced cost
WH	5.1852	0	4.8961	0	4.4829	0	2.6165	0
AL	0	−2,131.3001	0.0085	0	0.0727	0	0.4376	0
CO	0	−3,533.7723	0	−0.0004	0	−0.0003	0.3669	0
ME	0	−3,533.7723	0	−0.0004	0	−0.0003	1.5772	0
MA	0.6131	0	0.7733	0	0.651	0	0.1641	0
TO	0.2016	0	0.2972	0	0.3887	0	0.8374	0

The most profitable crop value is winter wheat. However, as the amount of water available becomes scarcer, the winter wheat land area gets smaller so net return decreases. Table 3 presents the shadow prices of water resources and other inputs as they become scarce with the different scenarios of water availability.

After describing the results from land area for the different crops in each scenario of water availability, now we present the results of the marginal shadow prices of each input and constraint in the model.

Table 3 shows the shadow price of water resource. It is the maximum price that the farmer is willing to pay for an extra unit of water-limited constraint. Water constraint has a marginal value of $3.620 for adding one more extra unit of water (mm) in his farm plan. This value implies that the farmer can increase his net return by this amount by using one extra unit of irrigation water (mm) in his optimal farm plan in the 100 % water availability case.

As we can see, the scarcer the water resource is for the farmer, more valuable the water is at the margin with values of $5.396, $6.2667, and $10.4588 for the 75, 50, and 25 % water available cases, respectively. The high shadow price is due to the water-limited availability. Therefore, it is profitable for the farmer in the area to purchase water at a price close or less than the shadow price so that he can increase his net return (Fig. 1).

Table 4 shows the shadow price of land resource. It is the maximum price that the farmer is willing to pay for an extra unit of land-limited constraint. As

Table 3 Value of additional 1 unit of water (mm) within a growing season for a representative farm

	100 % water available	75 % water available	50 % water available	25 % water available
Water value ($)	3.620	5.396	6.2667	10.4588

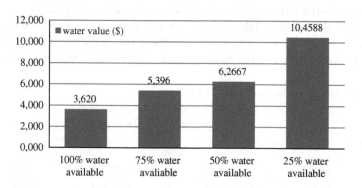

Fig. 1 Value of additional unit of water (mm) within a growing season for a representative farm

Table 4 Value of additional 1 ha of land within a growing season for a representative farm

	100 % water available	75 % water available	50 % water available	25 % water available
Land value ($)	3,533.772	0	0	1,332.318

shown on Table 4, the land constraint has a marginal value of $3,533.77 for adding one more hectare of land in his farm plan. This value implies that the farmer can increase his net return by this amount by using one extra hectare of land area in his optimal farm plan in the 100 % water availability case.

However, it is interesting that for the 75 and 50 % water availability scenarios, land has zero marginal value of adding one more hectare of land area in the farm plan. It is due to land resources are not binding in those cases. But, for the 25 % water available case, one more hectare of land area has a marginal value of $1,332.318 (Table 4 and Fig. 2).

Lastly, Table 5 shows the shadow price of labor and fertilizer inputs. Harvesting labor is not binding for the 100, 75, and 50 % water availability scenarios. But for the 25 % water available case, harvesting labor cost is a binding constraint. Fertilizer cost constraint is not binding for the 100, 50, and 25 % water availability scenarios. However, it is binding for a 75 % water available case. In this case, the cost of one more extra unit of manure fertilizer is $28.271.

Used resources amount in each of the four water scenarios is the last issue to be analyzed. Figure 3 presents the amount of each resource/input used in each of the four water availability scenarios within a growing season for a representative farm in Balkh region, Afghanistan. We can see from the figure that, as water resource used decreases, other inputs/resources used increase. Given land and fertilizer being scarce in different scenarios, harvesting labor is the only non-scarce resource and because of that, this is the most used input as water becomes scarcer at the farm level. It might explain the reason wheat activity is shifted to other crops with high harvesting labor requirements as water becomes scarcer.

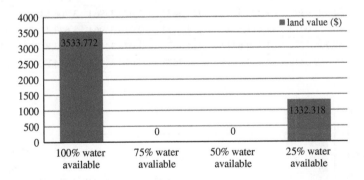

Fig. 2 Value of additional one hectare of land within a growing season for a representative farm

Table 5 Value of additional 1 h of harvesting labor and 1 unit of fertilizer (kg) within a growing season for a representative farm

	100 % water available	75 % water available	50 % water available	25 % water available
HV labor ($)	0	0	0	−5.2981
Fertilizer value ($)	0	28.271	0	0

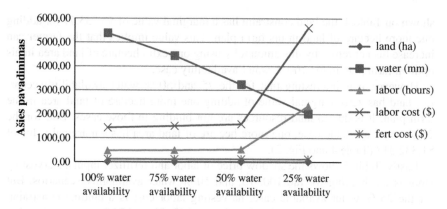

Fig. 3 Used resources within a growing season for a representative farm

7 Crop Net Returns for Each Water Availability Level

As part of the last economic assessment in this study, this section presents the net returns for each crop activity for the 100, 75, 50, and 25 % water shortage levels. Table 6 presents the results of the net returns by crop activity for the four water scenarios.

Table 6 Net return by crop given different water availability levels

% water	WH ($)	AL ($)	CO ($)	ME ($)	MA ($)	TO ($)
100	54,997.9895	−1,422.4	−714.3472	−2,070.34964	3,706.5454	−83.08113
75	46,324.0972	−1,411.564	−714.3472	−2,070.34964	3,672.43827	42.575553
50	37,275.3779	−1,350.6442	−714.3472	−2,070.34964	1,813.8172	−39.40372
25	18,669.644	−1,113.3659	−714.3472	−2,070.34964	2.98423233	138.80818

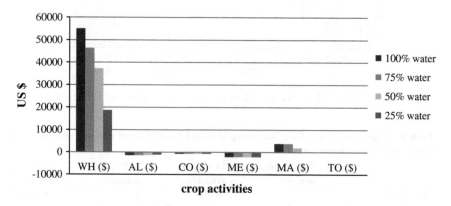

Fig. 4 Net return by crop given different water availability levels ($)

As seen, winter wheat has the most net return of all crop activities in a 6 ha land area. Winter wheat and maize are the crop activities with positive net returns in all scenarios. Tomato has positive net returns only in the 75 and 25 % scenarios. Alfalfa, cotton, and melon show negative net returns in all the water scenarios (Fig. 4).

8 Conclusions

Water is the main limiting factor in crop production in the hot and dry summer period within arid and semiarid regions in Afghanistan. When water resources are a limiting factor in crop production, irrigation programs need to be applied to enable maximum production per unit of irrigation water. Deficit irrigation is one way of maximizing water-use efficiency through higher yields per unit of irrigation water applied (Kiziloglu et al. 2009) and also a way to ensure sustainability of water management in the long term.

The irrigation time and water amounts are important determinants of maximizing crop yields. Predicting yield response to water use of crops is important in developing strategies and decision-making for use by farmers, their advisors, and researchers for irrigation management under limited water conditions. In arid and semiarid climatic regions, however, little attempt has been made to assess the water-yield relationships and optimum water management programs for different crops.

Also, there is a need to interact the water–yield relationship, optimum water amounts with the most profitable use of water in crop production land areas at the farm level. But, in order to get the most profitable use of water in crop production, the farmer needs to have information about the water supply, crop production, and how net income they would get from each water supply available. The typical farmer does lack all this information due to lack of irrigation infrastructure in Balkh region. So the problem being researched is how to find ways to optimize the use of water and land given different shortages in water availability at the farm level in Balkh, Afghanistan. This research question is important due to the lack of information stated before.

Developing this information will be useful for the MAIL in addition to the MEW in Afghanistan to help alleviate poverty and to promote economic development and enabling the Ministry personnel with knowledge for future actions.

The data used for this study were collected from on-site surveys and from secondary data sources such as the AWATT Diagnostic Analysis (DA) Report and the AWATT Crop Budget Group (2009).

The main objective employed a LP model for maximizing net returns of a representative individual farmer choosing the best crop activity land area of a combination of six different crop activities in this study. Results of the empirical model were estimated, and some discussion of the economic analysis of the results employing the economic marginal analysis so that the main objective of this study would be fulfilled. Also, net returns of each of the six crops used in the model were presented in Table 6.

Given the research question of this study, the main findings of this study lead us to conclude that water is a scarce resource in all the four water scenarios and a binding constraint. The static marginal value of water resource increases as the water availability decreases as expected. This indicates the likely importance of water in crop cultivation under irrigation schemes in arid and semiarid regions such as Siagard Canal, Balkh region. Thus, it is economically rational for the local farmer to pay at least the shadow price given a certain level of water availability for one more extra unit of water (mm), as long as its benefits are also more.

Second, land is binding in the 100 and 25 % water cases and fertilizer is just binding in the 75 % water scenario.

Third, the only non-binding input is the harvesting labor in all scenarios. It is not worth for the farmer to invest in labor, especially in the 25 % water case where an additional hour of harvesting labor would decrease net returns for the farm plan.

As the input/resources used are becoming scarcer as the water availability scenario decreases, we might conclude that harvesting labor is increasingly being used to its abundance. It leads to conclude why the crop mix changes to more harvesting labor-intensive crops as the water availability scenarios decrease.

9 Future Directions and Recommendations

There are some limitations in this study. First, as stated before, the model is static and deterministic which means that it does not take into account inter-temporal decisions and the uncertainties embedded in the crop production process. Inter-temporal decisions take into account the trade-offs between a decision today and tomorrow, and in crop production, this is an important decision-making that is missing in the present model. The missing uncertainties are related to climatic parameters, which are highly stochastic, and water supply uncertainties, which are very present in this part of the world, are also not considered in the model.

Second, the model does not set a limit on the water amount for a certain type of crop activity when a water availability scenario changes. It would be more realistic to do so, given the assumption that the farmer would pursue to crop the activity with highest net return.

Third, water and production are exogenous in the model and all the inputs are fixed, except for crop activity land area. It means that we assumed perfect competition markets for output and input prices so that they do not vary. Depending on the crop activity, some prices vary and the role of international prices for some commodities also fluctuates frequently, leading to different results in the model.

Fourth, all the functions and constraints were assumed to be linear. We know that in more realistic scenarios, water and yield functions should have, for example, a quadratic shape in order to behave in a diminishing marginal productivity fashion leading to different results in the model.

Fifth, the model does not take into account fallow land. We assume the farmer uses the whole 6 ha to crop within the growing season. Also, there is no crop rotation constraints employed in model. Usually, farmers rotate crops and have different crop growing seasons. The inclusion of those assumptions would probably change the crop mix in the model significantly.

Sixth, we assumed just one kind of labor. The model is missing pre-harvesting operation labor, which is an important input. Also, there are different types of fertilizers employed for different types of crops, but we assumed just manure in the model. Those assumptions lead us to conclude net returns were overestimated. The inclusion of those variables in the model would result in more realistic net returns.

Seventh, the data set used was small and due to lack of infrastructure in Balkh region, there were a significant amount of missing data. More accurate data should be collected for future studies.

Future studies should employ all those missing assumptions and gaps in the model and a better data set in order to have more representative conclusions.

References

Ahmad MD, Bastiaanssen WGM, Feddes RA (2005) A new technique to estimate net groundwater use across large irrigated areas by combining remote sensing and water balance approaches. Rechna Doab, Pakistan. Hydrogeol J 13:653–664

AWATT Crop Budget Group (2009) Enterprise budgets for Balkh watershed Afghanistan: an AWATT decision tool component. AWATT, May 28

De Brito PRB (2010) Farm management in crop production under limited water conditions in Balkh, Afghanistan. Technical paper, Colorado State University, Fort Collins

Griffin RC (2006) Water resource economics: the analysis of scarcity, policies, and projects. Massachusetts Institute of Technology, Cambridge

Jha A, Pritchett J (2008) Afghanistan water, agriculture, and technology transfer (AWATT). Colorado Water Newslett 25(5):26–27

Kiziloglu MF, Sahin U, Kuslu Y, Tunc T (2009) Determining water–yield relationship, water use efficiency, crop and pan coefficients for silage maize in a semiarid region. Irr Sci 27:129–137

Mahmoodi SM (2008) Integrated water resources management for rural development and environmental protection in Afghanistan. J Dev Sustain Agric 3:9–19

Mannocchi F, Mecarelli P (1994) Optimization analysis of deficit irrigation systems. J Irrig Drainage Eng 120(3):484–503

McKinney DC et al (1999) Modeling water resources management at the basin level: review and future directions. SWIM paper 6. International Water Management Institute, Colombo

Moore MR, Negri DH (1992) A multi-crop production model of irrigated agriculture, applied to water allocation policy of the bureau of reclamation. J Ag Resour Econ 17(1):29–43

National Development Strategy (2005) Afghan water law. Article no. 114

Nicholson W (2005) Microeconomic theory: basic principles and extensions, 9th edn. Thomson South Western, Ohio, p 671

Rae AN (1994) Agricultural management economics: activity analysis and decision making. CABI, Cambridge

Simon CP, Blume L (1994) Mathematics for economists. Norton, New York

Ward FA, Torell G (2009) Improving water, farm, and food policy choices for Afghanistan. Report presented to the US agency for international development. Afghanistan, water, agricultural, technology transfer (AWATT) project, Nov 2009

Author Biography

Paulo R. Borges de Brito a native of Sao Paulo, Brazil, is currently a Distance Section Coordinator in the Online Professional MBA Program at Colorado State University, USA. He also teaches a variety of Economics courses at Front Range Community College and Colorado Community Colleges Online. Paulo has been also involved in a number of non-profit organizations in Brazil and in the USA. He holds a B.S. in Economics from Mackenzie University in Brazil, an M.S. in Environmental Science from the University of Sao Paulo in Brazil, and an M.S. in Agricultural and Resource Economics from Colorado State University.

Sustainability of Effective Use of Water Sources in Turkey

Olcay Hisar, Semih Kale and Özcan Özen

Abstract Water is the most essential natural resource for sustainable development of human society as well as the most vital source for viability of human and natural systems. However, natural water resources had been threaten by increase of temperature due to global warming and improper usage, causing health problems both for human and aquatic environment. On the other hand, global water consumption has increased because of growth in population and increase of the per capita water use. To make adjustments to the water utilization, the need is allocating limited water resources and increasing local water use efficiency. Sustainability is a relative concept that must be applied in an environment undergoing multiple changes that are occurring over different temporal and spatial scales. Contrary to the popular belief, Turkey is not a water-rich country. Turkey depends on its water resource systems for survival and welfare. Therefore, new studies have been forced in the rehabilitation and sustainable usage of water sources recently in the world. In this paper, information about their current state and future projections is given based on many published data.

Keywords Water resources · Sustainability · Global warming · Climate change · Turkey

1 Introduction

Sustainable water resources systems are designed and managed to best serve people living today and in the future. The actions society take now to satisfy our own needs and desires should depend not only on what those actions will do for us but also on

---

O. Hisar (✉) · S. Kale · Ö. Özen
Faculty of Marine Science and Technology, Canakkale Onsekiz Mart University,
Terzioglu Campus, 17100 Canakkale, Turkey
e-mail: o_hisar@hotmail.com

© Springer International Publishing Switzerland 2015
W. Leal Filho and V. Sümer (eds.), *Sustainable Water Use and Management*,
Green Energy and Technology, DOI 10.1007/978-3-319-12394-3_11

how they will affect our heirs. This consideration of the long-term impacts on future generations of actions taken now is the core of sustainable development. While the word "*sustainability*" can have different meanings for different people, it always includes a consideration of the welfare of those living in the future. While the debate over a more precise definition of sustainability will continue, and questions over just what it is that should be sustained may remain unanswered, this should not delay progress toward achieving more sustainable water resources systems.

The concept of ecological and environmental sustainability has largely resulted from a growing concern about the long-run health of our planet. There is increasing evidence that our present resource use and management activities and actions, even at local levels, can significantly affect the welfare of those living within much larger regions in the future. Water resource management problems at a river basin level are rarely purely technical and of interest only to those living within the individual river basins where those problems exist. They are increasingly related to broader societal structures, demands, and goals.

The containment of sustainability criteria along with the more common economic, environmental, ecological, and social criteria used to evaluate alternative water resources development and management strategies may identify a need to change how we commonly develop and use our water resources. We need to consider the impacts of the change itself. Change overtime is certain; however, how the change will be like is the challenge. These changes will affect the physical, biological, and social dimensions of water resource systems. An essential aspect in the planning, design, and management of sustainable systems is the expectation of change. These includes changes due to geomorphological processes, the aging of infrastructures, shifts in demands or desires of the changing society, and even increased variability of water supplies, due to the changing climate. Change is an essential feature for the development and management of sustainable water resources.

Sustainable water resources systems are designed and operated in ways that make them more adaptive, robust, and resilient to an uncertain and changing future. They must be capable of functioning effectively under conditions of changing supplies, management objectives, and demands. Sustainable systems, like any others, may fail, but when they fail, they must be capable of recovering and operating properly without undue costs.

Given the ambiguity of what future generations will want, and the economic, environmental, and ecological problems they will face, a guiding principle for the achievement of sustainable water resource systems is to provide options to the future generations. One of the best ways to do this is to interfere as little as possible with the proper functioning of natural life cycles within river basins, estuaries, and coastal zones. Throughout the water resources system planning and management processes, it is important to identify all the beneficial and adverse ecological, economic, environmental, and social effects—especially the long-term effects—associated with any proposed project (Loucks and van Beek 2005).

Being one of the most important key elements influencing social health, well-being, the preservation of ecosystem, and the economic development of a country, water is a natural, yet limited resource. Due to global warming effects and its adverse impact on climate, many countries of the world will face serious shortages on this limited resource. Thus, planning, management, and preservation of water on a basin-wide scale are essential (Eroğlu 2007).

2 Current Situation of Water Resources

Population growth, industrialization, urbanization, and rising affluence in the twentieth century caused in a substantial increase in water consumption. While world's population grew threefold, water use increased sixfold during the same period. Water crisis is observed such that over one billion people around the world do not gain enough access to healthy drinking water. Moreover, half of world population does not have enough water and infrastructure for waste water. So, unavoidable water crisis could be foreseen in the whole world. On the other hand, according to a relevant research, contaminated waters cause 80 % of illness in developing countries and death of approximately 10 million people every year (Anonymous 2011).

While the population was 75 million in 2011, in 2012, it is 74 million in Turkey. Fifty-nine percent of the total population of Turkey, which is currently around 74 million, is presently dwelling in urban centers, whereas the remaining 41 % is living in rural areas. The amount of this increase in population was higher in provincial and district centers, while a decrease have occurred in villages and towns. The migration from villages to districts is thought to be effective in the realization of the situation in this way (TURKSTAT 2013).

Turkey is located on the crossroad of Europe and Asia. It has a total surface area of 779,452 km^2, of which 765,152 km^2 is land area and the remaining 14,300 km^2 is water surface. The climate of Turkey is semiarid with extremities in temperature. Climate and precipitation amounts exhibit great variance throughout the country: in the higher interior Anatolian Plateau, winters are cold with late springs, while the surrounding coastal fringes enjoy the very mild-featured Mediterranean climate. Average annual precipitation is 643 mm, ranging from 250 mm in the southeastern part of the country, to over 3,000 mm in the Northeastern Black Sea coastal area (Eroğlu 2007).

Natural precipitation, groundwater resources, freshwater rivers, streams, rivulets and lakes, dams and reservoirs, and marine and estuarine water resources are sources of water resources. Natural precipitation is the key source of water that feeds all the other water resources. Therefore, a decrease in rainfall will affect all the other water resources in a harmful way (Zakari 2013).

The average annual precipitation amount for Turkey corresponds to an average of 501 billion m^3 of water per year (Fig. 1). However, gross water potential of Turkey totals 234 billion m^3 because of discharge groundwater and surface runoff into neighboring countries or the various seas surrounding Turkey, drained surface water in closed basins and evaporation (Eroğlu 2007).

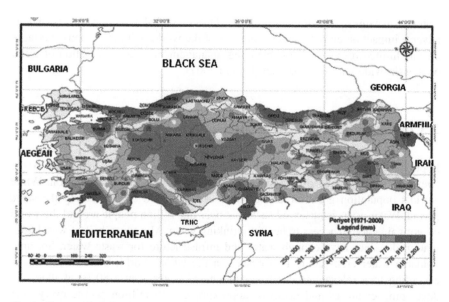

Fig. 1 Geographical distribution of mean annual precipitation (Şensoy et al. 2008)

Rivers are one of the main sources of freshwater. Seventy percent of total easily accessible water is provided by rivers. Moreover, 40 % of the world population depends for its freshwater on 214 transboundary rivers flowing through minimum two or more countries. For example, the Danube and Nile flow through 12 and 9 countries, respectively.

Turkey's water resources can be considered in 25 drainage basins. Figure 2 shows the water potential by drainage basins. The most important rivers are the Firat River (Euphrates) and Dicle River (Tigris), both of which are transboundary rivers originating in Turkey and discharging into the Persian (Arabian) Gulf. The Meric, Coruh, Aras, Arapcayi, and Asi Rivers are the other transboundary rivers (Bayazıt and Avcı 1997). Water potential and water consumption by basins are in Fig. 2.

The Euphrates and the Tigris are two of the most famous rivers in the world. The combined water potential of the two rivers is almost equal to that of the Nile River. Both rise in the high mountains of Northeastern Anatolia and flow down through Turkey, Syria, and Iraq and eventually join to form the Shatt–al–Arab 200 km before they flow into the Gulf. They account for about one-third of Turkey's water potential. Both rivers cross the Southeastern Anatolia region which receives less precipitation compared to other regions of Turkey. Therefore, during the 1960s and 1970s, Turkey launched projects to utilize the rich water potential of these rivers for energy production and agriculture. Turkey contributes 31 billion m^3 or about 89 % of the annual flow of 35 billion m^3 of the Euphrates the remaining 11 % comes from Syria. Iraq makes no contribution to the flow. As to the Tigris, the picture is entirely different. Fifty-two percent of the total average flow of 49 billion m^3 comes from Turkey. Iraq contributes all the rest. No Syrian water drain into the Tigris.

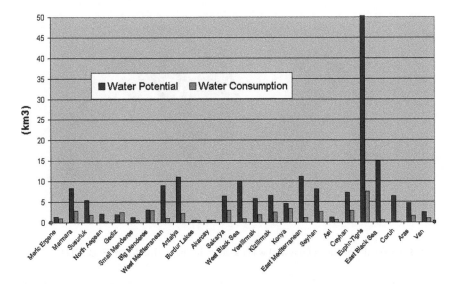

Fig. 2 Water potential and water consumption by basins (SHW 2009)

Southeastern Anatolia Project (GAP) is a regional integrated sustainable development project based on harnessing the water resources of the Euphrates and the Tigris rivers. It consists of dams, hydropower plants, and irrigation schemes and accompanying growth of agriculture, transportation, industry, telecommunications, health, and education sectors and services in this region. A total of 22 dams and 19 hydropower plants are to be constructed as components of GAP; 27,350 GWh/year will be produced (22 % of the country's hydropower potential), with an installed capacity of 7,500 MW and 1,700,000 ha will be irrigated (19 % of Turkey's economically irrigable land). Flow regulation and flood control will also be provided downstream (Anonymous 2011).

Swedish hydrologist Falkenmark (1989) points out that annual capitation of agricultural, domestic–urban, industrial water demand limit of minimum sufficiency is 1,000 m^3 in a country. So, under this limit means poverty in point of water. There are water famines especially in three regions of world at present time. These regions are Africa, the Middle East and South Asia. Similarly, covered by Water Basin Management Strategy in Turkey Aligned with the European Union and the Water Framework Directive, the countries that the amount of available water per capita exceeds 8,000–10,000 m^3 are defined "*water rich,*" less than 2,000 m^3 are defined "*have water scarcity*" and less than 1,000 m^3 are defined "*water poor*" (Akbulak 2011). The global distribution of physical water scarcity according to major river basin is presented in Fig. 3 (FAO 2011).

According to gross potential, water available per capita per year in Turkey as of 2009 was about 1,300 m^3. However, this amount is 3,000 m^3 in Asia, 5,000 m^3 in Western Europe, 7,000 m^3 in Africa, 18,000 m^3 in North America, 23,000 m^3 in South America, and 7,600 m^3 in overall the world (Türkkan 2006). Turkey is in

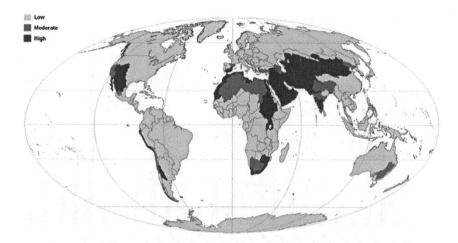

Fig. 3 Global distribution of physical water scarcity by major river basin (FAO 2011)

terms of per capita usable water potential in 182 countries 132nd ranks (Akbulak 2011). Actually, when capitation annual water amount is considered, the common aspect is that Turkey is not a rich country about water resources (Fig. 4).

The widespread nature of the risk of water scarcity may also limit the effectiveness of local solutions–such as acquiring more water from a neighboring country or basin–since many other localities will be trying to get control of the same resource (NRDC 2010). Turkey's policy regarding the use of transboundary rivers is based on the following principles:

- Water is a basic human need.
- Each riparian state of a transboundary river system has the sovereign right to make use of the water in its territory.

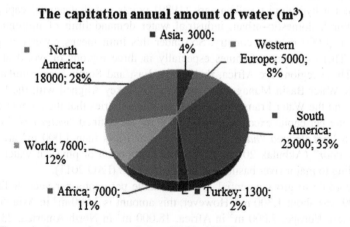

Fig. 4 The capitation annual amount of water (Türkkan 2006)

- Riparian states must make sure that their utilization of such waters does not give "significant harm" to others.
- Transboundary waters should be used in an equitable, reasonable, and optimum manner.
- Equitable use does not mean the equal distribution of waters of a transboundary river among riparian states.

As regards the Euphrates and the Tigris rivers;

- The two rivers constitute a single basin.
- The combined water potential of the Euphrates and the Tigris rivers is, in view of the Turkish authorities, sufficient to meet the needs of the three riparian states provided that water is used in an efficient way and the benefit is maximized through new irrigation technologies and the principle of "more crop per drop" at basin level.
- The variable natural hydrological conditions must be taken into account in the allocation of the waters of the Euphrates and the Tigris rivers.
- The principle of sharing the benefits at basin level should be pursued.

With respect to the utilization of the waters of the Euphrates and the Tigris rivers, Turkey has consistently abided by these principles and continued to release maximum amount of water from both rivers even during the driest summers thanks to the completed dams and the reservoirs in Southeastern Anatolia. For example, 1988 and 1989 as well as 2007–2008 water years were the driest years of the last half century. The natural flow of the Euphrates was around 50 m^3 per second. Yet, Turkey was able to release a monthly average of minimum 500 m^3 per second to Syria in conformity with the Article 6 of the Protocol signed by Turkey and Syria in 1987 (Bağış 1997). Article 6 reads as follows:

> During the filling up period of the Atatürk Dam reservoir and until the final allocation of waters of the Euphrates among the three riparian countries, the Turkish side undertakes to release a monthly average of more than 500 cubic meters per second at the Turkish–Syrian border and in cases where the monthly flow falls below the level of 500 cubic meters per second, the Turkish side agrees to make up the difference during the following month.

The motto of the Turkish government has always been that water should be a source of cooperation among the three riparian states. Turkey is eager to find ways of reaching a basis for cooperation, which will improve the quality of life of the peoples of the three countries. The point of departure should be to identify the real needs of the riparian states (Anonymous 2011).

In 2000s, total water supply (gross water consumption) is expected to be 3,973 km^3/year, and the net water consumption is expected to be 2,182 km^3/year (54 % of the water supply) in the world. This shows that water supply and net water consumption increased approximately threefold since 1950s. It is estimated that water supply and net water consumption reaches 5,235 km^3/year and 2,764 km^3/year, respectively. Nowadays, 59 % of the total water supply and 66 % of total net water consumption are carried out in the Asian continent, where major agricultural areas. It is expected that significant increase in water supply in the African and South American continent in the next year in spite of the decrease in water supply in the European and North American continent (Postel 1999; Shiklomanov 2000).

3 Use of Water Resources of Turkey

The trend of the water supply in Turkey is in Table 1. Quantity of water that sup-
plying/will supply for today and the future was calculated by using population
estimates (Table 2). Data related to the water consumption could not be reached.
Nevertheless, water consumption can be calculated agreeably that 40–50 % of the
supplying water lost in the distribution networks (Karakaya and Gönenç 2007).

When it comes to today, Turkey has water scarcity by the total potable water
potential, which is totally 112 billion m^3 and 1,500 m^3/person. On the other hand,
73 % of total utilization in this potential, reaching to 44 billion m^3, have realized
in the agricultural sector (SPO 2013).

Regarding to the utilization areas of water, irrigation is first rank (75 %),
household usage including drinking water is 15 % and industrial usage is 10 %.
Irrigation usage is 30 %, industrial usage is 59 %, and household usage is 11 %
in developed countries. However, these rates are different in the less developed
and developing countries. Irrigation usage is 80 %, industrial usage is 10 %, and
household usage is 8 % in these countries. Turkey is closer to less developed and
developing countries from the point of distribution of water utilization rates. When
looking for hydroelectric use, Turkey has 433 billion kWh gross hydroelectric
potential. It is estimated that the available part of this potential is 216 billion kWh
and 33 % is realized. The Euphrates and the Tigris rivers are in the first and second
rank in the available hydroelectric potential.

Table 1 The trend of water supply in Turkey (SPO 2001)

Years	Irrigation		Household		Industrial		Total Supply (10^9 m^3)
	Supply (10^9 m^3)	%	Supply (10^9 m^3)	%	Supply (10^9 m^3)	%	
1990	22,016	72	5,141	17	3,443	11	30,600
1992	22,939	73	5,195	16	3,466	11	31,600
1994	24,623	74	5,293	16	3,584	10	33,500
1997	26,415	74	5,520	15	3,710	11	35,645
2000	31,500	75	6,400	15	4,100	10	42,000
2030	71,500	65	25,300	23	13,200	12	110,000

Table 2 The trend of water supply for agricultural activities, industrial activities, and household
uses (SPO 2001)

Years	Irrigation (m^3/person/year)	Household (m^3/person/year)	Industrial (m^3/person/year)	Total (m^3/person/year)	Total (L/person/day)	Population (10^6 people)
1997	432	88	59	579	1,586	62,411
2000	482	98	62	642	1,758	65,300
2030	801	283	148	1,232	3,375	89,206

Table 3 Aboveground waters (billion m³) (Kıran 2005)

Flow	186.05
Consumable annual average amount of water	95.00
Actual annual consumption	27.50

Table 4 Underground waters (billion m³) (Kıran 2005)

Drainable annual water potential	12.20
Assigned amount	7.80
Actual annual consumption	6.00

Quantity of the aboveground and underground waters and annual water consumption of Turkey are in Tables 3 and 4.

According to these tables, the potential of the aboveground and underground water resources is approximately 200 km³. About 35 km³ of these resources is allocated for consumption. It is expected to the amount of consumption reaches 110 km³ in 2030 in consequence of the calculations considering the population estimates (SPO 2001).

4 Global Warming and Climate Change

All the countries and science world started to ponder about more productive use and development sustainability of available water resources because of *unconscious usage* of natural water resources and *global warming.*

The greenhouse gases in the atmosphere notably carbon dioxide (CO_2), nitrous oxide (N_2O), and methane (CH_4) prevent the heat radiated from the earth from being escaped into space. Human activities have led to an increase in the concentration of these greenhouse gases in the lower atmosphere which is resulting in global warming and its attended climate change. High solar radiation intensities and global warming, elevated air temperatures, reduced rainfall amounts and occurrence of droughts, unreliable and erratic rainfall events, poor rainfall distribution, extreme climate events–floods and storms, hurricanes and tornadoes are the indicators of the climate change (Zakari 2013).

Due to the human activities, there are now 40 % more greenhouse gases in the atmosphere and there were a few hundred years ago. The Earth has already warmed as the consequence of this, and scientists expect that the next 20–100 years, the world will warm a lot more (Fig. 5).

Global average temperatures are expected to increase by about 2–13 °F (1–7 °C) by the end of the century. According to this scenario, Turkey is among the risk group countries about the global warming (Fig. 6). It was also reported that the temperatures increase to 0.25 °C every 10 years in Turkey, there is fall average percentage 10 in rainfall, when a line is drawn from Samsun to Adana between 2071 and 2100 years, its west part will warm up 3–4 °C, its east part will warm up around 4–5 (Fig. 7). Daily rainfall amount will fall to 0.25 mm,

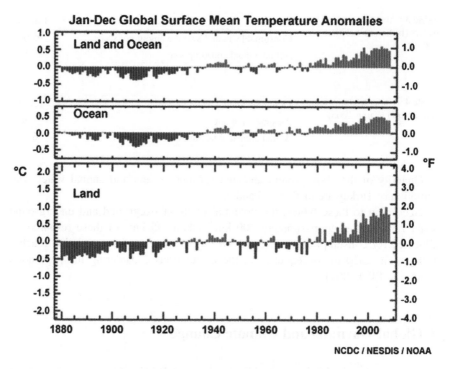

Fig. 5 Global surface mean temperature of land and ocean (NCDC/NOAA 2009)

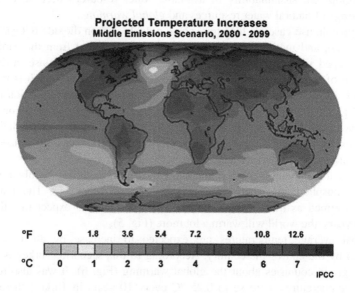

Fig. 6 Projected temperature increases (Gitay et al. 2002)

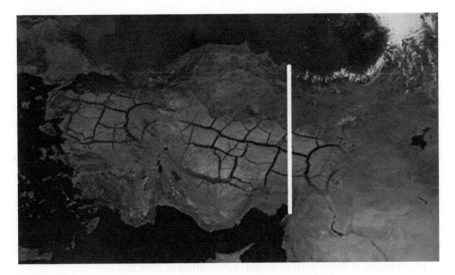

Fig. 7 Turkey map with a line from Samsun (in the North) to Adana (in the South)

vaporization and evaporation will increase, summer aridity will increase, there will be decline in fish species which live in interior waters depending on reducing in water resources (Atalık 2007). Again, it is made determined a lot of researches in parallel to report of IPCC (Bates et al. 2008); the negative effect of climate change to water resources will be pretty much in 10 years terms to come. Warming and sea level rise will continue and will probably occur more quickly than what we have already seen; even if greenhouse gases are stabilized, this will probably continue to occur for centuries, and some effects may be permanent.

One of the biggest problems of available resources is water pollution. Natural water resources are become dirty and unusable by industrialization, unplanned urbanization, agricultural activities, and polluting resources day by day.

5 Water Resources Management

Water resources management (WRM) is the wholeness that collects all the conditions and methods related to the determination and planning of need concerned with water resources, rational water use, detailed observation, and efficient protection under its framework.

In order to guaranty the supply of water in required place and at required time with sufficient amount and quality and to protect people and their activities from damaging effects of water resources, it is required to develop water resources development projects of different content and scope. A water resources project or system represent the group of measurements and activities that turn toward the aim of development or rehabilitation of water resources for serving into use of human beings and that contains structural or non-structural factors. Major targets of WRM of which the water resource is the basic elements can be listed as below:

- Determination of existing and future qualitative and quantitative characteristics of surface and groundwater resources, evaluation of supply possibilities,
- Determination, planning, and arrangement of community water demands,
- Formation of water balances, collection of factors that will provide continuity of these balances, and development of a long-term strategy for rational use of water resources,
- Monitoring of water resources in order to protect them from pollution and exhaustion,
- Planning water resources systems,
- Modeling of management,
- Designation of processes in water systems and operational conditions,
- Increase of assurance of water from quality and quantity point of views,
- Make it possible the multipurpose utilization of water resources, determination of priorities of these purposes, and reevaluation of allocations,
- Improvement of rational water use,
- Provide sustainability of natural potential of water resources and protect them,
- Provide effective utilization of technical elements (e.g., reservoirs and treatment plants) in order to protect communities from adverse effect of water resources,
- Benefit from managerial elements, economic instruments (e.g., pricing and penalties), laws, and regulations.

One of the most important WRM targets is planning and arrangement of community water demands after determination of existing and future qualitative and quantitative characteristics of surface and groundwater resources. Therefore, the extraordinary project of Turkey which related to inter-country transfers to make up for water deficits have been presented below as a case report.

6 The Northern Cyprus Water Supply Project

Cyprus is the third largest and third most populous island after the Italian islands of Sicily and Sardinia in the Mediterranean Sea (Carment et al. 2006). Also, it is the easternmost island in the Mediterranean Sea (Fig. 8). The island covers an area of 9,250 km^2 and a coastline of 648 km (Kleanthous et al. 2014). It is located in south of Turkey, west of Syria and Lebanon, northwest of Israel, and north of Egypt in the eastern Mediterranean.

As of 2010, requirement for domestic and drinking water of Turkish Republic of Northern Cyprus (TRNC) was 36 m^3/year and is expected to be 54 million m^3/year in 2035 (SHW 2011b). The population of TRNC was 294,906 in 2011. It is expected to be 349,650 in 2035 (TRNC-SPO 2013). Fifty-three percent of the population lives in the cities and 47 % lives in villages. The water is an indispensable resource for people and also it is important for developing social and economic activities (SHW 2011b) and critical to the functioning of human society and ecosystems (Alavian 1999).

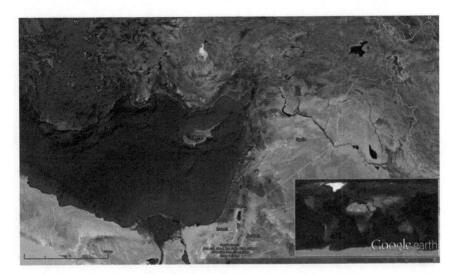

Fig. 8 Geographical position of Cyprus Island

Cyprus, politically divided into a Turkish northern region and Greek south-ern region, now has an even greater problem. Completely surrounded by the Mediterranean Sea, the island is facing a water crisis that has been aggravated by several years of drought and an increased need for water. Progressive climate change has seen a decrease in precipitation levels in Northern Cyprus by more than a quarter over the past 96 years. Four years of drought, which ended in 2008, have made Cyprus the first EU country to run out of water (Anonymous 2009). For years, TRNC has water shortage and this problem must be solved as soon as possible. To solving this problem, it is decided to the most suitable solution was uninterrupted water and electricity transferring from Turkey to TRNC (Fig. 9).

The Northern Cyprus Water Supply Project *(in Turkish: KKTC Su Temini Projesi)* is an international water derivation project designed to supply water for drinking, domestic, and irrigation from southern Turkey to Northern Cyprus via pipeline under the Mediterranean Sea. This pipeline project is original, unique, and the first of its kind project in the world.

TRNC has limited natural resources. Therefore, supplying drinkable, domes-tic, and irrigation water from Turkey to TRNC will significantly contribute to the development of Northern Cyprus. Seventy-five million m^3 of water to be taken based on a constant flow rate will be transmission annually, from Alaköprü Dam, built on Anamur (Dragon) Creek in the city of İçel (Mersin), by TRNC Water Supply Project. Of the 75 million m^3 water which will be transmission to TRNC through 107 km line in total length, 37.76 million m^3 (50.3 %) will be used for domestic and drinking purposes and the remaining part 37.24 million m^3 (49.7 %) will be allocated for irrigation (SHW 2011a).

Cyprus has the water shortage with regard to the groundwater and surface water due to drought. The project aims to supply Northern Cyprus with water from Turkey for a time period of 50 years. Following the realization of the project,

Fig. 9 A scheme for The Northern Cyprus Water Supply Project

irrigated farming at an area of 4,824 ha in Mesaoria Plains, one of the largest plains and the most valuable agricultural land of the TRNC, will help improve the standard of living in the region.

The project will be carried out by the Turkish State Hydraulic Works (SHW). It consists of the construction of a dam and a pumping station at each on both sides as well as a pipeline of 107 km running mainly under sea. The construction will have four stages: Turkey side, Land Line, Sea Crossing, and TRNC side.

6.1 Turkey Side

Alaköprü Dam is being built in Anamur, İçel (Mersin) Province on the Anamur Dragon Creek at 93 m elevation (Fig. 10). It will have a reservoir holding 130.5 million m^3 water. Thanks to this project, Alaköprü Dam will meet the water requirement in time of forest fire in the forestland. Also, the electricity energy will be produced in Alaköprü Dam. A hydroelectric power plant which has 26 MW power will be built on Alaköprü Dam and it will produce 111.27 million kWh of electricity, annually. In this way, the hydropower plant will contribute to the economy. In addition to this, it will contribute to the region in terms of fishing and tourism.

6.2 Land Line

A pipeline of 1,500 mm diameter and 23 km length will carry 75 million m^3 water of Alaköprü Dam to Anamurium Equalizing Chamber, which will have a

Fig. 10 A scheme for transmission line in Anamur, Mersin, Turkey

reservoir holding 10,000 m³ and connect to submarine pipeline in 1 km distance for transferring to Geçitköy Dam in TRNC side.

6.3 Sea Crossing

An 80-km-long submarine pipeline of 1,600 mm in diameter 250 m depth in Mediterranean Sea will transfer water from Anamurium Plant in Turkey to Güzelyalı Pumping Station in Northern Cyprus. The pipes will be made of high-density polyethylene (HDPE), a material commonly used to transport water. It will cross a channel as deep as 1,430 m, but the pipeline will be suspended 250 m below the sea surface and each 500 m section of pipeline will be fixed to the sea floor far below (Fig. 11). Earthquakes could destroy anchoring points or a tsunami could break the floating line. With all that, planning engineers considered potential hazards such as earthquakes, tsunami, and the high level of submarine traffic in the area.

6.4 TRNC Side

A pipeline of 3 km will elevate water from Güzelyalı Plant to the reservoir of Geçitköy Dam which is at 65 m elevation, close to the city of Girne. An 80-km-long submarine pipeline will be suspended in 250 m depth in Mediterranean Sea will transfer water from Alaköprü Dam in Turkey to Geçitköy Dam close to the

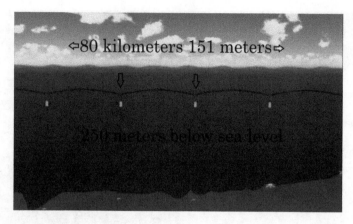

Fig. 11 A scheme for sea crossing of the project

city of Girne in Northern Cyprus (Fig. 12). Geçitköy Dam will have a reservoir holding 26.52 million m³ water. Also, Güzelyalı Pumping Station will have 5 MW power and Geçitköy Pumping Station will have 16.40 MW power (TSMS 2011).

The pipes will be laid on the sea floor and will be immersed to the seabed by filling with the seawater after combined over the sea by the vessels. The pipes will be suspended 134 pieces suspender under the sea. Such a suspended submarine pipeline of this size does not exist in the world (Anonymous 2013). The pipes to be settled under the sea are smart pipes and the pipeline will have sensors and transmitters mounted to signal any possible faults for repair. The results of experiments showed that the fatigue life of the pipeline system is 125 years and the creep life is more than 1,000 years (Fig. 13).

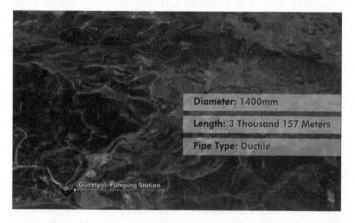

Fig. 12 A scheme for pipeline in Guzelyali Pumping Station and Geçitköy Dam in TRNC

Fig. 13 The HDPE pipes used in the project

Gruen (2000b) pointed out the preconditions for the successful importation of water. He stated that a plan for importation of water must meet four criteria to win general acceptance:

1. It must cause no appreciable harm either by reducing the supply to established users or by causing environmental damage.
2. It must prove to be technically feasible.
3. It must be politically acceptable. Facilities must be physically secure, and the agreement must be structured to insulate the scheme as much as possible from disruption or cancelation in case of political changes in the policies of the supplier country or of transit countries.
4. It must be economically viable.

The source of the water in Turkey, the Dragon River, has an annual capacity of 700 million m^3, about 1/10th or 75 million m^3 of which is to be piped to Northern Cyprus on completion (Anonymous 2013). So, the authors think that there is no risk for Turkey about reducing the resources and amount of supplying water to TRNC.

For solving the water shortage problem in Northern Cyprus, various measures and projects are planned and carried out to increase the amount of water supply and use it more efficiently. One of these projects have been planned to import water from Turkey by a tanker or through the use of large water bags. The purpose of this project is to import freshwater from Turkey by a tanker to meet the demand for potable water by households. The project does not aim to provide water for agricultural use or for recharging the aquifers which badly affected by the seawater intrusion. The first load of water was transported by water bags in 1998 by the Nordic Water Supply, the Norwegian producer of navigable plastic bags. Water bags of

10,000 m^3 capacity could potentially transport 3 million m^3 of water in the first year from Soğuksu River of Anamur in Turkey. Increasing of the water bags capacity to 30,000 m^3 would enable 7 million m^3 of water to be transported annually. This amount is the maximum that the system can allow to be pumped from Nicosia to Gazimagusa and then on to the main reservoirs. Bıçak and Jenkins (1999) explained in detail the total estimated costs for all parts of water import by water tanker are as follows: (1) The cost of water transportation was just US$0.4 per m^3. (2) It increased as US$0.79 per m^3 when substructure investments added. (3) Leakage of 30 % in the distribution system would increase total cost to US$1.13 per m^3 (Priscoli and Wolf 2009). This project aims to only freshwater importing but there is also a great need for agricultural use. So that this amount of the imported water is not enough for providing the water demands not only potable use but also agricultural use in Northern Cyprus due to the volume of water to be transported is limited by the capacity of the tankers. Therefore, when economic reasons were taken into account, the sustainability of this project was not possible.

Another important project purposes to prevent the excessive use of water by way of converting traditional irrigation systems to modern irrigation techniques. Through this project, a large amount of water will be preserved, salination will be prevented, and the quality and efficiency of agricultural harvest will be better. However, this measure will not supply water for potable use. Therefore, all of these measures are not adequate for sustainability of water sources.

There is another option for supplying demand for water desalination. Desalination is a general term for the process of removing salt from water to produce freshwater (Greenlee et al. 2009). Freshwater is defined as containing less than 1,000 mg/l of salts or total dissolved solids (Sandia 2003; Greenlee et al. 2009). Current desalination methods require large amounts of energy which is costly both in environmental pollution and in terms of money (Karagiannis and Soldatos 2008). We believe that there is variability in the cost of water desalination. The water desalination cost comprises of two main categories: The investment costs and the annual operating costs. The investment costs include all the costs related to the installation and appropriate method for the system such as drilling, desalination, and other equipment cost, buildings and installation and commissioning (Karagiannis and Soldatos 2007). Water desalination costs look like location specific and the cost per cubic meter ranges from installation to installation. Because, the water desalination cost depends upon a lot of factors which are the energy source, the desalination method, the capacity of the desalting plant, the level of feed water salinity, and other location-related factors. Karagiannis and Soldatos (2008) stated that the large desalination systems in many countries which could reach a production of even 500,000 m^3/day use mainly thermal desalination methods. In these cases, the produced freshwater cost ranges between US$0.50/m^3 and US$1.00/m^3. For medium size systems (12,000–60,000 m^3/day), the cost of seawater desalination shows the higher variability between US$0.44/m^3 and US$1.62/m^3. The cost can be higher which are between US$2.24 m^3 and US$19.11/m^3 in seawater desalination units having a capacity from a few cubic meters to 1,000 m^3. These small systems use mainly renewable energy sources and for this reason, as well as due to lower economies of scale the cost is so high. Seawater desalination plants have a cost which varies between US$0.44/m^3 and US$3.39/m^3 and only when the desalination unit is very small (2–3 m^3 daily production), the cost can increase to approximately US$6.9/m^3. These facts indicate that desalination requires great amounts of energy and this is high-priced way for supplying and sustainable use of water.

The water supply project was approved by Government decree No. 98/11202, dated May 27, 1998. It is being undertaken by a consortium of Turkish and European firms headed by Alsim-Alarko Holding of Istanbul, one of largest conglomerates and experienced construction companies of Turkey (Gruen 2000a). To maintain the sustainability of water import project, the impacts of each step should be evaluated and adverse effects should be mitigated as far as possible. For this objective, the environmental impact assessment (EIA) study is required as a decision aiding tool. For this project, there is also an EIA study carried out by Alsim-Alarko. Alsim-Alarko prepared the feasibility report and HSW approved it on 1999 (Rende 2007). The EIA study is a systematic procedure for classifying and estimating all potential impacts of the project (Lattemann and Höpner 2008). EIA report of this project shows that there is not a high-risk hazard and only a few moderate-risk hazards which are as low as can be implemented at a reasonable level. As a result, the EIA report pointed out the project is feasible. We would like to state that this project will continue to the Southern Cyprus if Southern Cyprus makes a request for supplying water.

Turkey has the capacity to contribute to the establishment of an enabling environment by realizing this project for socioeconomic development of the people in Cyprus which in turn could enhance peace and security in the region. We believe that this project will help to improved management of local demand.

The total investment cost of the project is budgeted at approximately US$320 million consisting of US$20.6 million for structures in Turkey, US$285.5 million for the submarine pipeline, and US$12.2 million for the structures in Northern Cyprus (Anonymous 2012). In case we are calculated the cost of water obtained using water pipeline, it costs US$4.2/m^3 water/a year. This value compared to whole desalination process is acceptable when the cost of maintenance is taken into account. Finally, a solution is still required to the water shortage problem in the island. 500 million m^3 of freshwater is annually flowing from southern of Turkey into the Mediterranean Sea (Bıçak and Jenkins 1999). Once the substructure is completed, it would be possible to export water to the Southern Cyprus firstly and then to the other Mediterranean countries. The benefits and impacts of water supply project should be considered on the scale of regional management plans. These facts indicate that the use of submarine pipeline between Turkey and Northern Cyprus is highly competitive to other methods of supply such as desalination and transfer by plastic bags. The authors point out the best solution is to construct a pipeline between Turkey and Northern Cyprus. Moreover, water shortage and worsening of water quality should be improved by successful water resources management strategies that require combining advanced technologies.

7 Conclusion

While water management and climate change adaptation plans will be essential to lessen the impacts, they cannot be expected to counter the effects of a warming climate. One reason is that the changes may simply outrun the potential for alternative such as modifying withdrawals, increasing water use efficiency, increased water recycling enhancing groundwater recharge, rainwater harvesting, and inter-basin or inter-country transfers to make up for water deficits. However, we use the isolated treatment of any component of water sources system results in suboptimal solutions. For that reason on integrated approach is inevitable for the rational management of water resources.

We believe that there is a great deal of effort in adopting and exercising an integrated approach to water resources management in Turkey. For this purpose, it has been taken into Water Frame Directive of European Union. The two main titles of this directive are as follows:

- "Usage of Sustainable Water" (80/68/EEC) topic (Efeoğlu 2005). To provide continuity of available resources is emphasized and to constitute necessary substructure about financial support is mandated.
- "Aquatic Ecosystem and Prevention of Waters." In other words to prevent pollution in available resources and to avert damage to nature stability is aimed.
- In addition, the two provisions are taken place in regulation of management of resource in 9th development plan of 2007–2013 years. These are as follows:
- "Environmental Protection and Development of Urban Groundwork" title of plan is 159 provision.
- 162 provision points out the agreement on the topic that "United Nations Climate Change Frame Engagement" was approved by the Turkish National Assembly in May 24, 2004.

Some foundations of Turkey unequivocally recognize the value of water resource information as a foundation for integrated management of resources. In this respect, they have lately had some initiatives and will hopefully achieve fruitful results in setting out to:

- implement better instrument for data collection;
- employ modern technology for data transmission, processing, and archiving;
- implement national water information system;
- take advantage of satellites and remote sensing applications for data transmission; analyze and present data using advanced computer models and Geographic Information Systems (GIS).

Acknowledgment The authors wish to thank anonymous reviewers for their insightful comments and constructive suggestions.

References

Akbulak Y (2011) The facts of the world and Turkey related to the available water resources. Retrieved 25 Oct 2013, from Dünya Newspaper: http://www.dunya.com/kullanilabilir-su-kaynaklarina-iliskin-dunya-ve-turkiye-gercekleri-118491h.htm

Alavian V (1999) Shared waters: catalyst for cooperation. J Contemp Water Res Educ 115(1):1–11

Anonymous (2009) Cyprus runs risk of desertification: geophysicist. Retrieved 10 Jan 2014, from Reuters: http://www.reuters.com/article/2009/01/12/us-cyprus-climate-idUSTRE5082PJ20090112

Anonymous (2011) Turkey's policy on water issues. Retrieved 24 Oct 2013, from Ministry of Foreign Affairs of Republic of Turkey: http://www.mfa.gov.tr/turkey_s-policy-on-water-issues.en.mfa

Anonymous (2012) Asrın Projesinde Kazma Vuruluyor. Retrieved 13 Jan 2014, from Gazete Vatan: http://haber.gazetevatan.com/asrin-projesinde-kazma-vuruluyor/486556/2/Ekonomi#.UHbLUvL1m2s

Anonymous (2013) Key challenge looms for longest undersea water pipeline. Retrieved 19 Dec 2013, from Bloomberg: http://www.bloomberg.com/news/2013-12-19/key-challenge-looms-for-longest-undersea-water-pipeline.html

Atalık A (2007) Effect of global warming on water resources and agriculture. In: Symposium of World Food Day, Ankara

Bağış (1997) Turkey's hydropolitics of the euphrates-tigris basin. Int J Water Resour Dev 13(4):567–581

Bates BC, Kundzewicz ZW, Wu S, Palutikof JP (eds) (2008) Climate change and water. Technical paper of the intergovernmental panel on climatic change. IPCC Secretariat, Geneva

Bayazıt M, Avcı I (1997) Water resources of turkey: potential, planning, development and management. Int J Water Resour Dev 13(4):443–452

Bıçak HA, Jenkins GP (1999) Costs and pricing policies related to transporting water by tanker from turkey to north cyprus. In: Brooks D, Mehmet O (eds) Water balances in the eastern mediterranean. IDRC Books, Ottawa

Carment D, James P, Taydas Z (2006) Who Intervenes? Ethnic conflict and interstate crisis. The Ohio State University Press, Columbus

Efeoğlu A (2005) E.U. water framework directive and related activities in Turkey. In: Water information systems of Turkey and Euro Mediterranean Water Information System (EMWIS), Ankara

Eroğlu V (2007) Water resources management in Turkey. In: International congress on river basin management. Antalya, pp 321–332

EU (European Union) (2000) Directive 2000/60/EC of the European Parliament and of the Council establishing a framework for Community action in the field of water policy. CELEX-EUR Official J 11(3):327

Falkenmark M (1989) The massive water scarcity now threatening Africa: why isn't it being addressed? Ambio 18(2):112–118

FAO (2011) Status and trends in land and water resources. In: FAO, The state of the world's land and water resources for food and agriculture (SOLAW)—managing systems at risk. Food and Agriculture Organization of the United Nations, Rome

Gitay H, Suárez A, Watson R, Dokken DJ (eds) (2002) Climate change and biodiversity. Technical paper of the intergovernmental panel on climate change. IPCC Secretariat, Geneva

Greenlee LF, Lawler DF, Freeman BD, Marrot B (2009) Reverse osmosis desalination: water sources, technology, and today's challenges. Water Res 43:2317–2348

Gruen GE (2000a) The politics of water and middle east peace, american foreign policy interests. J National Committee Am Foreign Policy 22(2):1–21

Gruen GE (2000b) Turkish waters: source of regional conflict or catalyst for peace? J Water, Air Soil Pollut 123(1–4):565–579

Karaaslan Y (2013) Water quality management in Turkey water resources management. SUEN, Istanbul

Karagiannis IC, Soldatos PG (2007) current status of water desalination in the Aegean Islands. Desalination 203:56–61

Karagiannis IC, Soldatos PG (2008) Water desalination cost literature: review and assessment. Desalination 223:448–456

Karakaya N, Gönenç İE (2007) Water consumption in turkey and the world. IGEMPortal, Istanbul

Kıran A (2005) Water in the middle east: a field of a conflict or compromise. Kitap, Istanbul

Kleanthous S, Vrekoussis M, Mihalopoulos N, Kalabokas P, Lelieveld J (2014) On the temporal and spatial variation of ozone in cyprus. J Sci Total Environ 476–477:677–687

Lattemann S, Höpner T (2008) Environmental impact and impact assessment of seawater desalination. Desalination 220:1–15

Loucks DP, van Beek E (2005) Water resources systems planning and management: an overview. In: Loucks DP, van Beek E (ed), Water resources systems planning and management an introduction to methods, models. UNESCO, Italy, pp 3–34)

NCDC/NOAA (2009) National Climatic Data Center. Retrieved 21 Oct 2013, from National Oceanic and Atmospheric Administration: http://www.ncdc.noaa.gov/

NRDC (2010) Climate change, water, and risk: current water demands are not sustainable. Retrieved 21 Oct 2013, from Natural Resources Defense Council: http://www.nrdc.org/global-Warming/watersustainability/

Postel S (1999) Last Oasis: facing water scarcity. (FS Sozer, Trans.) Tubitak-Tema, Ankara

Priscoli JD, Wolf AT (2009) Managing and transforming water conflicts. Cambridge University Press, New York

Rende M (2007) Water transfer from Turkey to water-stressed countries in the middle east. In: Shuval H and Dweik H (eds) Water resources in the Middle East. Springer, Berlin, pp. 165–174

Sandia (2003) Desalination and water purification roadmap—a report of the executive committee. DWPR Program Report. #95. U.S. Department of the Interior, Bureau of Reclamation and Sandia National Laboratories, Albuquerque, New Mexico

Shiklomanov IA (2000) Appraisal and assessment of world water resources. Water Int 25:11–32

SHW (2009) Turkey water report. General directorate of state hydraulic works, Ankara. Retrieved 20 Oct 2013, from State Hydraulic Works: http://en.dsi.gov.tr

SHW (2011a) KKTC'ye Su Temin Projesi. Retrieved 19 Dec 2013, from State Hydraulic Works: http://www.dsi.gov.tr/projeler/kktc-su-temin-projesi

SHW (2011b) Asrın Projesi Gerçekleşiyor. Retrieved 19 Dec 2013, from State Hydraulic Works: http://www2.dsi.gov.tr/basinbul/detay.cfm?BultenID=258

SPO (2001) Eighth-Five year development plan. State Planning Organization, Turkish Grand National Assembly, Ankara

SPO (2006) Ninth development plan. State Planning Organization, Turkish Grand National Assembly, Ankara

SPO (2013) Tenth development plan. State Planning Organization, Turkish Grand National Assembly, Ankara

Şensoy S, Demircan M, Ulupınar Y et al (2008) The climate of Turkey. Retrieved 12 Oct 2013, from Turkish State Meteorological Service, Ankara

TRNC-SPO (2013) Turkish Republic of Northern Cyprus State Planning Organization, Lefkoşa. Retrieved 01 Dec 2013, from http://www.devplan.org/

TSMS (2011) Asrın Rüyası KKTC İçme Suyu Projesinde Çalışmalar Bütün Hızıyla Devam Ediyor. Retrieved 19 Dec 2013, from Turkish State Meteorological Service: http://www.mgm.gov.tr/kurumsal/haberler.aspx?y=2011&f=kibrisicmesuyu

TURKSTAT (2013) Population statistics. Retrieved 20 Oct 2013, from Turkish Statistical Institute: http://www.turkstat.gov.tr/

Türkkan M (2006) Water resources potential and significance of our country. Western Mediterranean Forest Research Institute, Antalya

Uslu O, Türkman A (1987) aquatic toxicology and control. General Directorate of Environmental Management of Prime Ministry of Republic of Turkey, Ankara

Zakari A (2013) Impact of climate change, desertification on agriculture and food security. 10th Regional (West Africa) Meeting of the African Caribbean Pacific-European Union (Acp-Eu) Parliamentary Assembly, Abuja

Authors Biography

Prof. Olcay Hisar is a researcher at the Department of Basic Sciences, Faculty of Marine Sciences and Technology, at Canakkale Onsekiz Mart University in Turkey.

Semih Kale is researcher at the Department of Fishing and Fish Processing Technology, Faculty of Marine Sciences and Technology, at Canakkale Onsekiz Mart University in Turkey.

Özcan Özen is researcher at the Department of Marine Technology Engineering, Faculty of Marine Sciences and Technology, at Canakkale Onsekiz Mart University in Turkey.

Moving Toward an Anthropogenic Metabolism-Based and Pressure-Oriented Approach to Water Management

Xingqiang Song, Ronald Wennersten and Björn Frostell

Abstract Effective and efficient water management systems require a comprehensive understanding of anthropogenic pressures on the water environment. Developing a broader systems perspective and extended information systems is therefore essential to systematically explore interlinks between anthropogenic activities and impaired waters at an appropriate scale. For this purpose, this paper identifies information dilemmas in contemporary water monitoring and management from an anthropogenic metabolic point of view. The European Drivers-Pressures-State of the Environment-Impacts-Responses (DPSIR) framework was used as a basis for classifying and discussing two approaches to water management, namely state/impacts-oriented and pressure-oriented. The results indicate that current water monitoring and management are mainly state/impacts-oriented, based on observed pollutants in environmental monitoring and/or on biodiversity changes in ecological monitoring. This approach often results in end-of-pipe solutions and reactive responses to combat water problems. To complement this traditional state/impacts-oriented approach, we suggest moving toward an anthropogenic metabolism-based and pressure-oriented (AM/PO) approach to aid in alleviating human-induced pressures on the water environment in a more proactive way. The AM/PO ideas can equally be applied to water-centric sustainable urbanization planning and evaluation in a broader context.

X. Song (✉) · B. Frostell
Division of Industrial Ecology, KTH Royal Institute of Technology,
100 44 Stockholm, Sweden
e-mail: xsong@kth.se

B. Frostell
e-mail: frostell@kth.se

R. Wennersten
Institute of Thermal Science and Technology, Shandong University, 250061 Jinan,
People's Republic of China
e-mail: wennersten.ronald@gmail.com

© Springer International Publishing Switzerland 2015   229
W. Leal Filho and V. Sümer (eds.), *Sustainable Water Use and Management*,
Green Energy and Technology, DOI 10.1007/978-3-319-12394-3_12

Keywords Anthropogenic metabolism · DPSIR framework · Pressure-oriented · Water monitoring · Water management

1 Introduction

There is a growing consensus that it is essential to shift the focus from a single/ sectoral approach to a more holistic approach to water management and planning. Integrated Water Resources Management (IWRM) and Integrated River Basin Management (IRBM), based on a systems approach to water management, have attracted wide international attention. One key principle of IWRM is to integrate both within and between the following two categories: the natural system (e.g., water availability and quality) and the human system (e.g., resource extraction, production, and waste management). Unfortunately, IWRM is not yet effectively implemented on a wider scale, for a number of reasons, but primarily due to the lack of systematic approaches to better address complex water resources systems (Castelletti and Soncini-Sessa 2007).

In recent years, there have been important advances in understanding causes of water problems from an interdisciplinary perspective. For example, the United Nations Educational, Scientific and Cultural Organization (UNESCO) called for thinking outside the conventional water box: for water issues to be linked to decisions on sustainable development and for drivers of water pressures to be handled in broader and interrelated contexts (WWAP 2009). The UNESCO call (re-) emphasizes the importance of achieving an improved understanding of causal relationships in considering both quantitative water degradation and qualitative water degradation. Another example is the "System of Environmental-Economic Accounting for Water (SEEA-Water)," published by the United Nations Statistics Division in 2012. In the SEEA-Water, an experimental water quality accounting approach is introduced. However, it addresses only the stocks of certain qualities at the beginning and the end of an accounting period, without further specification of the causes (UNSD 2012).

Moving toward improved water management depends on the availability of relevant, accurate, and up-to-date information, and decision makers often lack access to the critical information needed for effective decision making (Hooper 2005). To facilitate the early observation of water quality changes, for instance, the European Union approach has focused on the improvement of water quality monitoring systems and ecological outcomes (EC 2003). However, Destouni et al. (2008) reveal that waterborne loads of nitrogen, phosphorous, and organic pollutants traveling from land to the Baltic Sea might be larger from small, unmonitored areas than from the main rivers that are subjected to systematic environmental monitoring. Developing a broader systems perspective on water monitoring and accounting approach is therefore essential to systematically explore interlinks between anthropogenic activities and impaired waters at an appropriate scale.

To our knowledge, there has, as yet, been no systematic examination of cause–effect relationships between anthropogenic metabolism and water quality degradation, while alleviating human-induced pressures on waters at their sources. As two forerunners in the development of metabolic thinking, Baccini and Brunner (2012) provide concrete approaches to accounting society's physical metabolism. In fact, the real strength of a metabolic approach is that it does not discriminate between inflows of resources and outflows of emissions. Instead, it sees both phenomena as linked and thus represents an improved systems approach to ecological sustainability, which could be applied to water resources management.

This paper aims to identify dilemmas in contemporary water monitoring and management approaches from an anthropogenic metabolic point of view. For this purpose, the European Environment Agency (EEA's) so-called Drivers-Pressures-State of the Environment-Impacts-Responses (DPSIR) framework is used as a basis for the classification of water management approaches. Furthermore, the paper recommends moving toward an anthropogenic metabolism-based and pressure-oriented (AM/PO) water management approach, the necessity for which is discussed.

2 The DPSIR Framework and Classification of Water Management Approaches

2.1 The DPSIR Framework

The DPSIR framework (Fig. 1) in general intends to provide a basis for describing environmental problems by identifying the cause–effect relationships between the environment and anthropogenic activities. In terms of the DPSIR framework, socioeconomic development and sociocultural forces function as drivers (D) of human activities that increase or mitigate pressures (P) on the environment. Environmental pressures then change the state of the environment (S) and result in impacts (I) on human, ecosystems, and the economy. Those changes in environmental conditions and the corresponding impacts may lead to societal responses (R) via various mitigation, prevention, or adaptation measures in relation to the identified environmental problems (Smeets and Weterings 1999). In practice, the DPSIR framework has been widely employed as an environmental reporting approach, e.g., in the EEA's State of the Environment Reports.

Although the DPSIR framework has been frequently used to aid in addressing various environmental problems, it has received a lot of criticisms. From the perspective of researchers, for example, typical criticisms are that (i) it employs static indicators without considering system dynamics; (ii) it fails to clearly illustrate specific causal relationships of environmental problems under study; (iii) it suggests only linear causal chains for complex environmental issues; and (iv) it has shortcomings to establish good communication between researchers and stakeholders (Rekolainen et al. 2003; Svarstad et al. 2008). Moreover, Friberg (2010) claims that "the DPSIR framework is seldom used by applied scientists, who often use 'stress' and 'stressors' rather than 'pressure'." Typically, a stress-based approach in stream ecology focuses on how point source and diffuse pollution affect ecosystems at various levels of organization (Friberg 2010).

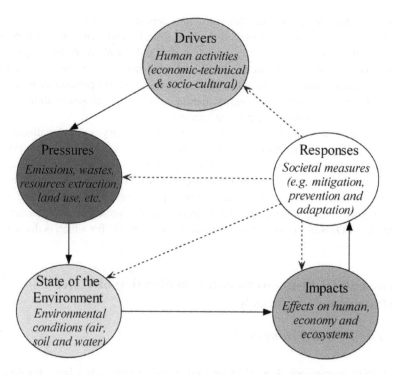

Fig. 1 The European DPSIR framework (after Gabrielsen and Bosch 2003)

On the other hand, Carr et al. (2007) hold the view that "DPSIR is not a model, but a means of categorizing and disseminating information related to particular environmental challenges." These authors further argued that the original goal of the framework is to identify appropriate indicators for framing particular environmental problems, rather than the elaboration of their cause–effect relationships, aiming to make appropriate responses. Referring to recent applications of the approach, Atkins et al. (2011) argue that "an expert-driven evidence-focused mode of use is giving way to the use of the framework as a heuristic device to facilitate engagement, communication and understanding between different stakeholders." In addition, Tscherning et al. (2012) highlight the usefulness of the application of DPSIR in research studies by providing policy makers with meaningful explanations of cause–effect relationships.

In this paper, the DPSIR framework is employed as a basis for identifying dilemmas in contemporary water monitoring and management systems. Furthermore, it is used to aid discussions about the necessity of moving from a state/impacts-oriented approach to an AM/PO approach to managing water resources.

2.2 The DPSIR-Based Classification of Water Management Approaches

Based on the European DPSIR framework, two classifications, namely state/impacts-oriented approach and pressure-oriented approach, are made for water management approaches and the derivation of information systems (Fig. 2). In simple terms, the state/impacts-oriented approach includes societal responses to changes in water environmental state and their impacts in terms of information from environmental and/or ecological monitoring networks. On the other hand, the pressure-oriented approach refers to management efforts focusing on drivers, pressures, and responses to anthropogenic metabolism.

The first classification, the state/impacts-oriented approach, requires pollutant-oriented environmental information and species-oriented ecological information. Its information systems focus on the ambient water environment and ecosystems. In a broader sense, the atmospheric system may also be referred to in relation to vapor flows and air pollutant concentrations. In other words, the main concern of a state/impacts-oriented approach is hydrophysical and biogeochemical changes in the water environment and the net effect is often reactive responses to combat water problems (because of late recognition of pollutant accumulation, for instance). Regarding water quality management, the main focus is usually on monitoring and controlling pollutants discharged to the natural recipient, e.g., by means of constructing monitoring networks and wastewater treatment methods (either natural, physical–chemical, and/or biological).

The second classification, the pressure-oriented approach, derived from a Drivers-Pressures-Responses (DPR) model, is based on the underlying principle that anthropogenic metabolism determines human-induced pressures on the water environment. According to Graedel and Klee (2002), metabolism of the anthroposphere "represents the metabolic processes of human-technological and

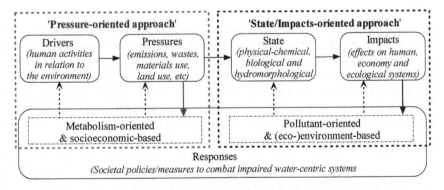

Fig. 2 The pressure-oriented and state/impacts-oriented approaches, where the *red arrow* shows the link emphasized in this paper to address the root causes of human-induced water problems (Song 2012)

human-social systems at all spatial scales, broadening the basic principles of urban metabolism to include all of the technical and social constructs that support the modern technologically-related human." In essence, the pressure-oriented approach includes the following: (i) an inventory analysis of water-related environmental loads (inflows of resources and outflows of emissions) of the system under investigation and (ii) assessing the water environmental consequences of the quantified environmental loads in the inventory analysis.

With respect to the pressure-oriented approach, accounting for the metabolism of the anthroposphere is necessary for identifying human-induced pressures on the water environment. Comprehensive water-related anthropogenic metabolic information system to a large extent could help optimize the allocation of limited resources in society for making proactive societal responses to water degradation. In this context, developing measures for combating environmental degradation would begin with investigating human-induced pressures exerted by production and consumption at their sources.

3 Dilemmas in Contemporary Water Monitoring and Management Systems

The following two examples are used as a basis for identifying dilemmas in demand for information in contemporary water management systems. However, a complete review of water monitoring techniques and indicators used in water management systems worldwide is beyond the scope of this study (it will be addressed in a follow-up study).

3.1 A Brief Conceptual Framework of Contemporary Water Quality Analysis

In recent decades, the focus of water quality management policy has gradually shifted from effluent-based (control of point pollution sources) to ambient-based (control of non-point pollution sources) water quality standards (National Research Council 2001). Generally speaking, water quality research is mainly driven by the following four needs: (i) toward scientific understanding of the aquatic environment; (ii) qualifying water for human uses; (iii) aiding in managing land, water, and biological resources; and (iv) identifying the fluxes of dissolved and particulate material through rivers and groundwater as well as from the land to the ocean (Meybeck et al. 2005).

According to Zhang et al. (2005), the current water quality approach "emphasizes the overall quality of water within a waterbody and provides a mechanism through which the amount of pollution entering a waterbody is controlled by the intrinsic conditions of that body and the standards set to protect it." This point is reflected in the two main streams of water quality management strategy: (i) setting water quality objectives (WQOs) and (ii) setting emission limit values (ELVs).

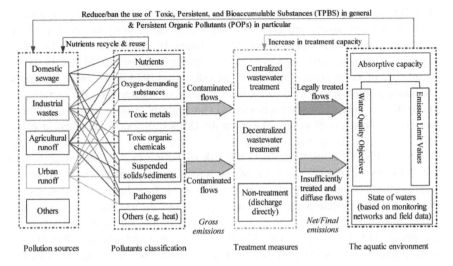

Fig. 3 A simplified conceptual framework of contemporary water quality research and management (after Song 2012)

In summary, Fig. 3 presents a simplified contextual framework showing some most frequently discussed issues in contemporary water quality research and management.

As shown in Fig. 3, discussions to date about causes of water quality degradation have mainly focused on substance and element fluxes (both point and non-point) from the anthroposphere to the environment. In fact, this point is also reflected in the two general objectives of water quality systems analysis (Karamouz 2003), i.e., to identify (i) major pollutant categories (e.g., nutrients, toxic metals/organic chemicals, pathogens, suspended solids, and heat) and (ii) principal sources of pollutants (e.g., domestic sewage, industrial waste, agricultural runoff, and urban runoff). Take, for example, the case of Helsinki Commission (HELCOM) Baltic Sea Action Plan (BSAP). The HELCOM BSAP, adopted in 2007, aims to restore good ecological status to the Baltic Marine Environment by 2021. One goal of the HELCOM BSAP is to have a Baltic Sea unaffected by eutrophication by means of cutting the nutrient (phosphorous and nitrogen) load from waterborne and airborne inputs (Backer et al. 2010).

3.2 Information Demand of the EU WFD Regarding Analysis of Pressures and Impacts

Regarded as a model framework for employing integrated approaches to water management, the European Water Framework Directive (WFD) came into force in December 2000 (EC 2000). To aid in the application of the WFD, the guidance document for pressures and impacts analysis (IMPRESS) was issued in 2003

(EC 2003). In the WFD, causal relationships are used to identify significant anthropogenic pressures and assess the impacts on the quantity and quality of surface water and groundwater (Article 5 and Annex II). In Annex VII of the WFD, significant pressures and impacts of human activities are presented as follows: (i) estimation of point source pollution; (ii) estimation of diffuse source pollution, including a summary of land use; (iii) estimation of pressures on the quantitative status of water including abstractions; and (iv) analysis of other impacts of human activity on the status of waters. Based on the IMPRESS, examples of cause–effect relationships are represented in Table 1.

The IMPRESS aims to evaluate the risk of failing to meet the objectives of the WFD by comparing the state of the aquatic environment with corresponding threshold values. In particular, the following four kinds of pressures are considered: (i) pollution pressures from point and diffuse sources, (ii) quantitative resource pressures, (iii) hydromorphological pressures, and (iv) biological pressures (EC 2003). In the IMPRESS, three prerequisites are identified for appropriately and successfully identifying pressures and assessing impacts, i.e.,

Table 1 Examples of driving forces, pressures, and impacts identified in the EU WFD (after EC 2003)

Type of pressures	Driving forces	Direct pressures	Possible impacts and change in environment
Diffuse source pollution	Agriculture	Nutrients (e.g., P and N) loss	Nutrients modify ecosystem
		Pesticide loss	Toxicity; water contamination
	Atmospheric deposition	Deposition of compounds of nitrogen and sulfur	Eutrophication; acidification of waters
Point source pollution	Industry	Effluent discharged to surface water and groundwater	Organic matter alters oxygen regime; increased concentration of suspended solids
	Thermal electricity production	Alteration to thermal regime of waters	Increased temperature; changes in biogeochemical process rates; reduced dissolved oxygen
Quantitative resource pressures	Agriculture and land use change	Modified vegetation water use	Altered groundwater recharge
	Water abstraction	Reduced flow or aquifer storage	Modified flow and ecological regimes; saltwater intrusion
Hydromorphological pressures	Physical barriers and channel modification	Variation in flow characteristics	Altered flow regime and habitat
Biological pressures	Fisheries	Fish stocking	Generic contamination of wild populations

(i) understanding of the objectives, (ii) knowledge of the water body and catchment, and (iii) use of a correct conceptual model. This strongly suggests that the proposed conceptual model for pressures and impacts analysis could describe both the quantitative nature and qualitative nature of the aquifer at a catchment scale and the likely consequences of pressures (EC 2003).

It is clear that the EU WFD focuses on discussing pressures of pollutants discharged into the water environment, in a form of either point or diffuse pollution. Indeed, the current water quality indicators—biological, physical–chemical, and hydromorphological—for determining the status of surface waters in the EU WFD are mainly state/impacts-oriented, while only water flow monitoring is partly pressure-oriented (Song and Frostell 2012).

Overall, the EU WFD and the above-mentioned HELCOM BSAP indicate the effort being made in Europe to improve the water environment. Although the main focus is on mapping and reducing emissions/wastes discharged to waters, they can be viewed as implementations of a semi-pressure-oriented approach. In fact, they show some promise of advancing toward an explicit pressure-oriented and proactive water management approach based on the metabolism of the anthroposphere.

4 Moving Toward an Anthropogenic Metabolism-Based and Pressure-Oriented Approach to Water Management

4.1 Facilitating a Broader Systems Perspective on Water Management

There is a growing consensus that a broader systems perspective is necessary for achieving improved environmental management in general and for improved water management specifically. This broader systems perspective could be regarded as a further clarification of the following opinion: "the former strategy of environmental management by controlling emission sources from industrial processes has to be replaced by a systematic approach that integrates all of the evaluations of environmental effects that can be assigned to a product" (Sonnemann et al. 2004). In the water domain, Biswas (2004) emphasizes that popular ways to address various national water problems "can no longer be resolved by the water professionals and/or water ministries alone." Moreover, Falkenmark (2007) claims that a shift in thinking about the focus of water management (like "blue" vs. "green" water) is needed because of past misinterpretations and conceptual deficiencies.

Here, we argue that a comprehensive understanding of the metabolism of the anthroposphere is essential for analysis and assessment of various water-related interactions between anthropogenic (human-made) systems and their environment(s) in a more integrated and proactive way. This often begins with using improved accounting for material and energy flows throughout the anthroposphere, followed by assessing the environmental impacts of resources used and waste/emission produced. Such a pressure-oriented approach aims to provide a basis for decisions and planning for more sustainable environmental performance, e.g., at an individual/household, company/industrial, or municipal/regional level.

Environmental changes and their pressures can only be properly understood if they are discussed in the context of the human activities or driving forces giving rise to them (EEA 2007). In particular, the root causes of human-induced water problems should be traced and analyzed from a metabolic point of view. Although WQOs and ELVs have been widely implemented for years, the traditional system of permits and enforcement is not leading to the required pollution abatement and is not effectively dealing with the sources of diffuse pollution (Van Ast et al. 2005). The traditional risk assessment-based approach does not fully capture the connections and interactions among individual existing environmental problems and drivers of environmental impacts (Bauer 2009).

Along with the suggested pressure-oriented approach, the DPR model (cf. the European DPSIR framework) should be promoted in order to effectively respond to emissions/wastes initially produced in the anthroposphere. A basic premise of the pressure-oriented approach to water management is that "the amount of resource flow into the economy determines the amount of all outputs to the environment including wastes and emissions" (EEA 2003). Here, it is worth emphasizing that the pressure-oriented approach includes, but is not limited to, input–output analysis. Theoretically, the pressure-oriented metabolic accounting approach could produce more pertinent water-related information with regard to the inputs of material/energy and the outputs of emission discharge including their interim transformation.

In principle, the metabolism-based pressure-oriented approach and the derivation of information systems could aid in effectively addressing water environmental degradation by means of avoiding/reducing various pressures exerted by human activities at all scales. The target information users include water researchers, water policy and decision makers, water-related socioeconomic decision makers, and other stakeholders involved in water-centric planning and decision making. In a broader sense, the AM/PO information could provide a basis for developing sustainable urban cycles (e.g., on water use, carbon flows, nutrient flows, and energy use) toward ecological sustainability.

4.2 Calling for a Transition to the Pressure-Oriented Approach

The state/impacts-oriented approach has now been employed not only in water planning and management, but also in environmental management. Very often, an accurate assessment of the state of the environment in relation to water, air, and soil is regarded as a prerequisite for policy makers and their scientific advisor committees to identify problems and take action for improvement (Kim and Platt 2008). In many cases, this holds true for indicator selection. Referring to the DPSIR framework, for instance, Bell and Morse (2008) emphasized that "impact and state sustainability indicators (SIs) are the primary measure applied to sustainability projects, but that drivers, pressure and response SIs may be developed at a later stage by the project team in order to help the team understand what the state SIs are describing—and thus to explain exactly what influences and drives the state and impact SIs." Here, according to Bell and Morse (2008), impact and state SIs (related to the state of a variable) should largely describe project impacts, while drivers, pressure, and response SIs (related to control, process, etc.) are more exploratory and analytical.

The current state/impacts-oriented approach is largely based on the concept of carrying capacity of environment and ecosystems. Under these circumstances, emissions and wastes would not be paid adequate attention until negative changes in water environment and ecosystems are monitored. As pointed out by Beder (2006), the implementation of environmental carrying capacity often "depends on value judgments about how much pollution a community is willing to put up with." In this context, societal measures often relate to pollution control, e.g., by means of increasing the extent of pollutant collection and treatment before being discharged into the ambient water environment. On the other hand, there is a time lag of about a decade, at a minimum, between nutrient concentration changes in a river basin and ecological and water quality response in waters (National Research Council 2009). In order to achieve better proactive water planning and decision making, the suggested pressure-oriented approach is one necessity in many ways.

Figure 4 briefly illustrates the state/impacts-oriented and pressure-oriented approaches, focusing on information flows. In contrast to the state/impacts-oriented approach, the pressure-oriented approach begins with exploring driving

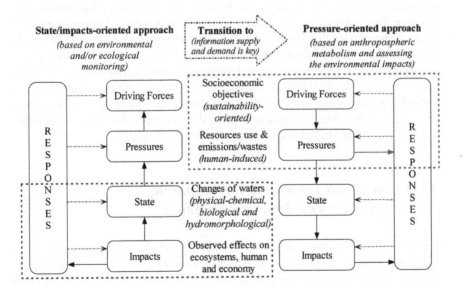

Fig. 4 Facilitating a transition from the state/impacts-oriented to the pressure-oriented water management approach (after Song 2012)

forces (various socioeconomic activities) and accounting for anthropogenic pressures (caused both by resource depletion and by pollution) on the environment. Thereafter, corresponding societal responses can be suggested, aiming to design an environmentally friendly anthropogenic metabolism in society at large. Most importantly, analysis of environmental pressures and impacts of socioeconomic development objectives could be comprehensively made beforehand by the use of the suggested pressure-oriented accounting approach as well as tools of integrated environmental assessment.

In order to trace the origins and pathways of pollutants in an area, one useful tool is material flow analysis (MFA), including substance flow analysis (SFA). MFA is a systematic assessment of flows and stocks of materials within a spatial or temporal system boundary by connecting the sources, the pathways, and the intermediate and final sinks of materials (Brunner and Rechberger 2003). When discussing the potential use of MFA for environmental monitoring, Brunner and Rechberger (2003) state that a well-established MFA of a region could replace traditional soil monitoring programs that are costly and limited in their forecasting capabilities by the use of statistics. On the other hand, Binder et al. (2009) argue that the efforts in MFA and SFA so far have been mainly academic and their actual impact on policy making is not clear. This is probably because most MFA studies are about material flows and stocks in a given area, while very few are accompanied by further discussion about their pressure-oriented contributions to environmental degradation at different scales from a broader systems perspective (Song 2012). In other words, facilitating the practice of the pressure-oriented approach largely depends on providing pertinent information by means of pressures and impacts analysis of anthropogenic metabolism.

In facilitating a transition to the pressure-oriented water management approach, it is essential to set an appropriate system boundary for monitoring, documenting, and reporting. The traditional socioeconomic statistics usually use an administrative boundary. In the water domain, a hydrological boundary has been suggested for IWRM at the scale of a river basin. In this context, alternative system boundaries suitable for pressure-oriented water systems analysis and management may be a hydrological boundary such as introduced in EU WFD, an administrative region, or a combination of these. In particular, an administrative approach on land (the socioeconomic system, e.g., companies, organizations, municipalities, provinces) should be used for data collection first, and then, the data are transformed to suit the hydrological system.

4.3 A Conceptual Framework for Accounting for Anthropogenic Pressures on Waters

Regarding water quality monitoring and management, a preliminary conceptual framework (Fig. 5) is developed as a brief demonstration of interlinks among the atmosphere, the natural water system, and the human-oriented system within an expanded systems boundary. Comprehensively identifying those links is a prerequisite to quantifying potential anthropogenic pressures on the water environment.

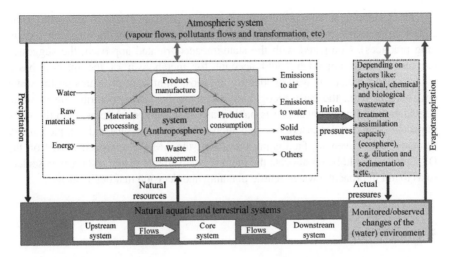

Fig. 5 A conceptual framework for a brief illustration of linkages between atmospheric, human-oriented, and natural water systems, which is a prerequisite to quantifying anthropogenic pressures on waters (after Song 2012)

In order to promote the use of a pressure-oriented water management approach in practice, a large amount of anthropogenic information needs to be produced by means of a metabolic approach. A metabolic approach also fosters different mass and energy balances, and thus, mass and energy accounting approaches are required to keep track of progress and deterioration of systems function. Same as any accounting approaches, the success of accounting for and analyzing anthropogenic pressures on the water environment depends on the extent of available information of relevant flows/stocks of materials and emissions throughout the human-oriented system (technosphere). A good documentation of this type of metabolic information is essential to effectively address the underlying drivers of both point source and non-point source of pollutants and human-induced water quantitative problems.

In principle, a comprehensive anthroposphere metabolic accounting process needs to be developed and implemented in order to systematically trace both input-related categories (e.g., resource depletion) and output-related categories (e.g., pollution) related to water degradation. Producing pressure-oriented metabolic information could complement the traditional water information systems and management by means of tracing the root causes of human-caused water problems back to the anthroposphere. To begin with, information on pressure-oriented water monitoring and accounting could theoretically (and later in practice) be achieved by the use of environmental systems analysis tools such as MFA/SFA, life cycle assessment (LCA), and environmental input–output analysis (IOA) over agreed system boundaries (Song 2012). Finally, the inventory results of emissions/wastes could be aggregated and assigned to different impact categories, such as eutrophication and toxicity. Then, the significant potential water quality/quantity pressures and their root sources could be determined.

The development and promotion of such a pressure-oriented approach could significantly assist proactive water policy and decision making (basically, planning practices). Compared with the state/impacts-oriented approach, the suggested pressure-oriented approach could account for metabolism of the material-based industrialized society as well as identify the most significant pressures on the water environment (e.g., the early recognition of metabolic factors contributing to water degradation). Moving toward sustainable water management systems, both the state/impacts-oriented and pressure-oriented approaches are necessary and complementary in many ways. Even so, facilitating the use of the AM/PO approach (based on the DPR model) could better allocate the majority of available scarce resources in society so as to aid in proactive water-centric planning and decision making in society.

5 Conclusions

Using the European DPSIR framework as a basis, this paper argues that the current water management approaches and associated information systems are mainly state/impacts-oriented, while very little is pressure-oriented. The state/impacts-oriented approach focuses mainly on physical and biogeochemical state changes in recipient waters, which often results in reactive responses to combat water problems (owing to the late recognition of contributing factors). To complement those traditional water management approaches, an AM/PO approach to water management is suggested at the conceptual level. The AM/PO approach is characterized in general by accounting for input of resources and output of wastes/emissions through the anthroposphere as contributing factors to water degradation. In principle, the produced metabolic information could help water-related planners and decision makers take proactive measures to address human-induced pressures on the water environment at their sources.

The suggested AM/PO approach, derived from a DPR model, shows a promising shortcut to effectively alleviating human-induced pressures (initially on land) with a focus on accounting for the anthropogenic metabolism. In order to cope with complex water problems in a more proactive way, it is not enough to focus only on water bodies (surface water and groundwater), e.g., either from the perspective of ecohydrology, biogeochemical monitoring and modeling, climate change, and/or adaptive water management. From a metabolic point of view, the root causes of human-induced water degradation are embedded in anthropogenic activities. A comprehensive understanding of the metabolism of the anthroposphere should be achieved and used as a basis for accounting for various pressures on waters and assessing their environmental impacts at their sources.

This paper only presents results of the first stage of this research at the conceptual level. Further studies will focus on developing pilot implementation projects. However, we hope that the preliminary results will stimulate interdisciplinary scientists and decision makers to rethink their individual preferred perceptions and approaches to environmental management in general and water management specifically. Concerns relate to, but are not limited to, guiding principles of water management (reactive vs. proactive), water-related data documentation (both socioeconomic and environmental), how to better use available scarce resources in water monitoring/accounting, and the water-centric planning process as well. In our opinion, regarding ecological sustainability in particular, the important issue is not only to make science-based decisions, but also to make decisions to address the "right" problems in an effective and efficient way. In a broader context, achieving a comprehensive understanding of human-induced pressures on the environment is essential to design or envision any "sustainable" society from a systems perspective.

References

Atkins JP, Gregory AJ, Burdon D, Elliott M (2011) Managing the marine environment: is the DPSIR framework holistic enough? Syst Res Behav Sci 28:497–508

Baccini P, Brunner PH (2012) Metabolism of the anthroposphere: analysis, evaluation, design, 2nd edn. The MIT Press, Cambridge

Backer H, Leppänen JM, Brusendorff AC, Forsius K, Stankiewicz M, Mehtonen J, Pyhälä M, Laamanen M, Paulomäki H, Vlasov N, Haaranen T (2010) HELCOM Baltic Sea action plan—a regional programme of measures for the marine environment based on the ecosystem approach. Mar Pollut Bull 60:642–649

Bauer D (2009) Environmental policy: a growing opportunity for material flow analysis. J Ind Ecol 13:666–669

Beder S (2006) Environmental principles and policies: an interdisciplinary introduction. University of New South Wales Press, Sydney

Bell S, Morse S (2008) Sustainability indicators: measuring the immeasurable?. Earthscan, London

Binder CR, Van Der Voet E, Rosselot KS (2009) Implementing the results of material flow analysis. J Ind Ecol 13:643–649

Biswas AK (2004) Integrated water resources management: a reassessment. Water Int 29:248–256

Brunner PH, Rechberger H (2003) Practical handbook of material flow analysis. CRC Press, Florida

Carr ER, Wingard PM, Yorty SC, Thompson MC, Jensen NK, Roberson J (2007) Applying DPSIR to sustainable development. Int J Sust Dev World 14:543–555

Castelletti A, Soncini-Sessa R (eds) (2007) Topics on system analysis and integrated water resources management. Elsevier, Amsterdam

Destouni G, Hannerz F, Jarsjö J, Shibuo Y (2008) Small unmonitored near-coastal catchment areas yielding large mass loading to the sea. Global Biogeochem Cycles 22:GB4003

EC (European Commission) (2000) Directive 2000/60/EC of the European Parliament and of the Council of 23 October 2000: establishing a framework for Communities action in the field of water policy. OJEC L327:1–72

EC (2003) Analysis of pressures and impacts. Common implementation strategy for the water framework directive (2000/60/EC). Guidance document no. 3. Office for Official Publications of the European Communities, Luxembourg

EEA (European Environment Agency) (2003) Europe's environment: the third assessment. European Environment Agency, Copenhagen

EEA (2007) Europe's environment: the fourth assessment. European Environment Agency, Copenhagen

Falkenmark M (2007) Shift in thinking to address the 21st century hunger gap: moving focus from blue to green water management. Water Resour Manage 21:3–18

Friberg N (2010) Pressure-response relationships in stream ecology: introduction and synthesis. Freshwater Biol 55:1367–1381

Gabrielsen P, Bosch P (2003) Environmental indicators: typology and use in reporting. European Environment Agency, Copenhagen

Graedel TE, Klee RJ (2002) Industrial and anthroposystem metabolism. In: Douglas I (ed) Encyclopedia of global environmental change: causes and consequences of global environmental change. Wiley, Chichester

Hooper B (2005) Integrated river basin governance: learning from international experiences. IWA Publishing, London

Karamouz M, Szidarovszky F, Zahraie B (2003) Water resources systems analysis. CRC Press, Washington, DC

Kim YJ, Platt U (2008) Advanced environmental monitoring. Springer, Dordrecht

Meybeck M, Peters NE, Chapman D (2005) Water quality. In: Anderson MG, McDonnell JJ (eds) Encyclopedia of hydrological sciences. Wiley, Chichester

National Research Council (2001) Assessing the TMDL approach to water quality management. National Academy of Sciences, Washington, DC

National Research Council (2009) Nutrient control actions for improving water quality in the Mississippi river basin and northern gulf of Mexico. The National Academies Press, Washington, DC

Rekolainen S, Kämäri J, Hiltunen M (2003) A conceptual framework for identifying the need and role of models in the implementation of the water framework directive. Intl J River Basin Manage 1:347–352

Smeets E, Weterings R (1999) Environmental indicators: typology and overview. European Environment Agency, Copenhagen

Song X (2012) A pressure-oriented approach to water management. Doctoral thesis in industrial ecology. KTH Royal Institute of Technology, Stockholm

Song X, Frostell B (2012) The DPSIR framework and a pressure-oriented water quality monitoring approach to ecological river restoration. Water 4:670–682

Sonnemann G, Castells F, Schuhmacher M (2004) Integrated life-cycle and risk assessment for industrial processes. CRC Press, Boca Raton

Svarstad H, Petersen LK, Rothman D, Siepel H, Wätzold F (2008) Discursive biases of the environmental research framework DPSIR. Land Use Policy 25:116–125

Tscherning K, Helming K, Krippner B, Sieber S, Paloma SG (2012) Does research applying the DPSIR framework support decision making? Land Use Policy 29:102–110

UNSD (United Nations Statistics Division) (2012) System of environmental-economic accounting for water (SEEA-Water). United Nations, New York

Van Ast JA, Blansch KL, Boons F, Slingerland S (2005) Product policy as an instrument for water quality management. Water Resour Manage 19:187–198

WWAP (World Water Assessment Programme) (2009) The United Nations world water development report 3: water in a changing world. UNESCO, Paris and Earthscan, London

Zhang HX, Corporation P, Fairfax V (2005) Water quality management and nonpoint source control. In: Lehr JH, Keeley J (eds) Water encyclopedia, vol 2. Water quality and resource development. Wiley, New Jersey

Authors Biography

Xingqiang Song and Björn Frostell work at the Division of Industrial Ecology at the Royal Institute of Technology (KTH) in Stockholm, Sweden.

Ronald Wennersten works at the Institute of Thermal Science and Technology, Shandong University, China.

Sustainable Water Management Defies Long-term Solutions

Kristan Cockerill and Melanie Armstrong

Abstract The popular and academic media are rife with calls to sustainably manage our water resources and to 'solve our water problems.' Yet, evidence suggests that throughout history, our efforts to 'solve' water problems have simply generated new problems. Humans have drained swamps to solve problems of disease and land shortage. This subsequently reduced water supply and increased flooding in many areas. Humans dammed rivers to solve problems related to energy and irrigation, thereby reducing ecosystem resiliency. Humans established a water-based sewerage system to solve problems of aesthetics and health and as a result increased water consumption and created a dependency on massive infrastructure. The insistence on solutions may exacerbate rather than alleviate negative conditions, in part, because it discourages decision-makers and citizens from accepting long-term responsibility for managing water to sustain ourselves. The authors argue that addressing water problems requires a cognitive shift to recognize the concept of 'wicked problems' and to subsequently change discourse about water to resist the idea of solutions.

Keywords Wicked problem · Water management · Historical context · Environmental discourse

K. Cockerill (✉)
Department of Cultural, Gender and Global Studies,
Appalachian State University, ASU Box 32080, Boone, NC 28607, USA
e-mail: cockerillkm@appstate.edu

M. Armstrong
Department of Geography, University of California,
Berkeley and U. S. National Park Service, Canyonlands National Park,
PO Box 580, Moab, UT 84532, USA
e-mail: melnps@gmail.com

© Springer International Publishing Switzerland 2015
W. Leal Filho and V. Sümer (eds.), *Sustainable Water Use and Management*,
Green Energy and Technology, DOI 10.1007/978-3-319-12394-3_13

1 Introduction

"The only thing that makes life possible is permanent, intolerable uncertainty; not knowing what comes next."

Ursula K. LeGuin.

The popular and academic media are rife with calls to sustainably manage water resources and to 'solve' water problems. A Google search on 'solve water problem' finds 179,000 hits, and 'water problem solution' garners 255,000 hits. But what does 'solving water problems' imply? If the 'solution' includes becoming 'sustainable,' what does this mean? The verb 'to sustain' carries several meanings, including to support, to buoy up, and to prolong (Merriam-Webster online). Within contemporary rhetoric, sustainability encompasses a positive relationship between ecologic, economic, and social conditions. When applied to water, sustaining implies ensuring that water will support something (e.g., people, habitat) for some prolonged period of time. When water's ability to support is threatened, citizens, interest groups, and politicians declare the situation a problem (or even a 'crisis') that should be solved. 'Solving' typically means bringing something to an end, and if something is truly solved, the implication is that it no longer requires attention. Water, however, has always and will always require human attention if the goal is to sustain ourselves. The history of human development is also the history of harnessing water to serve us. The first wells date to 7,000 BCE, and the great ancient societies (e.g., Egypt, Mesopotamia, Greece, China) developed extensive waterworks (Solomon 2010, Fagan 2011).

While the details of human water use have adjusted to address changing wants and needs, water management has remained remarkably consistent. Fagan (2011) writes, 'I was struck by how little most people's relationship with water changed over the thousands of years...' Reuss (undated) notes that even the basic technologies have changed little, '...lessons learned 2,000 years ago in ancient China or 200 years ago in Napoleonic France may well be equally valid for current water management and engineering.' Even when the word sustainable has not been used, the intent of water management has been to sustain something. This has generated a consistent pattern of identifying a problem (i.e., a threat to water's ability to sustain), then using knowledge and technology to 'solve' that problem.

Developing furrow irrigation was a 'logical step for villagers grappling with irregular rainfall, potential crop failures, and long dry seasons' (Fagan 2011). By the sixteenth century, gravity-driven water systems were no longer sufficient, invoking more intense technology to acquire and move water, and then, the industrial revolution changed 'the entire water equation for humanity' (Fagan 2011). Boiling water for steam power enabled water to be pumped from deep underground or moved far from its source. Solomon (2010) observes that 'It was also a common pattern of history that expansions driven by intensified use of water and other vital resources were followed by population increases that in turn so increased consumption that they ultimately depleted the further intensification capacity of the society's existing resource base and technologies. Such resource depletions thus presented each society with a moving target of new challenges requiring perpetually new innovative response to sustain growth.' This pattern has long been recognized, as Ludlow wrote in 1884 that 'Sooner or later all cities are brought face to face with the water problem, and even when it has been thought that a solution has been reached, the development of industries and the growth of population out-run the provision which it was believed would suffice for long periods, and call for constant watchfulness and care to meet the growing demands.'

Today, sustainability proponents might classify low-technology gravity-driven furrow agriculture as being 'sustainable,' but the impetus that drove developing that technology is the same impetus driving large dam construction. The problems of irregular rainfall, drought, increased population, and changing wants and needs have been repeatedly 'solved' by holding water where society wants it and moving it from one place to another. Each perceived solution has generated new problems, sometimes creating a new version of the initial problem. While the rationale for managing water has remained consistent, the scale of activity has changed as population growth has catalyzed technology development to meet demands, which has then increased the scale of subsequent problems.

Water management dilemmas extend beyond human survival, and definitions of sustainability have increasingly brought non-human subsistence into conversations about water systems. Humans also manage water for fish and wildlife, whether for food or recreation. Technologies as simple as rock walls can be used to manage wetlands and have long been used to create fish habitat. In Polynesia, people carried stones many miles to the ocean to enclose bays and regulate flows of sea- and freshwaters for aquaculture (e.g., the Kaloko fishpond in Hawai'i). The archaeological record shows that Native Americans used mounds and terraces in estuaries, along with prescribed burning, to cultivate water plants and habitat for waterfowl (Deur 2002, Bovy 2007). The popularity of waterfowl hunting prompted enhancing and developing wetlands for sport birds (Dolin and Dumaine 2000). Even cultural taboos governing fishing and hunting activities, or managing human waste, had a historic effect in preserving water quality and promoting species survival to benefit both humans and wildlife (Jianchu et al. 2005).

While benefits to wildlife often fall second to human needs, historic water management technologies illuminated the human ability to impact other species. From the prairie farmer who deliberately overflows a well to create a habitat for wildlife to the dam builder who installs ladders for migrating fish, humans recognize that technologies created in response to water problems do create problems for wildlife. This further complicates the problem–solution paradigm and inevitably challenges people to evaluate complex and competing needs in formulating water policies. Just as the history of water development reveals expanding water management systems to meet demands of growing populations, new scales of industry also changed the scope of wildlife and habitat management. Environmental legislation in the mid-twentieth century that mandated protecting endangered species forced people to reconsider aquatic and associated terrestrial environments as spaces that sustained wildlife. Debates about climate have further shifted the scale to problems as large as sea-level rise and global drought, encompassing the human and non-human in a wicked problem in which the unintended consequences of development threaten all species, including *Homo sapiens*.

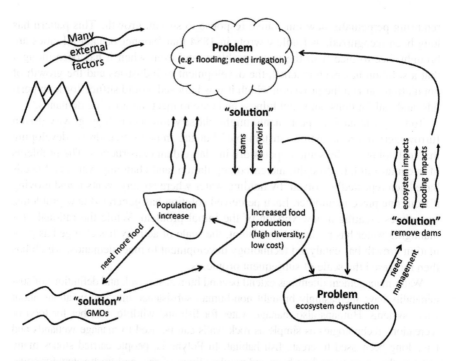

Fig. 1 Shows the cyclical nature of problems generating solutions, which then have unintended consequences that catalyze new problems, which are sometimes new forms of an old problem

Given this history, there seems to be a gap between the perceived need to 'solve' water problems and the reality of managing water systems. Indeed, there is ample evidence that 'the chief cause of problems is solutions,' an aphorism usually attributed to American journalist Eric Sevareid. This chapter presents several examples of this phenomenon, highlighting that solutions become problems within a sustainability paradigm because what 'solves' a perceived economic or social problem often generates new environmental problems, or solving an environmental issue may generate social concerns (Fig. 1). The authors then offer an analysis suggesting that sustainable water management requires recognizing the power of language in framing problems and that water issues defy solution. Given the consistent and uncertain conditions under which humans have always managed water, any approach to sustainable water management must resist the idea that water problems can be solved and change language use accordingly.

2 Dams as Problematic Solutions

Few water-related topics have generated as much attention as dam building. Retaining water for human use has been part of the water management portfolio for centuries. Installing dams has 'solved' all kinds of problems including a need for irrigation, water supply, flood control, and navigation. These same dams have simultaneously generated multiple problems, including negatively affecting ecosystems, displacing people, and perpetuating an unrealistic sense of security for communities reliant on services the dams provide. In response to growing controversy surrounding large dams, the World Commission on Dams (WCD) was initiated in 1997 to assess impacts and develop guidelines for planning, constructing, and operating dams. The WCD spent two years completing a comprehensive assessment for its 2000 report, which was 'intended to increase the chances of win-win solutions through more comprehensive and inclusive planning processes' (Steiner 2010). A decade later, the conflict centered on large dams had not significantly abated and the journal, Water Alternatives, focused a special issue on the WCD legacy to 'help galvanise renewed interest in jointly creating effective solutions for water and energy development' (Moore et al. 2010). Ten years after the WCD report, the debate about dams had not lessened, yet the call to 'solve' the problem(s) remains. This offers a classic example of being caught in the perpetual solutions as problems cycle. The specifics of this debate are less relevant than the persistence of debate. This consistency suggests that issues of planning, constructing, and operating dams are perpetual and will not be 'solved' in any finite sense.

Bringing the dam argument full circle is the increased attention to removing dams as a water management approach. Within the span of a century (a relatively short time frame), human society has transitioned from envisioning large-scale dams as a long-term solution for water management, to realizing the dams themselves cause their own long-term problems. In contemporary discussions of river management, removing dams is often perceived as an ecological solution. In the USA, a National Park Service brochure describing dam removal on the Elwha River promises that 'as the dams come down, the salmon can return, bringing with them the promise of a restored ecosystem and a renewed culture.' The presence of charismatic mega-fish in this river signals a 'solved' environmental problem, and the New York Times reported the presence of a dozen steelhead salmon in the river upstream of the dam sites as evidence of 'a river newly wild' (Johnson 2012). Similarly, the return of fish after removing a dam on Maine's Kennebec River or opening gates on the Pak Mun Dam on Thailand's Mun River has been cited as proof that 'rivers can heal' (Postal and Richter 2003). Healing implies, as does much language surrounding dam removal, that the work of the river will erase the dam's mark and 'health' will return. The prefix 're-' modifies verbs throughout reports on dam removal, promising that the projects will 'replenish,' 'restore,' 'reestablish,' 'revitalize,' 'renew,' and 'redeem' these rivers. Clearly, eradicating dams does not erase years of changes in the lives of people, plants, and animals, yet the ways people talk about dam removal persistently look back in time as if to suggest that a river without dams will be some imagined river of the past, but without the problems that led to constructing the dams in the first place.

Dam removal creates social and economic impacts related to changes in flooding dynamics, access to recreation, and water available for irrigation or municipal supply. Notably, the decision to remove the dams on the Elwha may have had as much to do with declining electrical production and structural weakness in the dams as it did with ecosystem health. Before the removal commenced, planners calculated the impacts upon human developments downstream and built dikes and levees to provide flood protection to private properties. The ecological impacts of these actions must still be accounted for, undermining claims that the river is now 'free' and 'wild.' Issues of temporal scale are pertinent here and offer additional complexity to the cyclical nature of problems and solutions. On a human time-scale, the impacts of dam removal can be significant. At Elwha, the most notable short-term impact is a rush of sediment pouring downriver, raising the riverbed up to two feet and clouding the water the length of the river and into the ocean, potentially damaging fragile marine and fisheries habitats (Johnson 2012). Over a longer time, ecosystems, including human systems, will adapt and create something new, with concomitant problems.

3 Solving Floods on Inland Waterways

Decisions to develop inland waterways in the USA offer another example of perceived solutions generating new problems. A key motivation for developing these waterways was economic, as the US officials perceived the nation to be at a competitive disadvantage with Argentina, Australia, and other countries in getting goods to Europe. Secretary of Commerce, Herbert Hoover (1928), wrote, 'we must find fundamentally cheaper transportation for our grain and bulk commodities which we export and the raw materials which we import into the Mid-West.' Hoover continued, 'In any examination of our country for remedy, we have naturally turned to a consideration of the magnificent natural waterways which Providence has blessed us with.' This example highlights that economic sustainability is directly connected to water management. The 'solution' to a perceived economic disadvantage was to more intensively manage waterways, which has had diverse social and environmental consequences.

Flooding posed a significant barrier to improved navigation on US rivers and was a focal point for waterway development. Flood management on the Mississippi River is a textbook case of solution hopping with perceived solutions generating new problems prompting yet more solutions. Prior to a massive flood event on the Mississippi River in 1927, US policy promoted levees as the single best way to manage floods. The proposed solution to preclude another devastating flood was to move away from 'levees only' to employing more intense structural design and engineering techniques.

The Committee on Mississippi Flood Control was commissioned to assess the need for and appropriate methods of flood control. Their final report focused on developing '"a program which will insure, so far as is humanly possible, a permanent solution" of floods of the Mississippi River' (Delano 1928). Further, the Committee concluded that, '… securing the judgment of the highest engineering talent in the country, both governmental and civilian,—federal, state and local,—the committee is impressed with the unanimity of opinion that adequate control of the Mississippi River is practicable.' Gifford Pinchot (1928), a Governor of Pennsylvania and first head of the US National Forest Service, weighed in stating that 'every useful and available means of establishing [control of streams], including levees, spillways, soil conservation, forest conservation, storage reservoirs, and any others should be considered and made use of to the fullest practicable extent.'

These arguments succeeded in convincing the federal government to implement an extensive engineering approach to control flooding along the Mississippi and other US rivers. The dams, spillways, and reservoirs constructed generated numerous subsequent problems, including exacerbating flooding in some areas; radically altering riparian and floodplain ecosystems; and creating a false sense of security that motivated people to build within the floodplain under the presumption that flood control would protect them (Pinter 2005, Frietag et al. 2009). The problem–solution cycle continues in debates surrounding river management as non-structural options, ranging from re-establishing wetlands to 'flood proofing' buildings to reduce damage when floods do happen, have become more prominently presented as solutions to address flooding. While this recognizes the inevitability of flooding, implementing non-structural approaches will undoubtedly present unintended social, economic, and/or environmental consequences.

4 The Accidental Problem of the Salton Sea

Growing populations demand greater food supplies, and water development in the western USA capitalized on new technologies to expand agricultural lands to feed a growing nation. The State of California has diverted water from every direction into its Central and Imperial valleys to grow cash crops such as lettuce and oranges. While flooding was not a primary concern in developing water sources for agriculture in the desert, the flood that created the Salton Sea exemplifies water's potential to reshape landscapes and establish new cultural relationships to local water.

In a 1905 flood event, the Colorado River breached irrigation canals built to carry water into California's Imperial Valley. The overflowing river followed an ancient waterway to the Salton Sink, where it formed a large lake with no outlet in one of the lowest valleys on the continent. The geologic record shows that the Salton Sea has existed many times in the past, for as the Colorado River pushed its historically heavy loads of sediment into the Gulf of California, it regularly clogged its own outlet, diverting water into the Salton Sink. As in the geologic past, an abundance of sediment created this modern incarnation of the Salton Sea, but the 1905 event stood apart from these historic floods because the river's sediment overcame the technologies created to manage the water. Unlike lakes created by dams, the Salton Sea does not exist where humans desire it. A social and economic impact of this event was that it submerged more than half of the Torres Martinez Desert Cahuilla Indian Reservation, land these native people historically and currently depend upon for survival. For a brief time, the sea in the desert drew celebrities and tourists to its shores. With no major water input source, however, the sea began to shrink and grow salty, and people abandoned their resorts on a shoreline littered with dead fish. The lake is sustained by runoff from the agricultural industry, which is rich with pesticides and fertilizers and increases the lake's salinity by about one percent each year (Bali 2010).

This pollution threatens the survival of more than 400 species of birds that depend upon this habitat. As agriculture appropriated historic swamps and waterways, migratory bird populations flocked to the Salton Sea. The 'problem' of the Salton Sea now centers on the quality of the water, which generates unpleasant smells for tourists and is depleting food sources for birds. The 1990s saw major bird die-offs, including 150,000 Eared grebes and 9,000 American white pelicans, which represents 15–20 % of the white pelican population and is the largest reported die-off of an endangered species (Friend 2002). Eliminating the sea to restore the landscape to its pre-1905 dryness raises the question of habitat for hundreds of thousands of birds and prompts managers to weigh non-human species survival as part of the equation.

When modern advocacy groups take up a call to 'Save the Salton Sea' or 'Restore the Salton Sea,' the question follows: Restore to what state? Dry land (pre-1905) or non-polluted sea (early 1900s)? Or, can (should?) the sea be managed to create a viable water system unlike any that has existed there in the past, for the benefit of humans and wildlife? Are tribal rights to the land to be honored? Proposed alternatives range from connecting the sea to the ocean to blend polluted lake waters with the less-salty ocean water; building a dam to divide the lake into a marine environment and a brine sink with saline habitat; or instigating a large-scale desalination project. If the goal is sustainability, are any of these options sustainable and over what temporal scale? Is it possible to balance economic, social, and ecological issues in managing the Salton Sea?

California established the Salton Sea Authority in 1993 to bring together relevant local and tribal governments to evaluate the Salton Sea. They advocated a technological fix that separates the sea into two bodies and manages the water quality along with fish and bird habitat. Notably, the Authority sees the restoration process as an endgame, lasting 30–40 years. 'The Authority's Plan would provide a *restored Sea* along the current shoreline coupled with the development of habitat areas that could stimulate development and improve the economic conditions for the Tribe and Imperial and Riverside counties' (Salton Sea Authority Plan for Multi-Purpose Project 2006, italics added). 'Once restored' the plan envisions a future lake 'with a stable shoreline, rich wildlife and a growing number of visitors.... Property values along the shoreline will stabilize and undoubtedly increase.' The accidental sea has social benefit, including agriculture and tourism, and managing its waters hinges upon a debate over which benefits can prevail.

The US Bureau of Reclamation (2007) has also evaluated six alternative actions for the Salton Sea, including four variations of dam projects; one 'habitat enhancement without marine lake'; and no action. The Bureau's report acknowledges 'substantial risk and uncertainties' with each alternative, primarily due to lack of data, but recognizes that even a 'detailed evaluation would not resolve the hydrologic and biologic uncertainties.' Ultimately, the Bureau declined to select a preferred alternative primarily due to the cost of each (their alternatives ranged from $3.5 to $14 billion), proposing instead further study of the land and restoration mechanisms. This plan reflects Pahl-Wostl's (2008) idea that 'In recent years the command and control paradigm has been replaced by a paradigm based on the notion of "living with water." In this paradigm the limits of control and the importance of uncertainties are clearly acknowledged. Acceptable risks and decisions are negotiated. This cultural framing supports integrated solutions and the implementation of multifunctional landscape with an increased adaptive capacity of the system.' Planners for the Salton Sea seek to build such a multifunctional landscape out of a polluted floodplain, envisioning a space with capacity far exceeding its pre-1905 use, incorporating traditional tribal land claims, large-scale industrial agricultural, tourists seeking refuge, and millions of migratory birds.

Questions of jurisdiction, authority, and funding continue to bind water managers and impede efforts to take action on the Salton Sea. In 2013, yet another environmental report certified a plan to build cascading ponds as the preferred alternative for species conservation, and yet another bill passed the California legislature attempting to delineate a new structure of governance to collectively plan the fate of the largest lake in the state.

For a hundred years, the Salton Sea has sat waiting, evaporating, heedless of the debates raging and technological fixes being promoted in the cultural realm. The social and ecological consequences of any action upon this landscape are unknown, but no action will 'solve' the flooding event that was caused by efforts to solve the problem of irrigating the desert. Indeed, large dam projects upstream have greatly reduced flooding as a social problem in the area. 'Solving' the flooding problem, however, has not eliminated water management issues in the Imperial Valley, and current debates over the future of the Salton Sea highlight the cycle in which acting upon an environmental problem generates social issues and mitigating social impacts creates more environmental concerns.

5 The Power of Water-Based Sewerage

In another example demonstrating that 'solving' water problems can have far-reaching social, economic, and ecologic consequences, Broich (2007) offers a fascinating analysis of water management systems as a tool in extending British colonial authority. He argues that employing gravitational systems to provide water and remove waste was promoted in Britain and in the colonies not only as a physical issue, but also as a way to address social and moral concerns as well. 'This water system was a practical solution to the challenge of providing growing urban populations in Britain with water, but it represented more than that to the advocates of improvement in cities such as Bradford, Glasgow, and Manchester. For them, the design was ideally suited to accomplish goals that combined physical and moral amelioration' (Broich 2007).

Broich discusses how changes in water availability affected individual behavior and set new norms for cleanliness. This transition posited that being clean could help avoid moral failings such as alcoholism. The private water companies serving slums often did not have enough pressure to get water to upper floors, thereby perpetuating the perception that poverty and filth were connected. 'A writer in Dickens's weekly journal, *Household Words*, anticipated that once there was "a constant supply of water at high pressure within reach of every housewife's thumb...we shall have advanced also in the moral and mental discipline of urban life to a better state"' (Broich 2007). At the same time, the germ theory of disease refuted the link between disease and moral failing. Instead, it located the source of illness outside the body, making it something that individual actions could control. Clean environments could be achieved through civic sanitation systems, in part, but also by care for the individual body and domestic space. Tomes (1990) argues that the late-nineteenth-century cult of domesticity created the moral imperative for homemakers to maintain high standards of cleanliness, primarily by consuming goods such as ceramic toilets, water filters, and chemical disinfectants. Before municipal water systems were in place, the ability to install a private filtration system was both a mark of affluence and a material deterrent of disease.

Because water management was credited with 'solving' a multitude of social ills, Broich (2007) notes that Britain experienced a 'water reformation,' beginning in the 1840s as municipalities purchased or 'municipalized' private water companies. The municipalities then began developing new infrastructure and locating new water sources. This solution to physical, social, moral ills flowed from applications in Great Britain to the British colonies. Broich gives examples of the British imposing gravitational water systems in places where it was not wanted (i.e., from the local perspective, there was no problem to be solved) and where it was physically too different from England to work appropriately. 'At home and abroad, rulers and reformers identified the same practical problems, the unhygienic habits of the working class or native city dweller, and the same abstract predicament, the moral degeneration of townspeople living among "filth," and applied the same environmental solutions.' The British closed wells and tanks to prevent Indian townspeople from using traditional sources and to catalyze behavior change that embraced the new 'official water supplies.' This served to make the 'subjects more dependent on colonial authorities; they centralized control of the most critical element in the hands of the British, when the availability of water had formerly been decentralized' (Broich 2007).

Water imperialism removed local control and changed local habits, and therefore, as a 'solution,' it defies popular contemporary thinking that managing water locally is more sustainable than centralized efforts. This imperial approach generated multiple repercussions, including contributing to general offense at colonial power. More relevant to this chapter, it offers yet another example of a solution generating long-term problems. This is a textbook case of the idea of path dependency, as technology was thoroughly implemented such that it now represents massive infrastructure, precluding any easy shift to different, potentially more water-efficient methods. While this approach did seem to solve an immediate problem by pushing waste away from human habitation, the 'away' was only downstream. As downstream populations have increased, some of those downstream places are now at the center of environmental alarm.

Gravitational water technology allowed for increased water use in many instances and has often been perceived as an unalloyed 'good.' It is now accepted that ensuring a water supply is a 'legitimate, beneficial activity of the modern state' (Broich 2007), but sanitation also creates an entry point into private life. In the USA, people had to be persuaded to allow boards of health into their homes to inspect their plumbing, and laws governing sanitation required social acceptance of such practices as necessary for human health (Tomes 1990). Declining mortality in the 1870s buoyed public health campaigns, and the fact that people with means suffered less mortality stood as popular evidence that good sanitation had positive health impacts. Managing private space for public good forms the basis of public health in the modern era, swaying debates over vaccination, fluoride in water, and even compostable toilets. State actions to manage water for the good of the population are accepted as solutions to health problems, often without scrutinizing the broader social and economic outcomes that may unequally impact members of society.

Water-based sewerage has encouraged water habits that are increasingly problematic as human population grows and climate change affects water flow. As a 'solution,' sanitation well fits the documented cycle as it lowers disease rates, which has contributed to population growth, which increases water demand. While it is highly unpopular to acknowledge, this presents another tension among the commonly cited core ideas of balancing social, economic, and ecologic conditions as inherent to sustainability. Within an ecological frame, disease and death 'solve' problems of resource scarcity. For *Homo sapiens*, this biologic reality is tightly linked with social and economic conditions, as it is those with lower status who disproportionately suffer. This does not change the physical reality that preventing disease to extend life span and subsequently increasing population will have consequences. While certainly stretching far afield from the primary focus of this chapter, the problem–solution chain extends to modern debates about health care and the role for government in protecting a population from disease. If sustainability is the focus, is protecting and lengthening human health compatible with ecosystem resilience and long-term viability?

6 Water Is Wicked

The notion of a feedback loop among problems and solutions is well recognized, yet the pattern persists. Contemporary scholarship simultaneously acknowledges that solving water problems is a misplaced concept, yet continues to raise the specter of a solution in addressing challenges.

As an example, water experts Vorosmarty and Pahl-Wostl (2013) recognize, 'It is ironic that many of today's water problems arise from the very solutions we administer. Proliferation of costly, so-called hard-path engineering, like centralized sewers and large dams, provide undeniable benefits, such as improved hygiene and stable water supply. But they also degrade waters with pollution, obliterate natural flow cycles and block the migration routes of fish and other aquatic life.' Yet, in an interview, the same Pahl-Wostl called for solutions to water problems: '"We need to move from problem identification to the co-design of evidence-based solutions"' (Pahl-Wostl quoted in Perez 2013).

Notice a similar sway back to solution-finding in Fagan's (2011) discussion of water issues: 'The Aswan Dam has not solved Egypt's water problems...' He then explains the complex issues relevant to water management among Egypt, Sudan, and Ethiopia. In this same paragraph, he writes, 'Many people along the river are starving. One solution may be irrigation, but any large-scale dam and storage schemes would have a serious impact on water supplies downstream.' He continues by noting that the only 'logical solution' is in establishing cooperative agreements among the effected countries. In one paragraph, it is made clear that a dam was not a solution, but that there are solutions, potentially including a dam—albeit with recognized consequences.

This phenomenon of simultaneously recognizing the problem–solution cycle *and* continuing to seek solution is a function of the limits of language and the nature of 'wicked problems.' Such problems are embedded in complex systems, involve multiple definitions, and are never solved, '[a]t best they are only resolved—over and over again' (Rittel and Webber 1973). This is in part because the '"solution" to one interested party is a "problem" for others' (Freeman 2000). Despite 40 years of recognizing that Rittel and Weber well captured the idea that trying to solve 'wicked problems' is itself a problem, the word 'solution' remains prominent in scholarly and popular debates about wicked problems. Even Rittel and Webber themselves use the word solution in discussing the idea of wicked problems. Freitag et al. (2009) explicitly state that long-term solutions can be worse than the problem and highlight the importance of language and the need to define terms in addressing floodplain management. Yet, they then continue to use the word 'solution' in discussing improved approaches. This includes their conclusion that the 'solution' to floodplain management is to work with the river, not against it. They promote adapting and accommodating floods, rather than seeking to control rivers. The premise of this chapter is that such an approach is not a solution, sensu stricto, because it does not 'fix' the problem. Rather it is a flexible strategy that recognizes the need for consistent attention. Acknowledging this difference between a solution and an ongoing management need requires rethinking how sustainability is conceptualized and how language perpetuates the problem–solution cycle.

As the introduction to this chapter notes, descriptions of sustainability typically link environmental, social, and economic conditions. What is often glossed over in discussing sustainability, however, are issues of scale, both spatial and temporal. Thinking about sustainable water management hinges on key questions: What is to be sustained over what period of time? Issues of scale have perpetuated the focus on solutions because in the short term, many solutions have worked (e.g., dams provide water for irrigation; water-based sewerage reduces disease). Only with the perspective of time, does the solution become recognizable as a new or reframed problem. Ignoring scale in water solutions is compounded by the inclination to apply the concept of sustainability to a single portion of a complex system. Discussing sustainable water management without recognizing its relationship to sustainable energy management, or sustainable agricultural practices, or sustainable population is problematic. If one part of a complex system is not sustainable, then the entire system is not sustainable. Additionally, promoting the idea of sustainability as an end goal perpetuates the quest to solve problems so that sustainability can be achieved. But sustainability is better framed as a process or discourse, not a goal (Dryzek 1997), and this requires non-solution-oriented discourse.

Language simultaneously illuminates problems and promises solutions, and this self-referential process will persistently be inadequate in opening new paradigms. Language is pivotal in creating perceptions of what constitutes a problem, a risk, a crisis, or a resolution and is core to how humans understand ideas of security (Bishop 1980, Buzan et al. 1998). Contemporary theorists of security recognize the power of language to imagine a future in which problems have been solved (Buzan et al. 1998), to constitute security through daily practice (Bigo 2002), and to entwine the logics of security with risk management practices (Methmann and Rothe 2012). In contrast, wicked problems that defy resolution, including water management, create an apocalyptic frame, characterized by a sense of uncertainty and doom. By integrating the logics of security and risk, social groups can act against a wicked problem. In turn, the ways people articulate the character of the environmental or other risk impact the type of actions taken to create security against the future threat. 'The actual political practices that result from any given act of securitization will depend on the particular way in which antagonism and security are discursively constituted' (Methmann and Rothe 2012). Thus, the articulation of a 'solution' directs political actions toward finding or creating just that for security depends upon 'solving' that which creates risk.

Buzan (1998) defines security as a speech act that follows a precise script, and therefore, conventions of environmental discourse illuminate understandings of what a solution actually is. Buzan's security script first creates an external enemy or threat and then identifies exceptional measures to handle threat. Dryzek (1997), however, writes 'the discourses of environmental problem solving recognize the existence of ecological problems, but treat them as tractable within the basic framework of the political economy of industrial society, as belonging in a well-defined box of their own. The basic storyline is that of problem solving, rather than heroic struggle.' The distinction here is in the shift from problem defining to problem solving. Dryzek recognizes that apocalyptic rhetoric exists and is used to create a script for environmental concern, but finds that within institutions that attempt to 'solve' problems, the discourse is not radical, and there are no large-scale attempts to address population control or end economic growth, for example.

This likely reflects an institutional understanding that solutions and blame-laying are interrelated. If modern environmental problems originate with people, they must be someone's fault and the language of 'solving' perpetuates this propensity to blame individuals or institutions for events that are often beyond their control. The need to assign blame consequently narrows people's ability to conceptualize complex, multisourced wicked problems. Chenhansa and Schleppegrell (1998) found that when grade school students studied environmental issues with no context or prior knowledge, they simply reversed their understandings of the causes to propose solutions, directly targeting people and their actions (i.e., 'People need to stop cutting down trees.'). When no human agent could be identified, they wrote statements like 'There is no solution. It was an accident.' Not only is finding a solution aligned with placing blame, but this thinking expresses an understanding that when accidents happen, or when a definitive actor and action cannot be identified, solutions do not exist. Solutions, then, are also aligned with explicitly human-caused problems. Additionally, when students were not exposed to creative or outside-the-box approaches, the solutions they suggested mirrored the problems they identified, a circular mode of thinking. Though in this study, the researchers' prompts to find a solution directed students' thinking, their processes exemplify the fierce bond between problem and solution in environmental studies, showing that the way a problem is presented to an audience will directly influence how they think about solutions.

Another study of solution-making explored how metaphor shaped solution generation, showing that the language used to describe an issue impacted which solutions people chose. In their research, Thibodeau and Boroditsky (2013) described crime as both a monster and a virus and found that respondents hearing the first metaphor favored an aggressive 'attack' to target crime, while those who heard crime described as a virus-favored internal 'healing' as a response. These discursive frames are seductive because they work to streamline information and offer relief from the complexity of political issues; however, the researchers pointed to limitations of using metaphors to explain issues because in multifaceted social situations, it can be difficult to identify the complexity that exists outside the metaphorical frame. In these examples, reducing complex processes through simplified language for the sake of mass communication directed individuals' responses in specific ways. In describing an issue as complicated as water system management, shortcutting, simplifying, and summarizing the character of the 'problem' will—through language alone—prompt a response that mirrors the issue as presented. Cockerill (2003) found this in testing how language about flooding impacts affected respondent support for specific policies. Language describing a flood positively as a natural event elicited more support for policy that allowed rivers to run freely, while those who read language describing the flood as 'devastating' were more likely to support stronger flood control measures. This work suggests that language shapes problem formation for readers and a problem (i.e., a devastating flood) requires a solution (i.e., more control), while the non-problem of floods being natural poses no risk and hence does not require a solution—the river just does what it does.

These theorizations of language strategies, when applied to mass media communications, further explain how broad conceptions of environmental problems are negotiated in the public sphere. Kensicki (2004) looked for language devices in news reporting, evaluating stories that presented cause-and-effect explanations, possibilities for citizens to directly combat issues, and including potential solutions in the news article itself. She found that audiences favored articles that reported a solution. Another study found that audiences were less satisfied with how environmental solutions were reported than with any other aspect of journalism and that the public sees media that are not reporting solutions as failing to do their job (Riffe and Reimold 2008). The possibility that solutions do not exist is not considered as an explanation for the absence of solutions in news reporting. Public expectation seems to hold that media will communicate both problem and solution in environmental issues. As an example, in 2010, an opinion poll asked experts: 'What are the technologies or changes in behavior which show the most promise for addressing water shortages over the next 10 years?' A headline in a news article reporting the responses was 'Experts name the top 19 solutions to global freshwater crisis' (Experts name the top 19 solutions to the global freshwater crisis 2010). This disconnect between posing a question using non-solution, non-crisis language, and the media headline using both solution and crisis perpetuates the problem–solution cycle.

Beyond media reporting on existing solutions, the value of communicating solutions has been studied as a persuasive mechanism and a strategy to rally individuals to support environmental causes. Communication scholars consider how framing issues 'explains who is responsible and suggests potential solutions' (Ryan 1991). Frames work to define problems, evaluate causes, and suggest remedies. Snow and Benford (1988) identify 'core framing tasks' for social issues, one that entails listing strategies, tactics and targets and proposing solutions. By this theorization, the solution is again part of the problem, or rather, problem articulation is partially dependent upon presenting possible solutions. This body of scholarship proposes that an organization that wants to recruit public support must be able to articulate solutions to the issue they are advocating. By this argument, to garner support for their actions, water policy makers and management agencies should be able to present solutions and predict likely outcomes of their practices. Again, this presumes a linear causality between actions and outcomes, ignoring unintended consequences and cyclical relations that characterize water systems.

This framing thesis exemplifies the cultural expectation that problems will have solutions, but also shows that consensus about whether a problem exists is linked to the presentation of a cohesive solution. Dardis (2007) argues 'there is an indication that the offering of solutions in relation to a specified problem may enhance individuals' acceptance or evaluations of a message, and thereby may lead individuals to agree more with the notions promoted by the message's source.' This reverses the linear logic that suggests that problems must be followed by solutions and shows the cultural power held by the idea of a solution. A desirable outcome has a powerful effect in framing the pre-existing problem and excluding contradictory evidence and complexities. Whether studying persuasive communications, media reporting, or environmental education, this combined scholarship shows that people of all ages experience discomfort when presented with problems without accompanying potential solutions and look for solutions in an ordered progression of temporality and cause and effect.

Consider the temporality in key terms of environmentalism: *sustain, restore, conserve, preserve.* These words are firmly tied to the past, creating double bind through language that refuses to look forward (Bowers 2001). This reflective language affirms culturally accepted norms that nature exists in the past; technologies move us forward. Our deeply entrenched ways of thinking about nature as a static, often primitive, state and technologies as mechanisms of change and linear progress shape a belief in interminable progression toward a solution. Within this logic, sustainable water management rhetoric often harkens back to more primitive systems that privilege one water use (often non-industrial agriculture) over another (e.g., urban development, recreation). By refusing to acknowledge the dynamic quality of water systems and the cyclical outcomes of technologies of control, modern thinkers persist in pursuing a singular solution. Trying to 'solve' water problems is an attempt to put the problem in the past.

As Stone (2001) well articulates, couching decisions in terms of solutions sets decision-makers up to fail and increases tension. She states that the idea of 'policy solutions' is misleading and writes that policy actions '…are really *ongoing strategies* for structuring relationships and coordinating behavior to achieve collective purposes' (emphasis original) and that 'policy is more like an endless game of Monopoly than a bicycle repair.' Attempts to promote sustainable water management will be better served by acknowledging that problems will not be solved, will not be removed to the past, but, rather, will require constant attention. This actually offers a path to better promote flexible, adaptable, social-learning-based approaches to water management, which are potentially more appropriate within a sustainability process.

Despite a significant body of work on wicked problems, there has been little attention focused explicitly on how wicked problems are constituted and why they resist 'solution.' Such work is essential to break the feedback loop that implies that 'solutions' are possible, and therefore, when 'solutions' fail (as they always will with wicked problems), it generates increased distrust and lack of confidence in the entities that initially developed the 'solution.' Shortly after Rittel and Weber first proposed the idea of 'wicked problems,' Churchman (1967) noted that trying to tame a wicked problem bordered on deception, as it created a false notion that taming is possible. Indeed, literature continues to reveal the conundrum between what constitutes a tame problem versus a wicked one. Returning to one example, many water scholars highlight the success of water-based sewerage as evidence of solving pressing environmental and human health problems. Indeed, the perceived tame problem of keeping pathogens out of household water supplies was solved. But, as the authors have already discussed, this has had ramifications socially, economically, and ecologically, thus connecting it to issues classified as 'wicked.' Conklin (2005) concludes that recognizing the nature of wicked problems and understanding that you are dealing with a wicked problem is necessary to find ways to collaborate effectively to address the problem.

7 Conclusion

The relationships between wicked problems, water management, and sustainability are deep and diverse. Embedded in these relationships are the artifacts of historical decisions, the connections between ecological, economic, and social conditions, and issues of discourse about water management. As depicted in Fig. 1, there is a cyclic relationship in which implemented solutions to water management problems create new problems demanding solution, which when implemented can sometimes catalyze some form of the initial problem. Examples from this chapter highlight these relationships, as dams, flood control, irrigation canals, and water-based sewerage were all implemented as solutions to economic and social problems. These solutions subsequently generated ecological problems and sometimes prompted new or reframed economic and social concerns. Reuss (undated) well captures a fundamental idea of these cyclic relationships: 'History... suggests the mutability of human vision; what seemed so obvious and relevant in one decade may seem outmoded in the next. In a constantly changing world, engineers and politicians must accept responsibility for both short and long-term consequences of their projects, and, like the country doctor, must be on the lookout for new ways to keep the body politic healthy and happy. *The challenge is continuous*' (emphasis added).

Despite evidence that water management is a wicked problem and hence presents continuous challenges, the idea of solving remains prominent in both academic and popular perceptions. This emphasis on solving water problems poses a real risk to human well-being, economic stability, and ecologic integrity. The insistence on solutions may exacerbate rather than alleviate negative conditions in part because it discourages decision-makers and citizens from accepting long-term responsibility for managing water to sustain ourselves. Although scholars/authors recognize the concept of a wicked problem, many continue to use the word solution and this perpetuates the self-referential notion that problems and solutions are parallel and enforces the idea that someone can, should, and will 'solve' the problem. If people believe that water management issues can be or should have been solved, the continued discourse about water management problems then creates a propensity to lay blame—someone is at fault for not solving this problem because the rhetoric emphasizes that solutions are possible. Changing this discourse is one step toward changing thought and behavior about how to sustainably manage water resources.

Readers could conclude that there is a bit of a circular argument embedded in this chapter, which can be read as implicitly proposing that the 'solution' is to stop thinking of 'solving' water problems. But the proposal to change discourse, to cease using the verb to solve, is not intended to be read as a solution. The authors recognize that changing discourse will have its own set of consequences. The core message, however, is that with a dynamic entity like water, solving is a misplaced frame. A more appropriate frame or discourse may be to allow language to be complex. For example, the answer is not to generate a new word or phrase to replace 'solution' but to use more informative, more accurate, more complicated language that is appropriate to the problem being addressed. Eliminating the word and complicating the problem by not framing it in terms of a solution can contribute to changing the thought process that demands solution, which in turn may contribute to behavior change that embraces adaptive, resilient approaches to water management.

References

Bali KM (2010) Salton sea salinity and saline water. imperial county agriculture briefs pp 10–12

Bigo D (2002) Security and immigration: toward a critique of the governmentality of unease. Altern Glob Local Polit 27:63–92

Bishop M (1980) The language of poetry: crisis and solution: studies in modern poetry of French expression, 1945 to the present. Rodopi, Amsterdam

Bovy K (2007) Prehistoric human impacts on waterbirds at Watmough Bay, Washington, USA. J I Coast Archaeol 2:210–230

Bowers C (2001) How language limits our understanding of environmental education. Environ Educ Res 7:141–151

Broich J (2007) Engineering the empire: British water supply systems and colonial societies, 1850–1900. J Br Stud 46:346–365

Buzan B, Wæver O, De Wilde J (1998) Security: a new framework for analysis. Lynne Rienner, Boulder

Chenhansa S, Schleppegrell M (1998) Linguistic features of middle school environmental education texts. Environ Educ Res 4:53–66

Churchman CW (1967) Guest editorial: Wicked problems. Manage Sci 14:141–142

Cockerill K (2003) Testing language: media language influence on public attitudes about river management. Appl Environ Educ Commun 2:23–37

Conklin J (2005) Wicked problems and social complexity, in dialogue mapping: Building shared understanding of wicked problems. Wiley, West Sussex

Dardis FE (2007) The role of issue-framing functions in affecting beliefs and opinions about a sociopolitical issue. Commun Quart 55:247–265

Delano FA (1928) The report of the committee on Mississippi flood control appointed by the united states chamber of commerce. Ann Am Acad Polit Soc Sci 135:15–24

Deur D (2002) Rethinking precolonial plant cultivation on the northwest coast of North America. Prof Geogr 54:140–157

Dolin EJ, Dumaine B (2000) The duck stamp story. Krause Publications, Iola

Dryzek JS (1997) The politics of the earth: environmental discourses. Oxford University Press, Oxford

Experts name the top 19 solutions to the global freshwater crisis (2010) Circle of Blue 24 May. Web accessed 20 Sept 2013

Fagan B (2011) Elixir: a history of water and humankind. Bloomsbury Press, New York

Friend M (2002) Avian disease at the Salton Sea. Hydrobiologia 473:293–306

Freeman DM (2000) Wicked water problems: sociology and local water organizations in addressing water resources policy. JAWRA 36:483–491

Freitag B, Bolton S, Westerlund F, Clark JLS (2009) Floodplain management: a new approach for a new era. Island Press, Washington

Hoover H (1928) The improvement of our mid-west waterways. Ann Am Acad Polit Soc Sci 135:7–14

Jianchu X, Ma ET, Tashi D et al (2005) Integrating sacred knowledge for conservation: cultures and landscapes in Southwest China. Ecol Soc 10:151–175

Johnson K (2012) A river newly wild and seriously muddy. The New York Times 2 August

Kensicki LJ (2004) No cure for what ails us: the media-constructed disconnect between societal problems and possible solutions. Journalism Mass Commun 81:53–73

Ludlow CW (1884) Surveys for future water supply. J Franklin Inst 117:453–459

Methmann C, Rothe D (2012) Politics for the day after tomorrow: the logic of apocalypse in global climate politics. Secur Dialogue 43:323

Moore D, Dore J, Gyawali D (2010) The world commission on dams +10: revisiting the large dam controversy. Water Altern 3:3–13

Pahl-Wostl C, Tabara D, Bouwen R et al (2008) The importance of social learning and culture for sustainable water management. Ecol Econ 64:484–495

Perez I (2013) WATER: Can we curb unlimited uses for a limited resource? ClimateWire 29 May in The NAEP National Desk 14 June 2013

Pinchot G (1928) Some essential principles of conservation as applied to Mississippi flood control. Ann Am Acad Polit Soc Sci 135:57–59

Pinter N (2005) One step forward, two steps back on U.S. floodplains. Science 308:207–208

Postal S, Richter B (2003) Rivers for life: managing water for people and nature. Island Press, Washington

Reuss M (undated) Historical explanation and water issues. UNESCO International Hydrological Programme to the World Water Assessment Programme

Riffe D, Reimold D (2008) Newspapers get high marks on environmental report cards. Newspaper Res J 29:65–79

Rittel H, Webber MM (1973) Dilemmas in a general theory of planning. Policy Sci 4:155–169

Ryan C (1991) Prime time activism: media strategies for grassroots organizing. South End Press, Boston

Salton Sea Authority Plan for Multi-Purpose Project (2006) Web accessed 26 Sept 2013

Snow DA, Benford RD (1988) Ideology, frame resonance, and participant mobilization. In: Klandermans B, Kriesi H, Tarrow S (eds) International social movement research, vol 1 JAI Press, Greenwich

Solomon S (2010) Water: the epic struggle for wealth, power, and civilization. Harper, New York

Steiner A (2010) Preface. Water Altern 3:1–2

Stone D (2001) The policy paradox: the art of political decision-making. W.W. Norton, New York

Thibodeau PH, Boroditsky L (2013) Natural language metaphors covertly influence reasoning. PLoS ONE 8:e52961

Tomes N (1990) The private side of public health: sanitary science, domestic hygiene, and the germ theory. Bull History Medicine 64:498–539

US Bureau of Reclamation (2007) managing restoration of the Salton Sea final report, U.S. Department of the Interior. Web accessed 26 Sept 2013

Vorosmarty CJ, Pahl-Wostl C (2013) Delivering water from disaster. The New York Times 10 June. Web accessed 26 Sept 2013

Authors Biography

Dr. Kristan Cockerill has an interdisciplinary background and 20 years of work experience to understand and improve the connections between cultural and scientific information related to developing environmental policy. Most recently, her work has focused on social and communication issues relevant to water management decisions, including developing collaborative models for water management; promoting community water education; and assessing public and decision-maker attitudes toward water management. She has taught a broad suite of courses at Appalachian State University, Columbia University's Biosphere 2 Center, and the University of New Mexico.

Dr. Melanie Armstrong is a postdoctoral researcher in geography at the University of California—Berkeley, where she studies new forms of nature and environment emerging from modern bioscience. She has also worked for more than ten years for the National Park Service, managing public lands throughout the western USA and witnessing how emerging ideas of nature shape the ways people interact with modern landscapes. Her research explores how the belief that humans can manage nature shapes how people respond to the crises of our day.

Sustainable Water Use: Finnish Water Management in Sparsely Populated Regions

Piia Leskinen and Juha Kääriä

Abstract In Finland, 1 million inhabitants of the population (5.4 million) live in sparsely populated areas. Since 2004, the Finnish legislation requires that every house outside the municipal sewer networks must have a water purification system that meets the minimum purification requirements for phosphorus, nitrogen, and organic matter. Existing dwellings were given an adaptation period of ten years, during which they would have to make the necessary investments. In our study, we focused on making research on the functionality of small-scale purification systems in 30 different households and on dissemination of information about the purification systems and the legislation to concerned property owners. The purification performance of the plants was monitored by traditional sampling and continuous on-line water quality sondes. The study was focused at determining how much the fluctuations in the incoming wastewater quality affect the purification performance. The main results showed that the small-scale purification systems function generally well if they are properly installed and regularly maintained. Unfortunately, this is not often the case. Several recommendations on how to prevent faults in installation of the systems and how to encourage property owners to maintain their systems were made.

Keywords On-site wastewater treatment · Single-house package plants · Wastewater management · Sparsely populated areas

1 Introduction

Eutrophication, caused by excess input of phosphorus and nitrogen nutrients into the water bodies, has been identified as a major threat to the quality of coastal water resources in Europe (European Environmental Agency 2001). In 2000, the European Union set a directive to improve the water quality in all member states (EU 2000).

P. Leskinen (✉) · J. Kääriä
Turku University of Applied Sciences, Sepänkatu 1, 20700 Turku, Finland
e-mail: piia.leskinen@turkuamk.fi

© Springer International Publishing Switzerland 2015
W. Leal Filho and V. Sümer (eds.), *Sustainable Water Use and Management*,
Green Energy and Technology, DOI 10.1007/978-3-319-12394-3_14

This so-called Water Framework Directive set the ambitious goal of having all water bodies in the European Union area in a good state, as defined using ecological classification system. Finland has a reputation of being a sparsely populated country with thousands of lakes and a clean nature. In fact, the country has 5.4 million inhabitants in 300,000 km^2 of land and 60,000 lakes of surface over a hectare. One-third of classified lakes and half of the coastal areas are in a poor or satisfactory state (Putkuri et al. 2013). One million of Finns live in rural areas, without connection to municipal water management systems. In addition, there are approximately half a million summer residences in Finland, most of which are located by a lake or the Baltic Sea. Traditionally, the summer cottages were modest cottages with dry toilets, but during the last 20 years, there has been a strong trend of upgrading the commodity standards of summer houses to the level of permanent residencies. As the nutrient loading from municipal water treatment facilities and industrial sources has diminished significantly due to strict regulations and investments in treatment technologies, the emissions from rural dwellings have become the second largest source of phosphorus after agriculture (Putkuri et al. 2013). In addition, there is a risk of contamination of drinking water wells by untreated wastewaters.

In order to reduce the nutrient loads and hygienic risks from rural dwellings, the Finnish legislation was modified in the early 2000s. The most important addition was the Government Decree on Treating Domestic Wastewater in Areas outside Sewer Networks (542/2003), which came into force in the beginning of 2004, and set minimum standards for wastewater treatment and the planning, construction, use, and maintenance of treatment systems in rural areas. The Decree does not make a difference between permanent and holiday residences. Instead, the level of sanitary and water facilities is considered, meaning that no wastewater treatment is required only if the house has a dry toilet and no water pipe. The requirements of the decree were applied to all new houses immediately, but the existing houses were given a transition period of 10 years.

The decree was not welcomed by residents of rural areas, and it started a wide ranging public discussion that went on from internet discussion groups, newspapers, and markets all the way to the parliamentary sessions. The main points of criticism were that the performance of small-scale purification is questionable, the investments are too expensive, and the limits for purification are too strict. Further, many property owners seemed to be unaware of what they were expected to do. In 2009, it was estimated that only 10–15 % of properties had done the required improvements in their wastewater systems (Tarasti 2009).

Due to the debate, the Government Decree on Treating Domestic Wastewater in Areas outside Sewer Networks (hereafter wastewater decree) was modified and the new decree (209/2011) came into force on March 15, 2011, with lower purification requirements and extended transition period for upgrading the treatment systems in old houses.

In this article, we describe the results on the functionality of different systems. Our aim was to answer the following questions raised by the public debate:

1. Is the average load and reduction percentages a good way of defining the purification requirements?

2. Are the purification results of single-house purification plants affected by variations in the daily load or occasional exposure to strong household chemicals?

In this article, we also evaluate the impact of public debate and the modification of
legislation on the willingness of property owners to comply with wastewater legislation.

2 The Purification Requirements and Measured Wastewater Quality

The modified Government Decree on Treating Domestic Wastewater in Areas
outside Sewer Networks in Finland (from 2011) requires that all properties must
remove 70 % of phosphorus (P), 30 % of nitrogen (N), and 80 % of organic matter
(BOD7) in their treatment systems. In especially sensitive areas, such as those near
water bodies or ground water areas, the requirements are 85 % P, 40 % N, and 90 %
BOD7. The purification requirements are counted from estimated average loading,
which is defined in the same decree, 2.2 g P, 14 g N, and 50 g BOD7 per person
per day. Purification requirements are defined as reduction percentages from initial
load in Finland and for example in Norway, whereas in some countries the purification requirements are expressed as maximum concentrations in outgoing wastewater. The actual wastewater concentrations that can be measured and reduction
percentage from average load can be compared using the following equation:

$$\text{concentration} = \left(1 - \left(\frac{\text{reduction requirement}}{100}\right)\right) \times \frac{\text{average load } \left(\frac{g}{d}\right)}{\text{water consumption } (L/d)}$$

For example, for phosphorus, the maximum allowed concentration in purified
wastewater can be calculated:

$$\text{concentration} = \left(1 - \left(\frac{70}{100}\right)\right) \times \frac{2.2\left(\frac{g}{d}\right)}{128\left(\frac{L}{d}\right)} = 5.2 \,\text{mg/L}$$

Above it can be seen that the water consumption needs to be known in order to
resolve the equation. Most properties in rural areas get their water from an own
well, and the water consumption is not measured, so the information on the actual
water consumption in rural areas is scarce. In centralized water distribution systems, the average water consumption rate of households in Finland was 128 L/day
in 2010 (Vesihuoltolaitosyhdistys 2012). During our studies, we measured water
consumption in a number of properties and the consumption rates varied from
70 L to 150 L/day, the average being 110 L.

To our knowledge, there is not much measured data on wastewater production in individual properties, and all estimates are based on data collected from
municipal wastewater treatment plants. The sampling of nonpurified wastewater
in an individual property is technically challenging, and the quality and quantity
of wastewater produced in an individual property have large daily and weekly

Table 1 The measured wastewater load in four properties compared to standard load in the decree

		BOD7 mg/L	Ntot mg/L	Ptot mg/L	Water consumption L/day/person
Property 1	Mean	322	101	16.3	74
	standard (n)	58.8 (17)	16.1 (17)	6.3 (17)	
Property 2	Mean	318	110	16.7	115
	standard (n)	89.7 (13)	22.4 (13)	5.4 (13)	
Property 3	Mean	520	122	23.3	80
	standard (n)	195 (10)	45.9 (10)	6.4 (10)	
Property 4	Mean	399	94.6	20.1	120
	standard (n)	107 (20)	19.2 (20)	6.25 (20)	
Decree standard		*391*	*109*	*17.2*	*128*

variations. Thus, estimating the average load is difficult. We attempted to address this question by installing sampling devices in four different properties. The unpurified wastewater samples were collected in a container in a 24-h period, and the samplers were equipped with a disintegrator. The samples of purified wastewater were taken from the inspection wells of the purification plants. The results are summarized in Table 1. According to our data, in these four properties, the decree standard values corresponded quite well to the measured wastewater load. Our results are in line with the few other studies (Lowe et al. 2009; Nieminen et al. 2013) where wastewater production in individual properties has been measured.

3 Treatment Systems Overview

Traditionally, rural dwellings in Finland get their drinking water from an own well and dispose their waste water into the environment, after one or two sedimentation tanks. In many properties, there are also septic tanks where either only toilet water or both washing and toilet waters are led. In a long term, this is a very expensive solution and the cost of the tanker truck visit may lead the property owners to emptying the tanks in the nature. Many of these systems have been installed 20–30 years ago, and not maintained since, apart from eventual emptying of the tanks.

In order to meet the requirements of the current wastewater legislation, a property must have an advanced purification system, including two to three sedimentation tanks followed by a filtering field or a small-scale purification plant. In properties that have a dry toilet or where toilet water is lead to a septic tank, a simple filtering system is sufficient for washing water treatment. The wastewater decree is based on the idea that no system is better than another, as long as the purification requirements are met. The choice of the treatment system depends on local conditions and the property owners should seek for advice from an expert in order to make the right choice (Lehtoranta et al. 2014). The water using habits and personal preferences of the system users should be taken into account when planning a system, as well as the soil type and dimensions on the property. For example, soil filtering systems can be only used in areas where groundwater is not near the soil surface, in properties that are large enough to allow construction of a filtering field of about 30 m^2.

If the distance to the neighboring houses is short, it is recommended that the possibility of putting up a shared or a community-based system is inspected first. If situated reasonably close to the cities, the communities may put up a cooperative for the construction of a water and sewage network that will then be connected to an existing municipal treatment plant. An own treatment plant can be put up by those community-based cooperatives that are situated in remote areas. Shared and community-based systems are recommended due to easier maintenance, better performance, steadier wastewater flow, and financial advantages. However, the Finns traditionally like to have their own space and typically houses are built far apart. In these cases, an individual system for the property is the only option.

A factory designed package plant with a combined active sludge and chemical treatment process is generally chosen in properties that are limited in space. In Finland, there are several manufacturers of such plants and new package plants came into the markets upon the enforcement of wastewater decree in 2004. Although some of the plants are working on a continuous flow principle, most of them have a sequencing batch reactor that start the purification process either at a certain time of a day or when the amount of wastewater reaches the preset level. The reactors typically have an aerobic mixing period during which a compressor feeds pressurized air into the reactor resulting in degradation of organic matter and nitrification of ammonium. This is followed by a settling period, during which the oxygen is rapidly consumed from the reaction tank creating anaerobic conditions that are favorable for denitrification (conversion of nitrates into elementary nitrogen gas). The phosphorus precipitation using aluminum or ferric salts is done either in the aerobic process tank or in a separate tank after the biological process. A part of the sludge from the process tank is used to maintain the process stability, but the excess sludge is stored in a separate container or a tank, from where it should be emptied regularly and transported for treatment to an authorized treatment plant.

4 Impact of Incoming Wastewater Quality on Treatment Efficiency of Single-House Package Plants

From the beginning, the Finnish wastewater debate raised many critical questions on the functionality of biological process of the package plants over cold winter periods and on their ability to deal with large fluctuations in wastewater quality. Several studies show that although the performance of on-site systems varies greatly, they generally work well if they are properly maintained (Hellström and Jonsson 2003; Vilpas and Santala 2007). Garcia et al. (2013) found that the purification results of aerobic on-site treatment systems were similar to those of municipal wastewater treatment plants. However, previous studies have been carried out by taking samples from purified wastewater only. In our study, we addressed the impact of varying wastewater quality on the purification results of the package plants by taking samples from the raw and purified wastewater in three different package plants during a period of about one month. All plants were in normal use during the study. The results are presented in Fig. 1. According to these results, it seems that the fluctuations in the quality of incoming wastewater are generally not reflected in the quality of outgoing wastewater. Rather, when the purification plant is functioning well, it can treat even high concentrations of nutrients.

In order to reveal short-term variance in the purification efficiency, we installed continuous sensors in eight different purification plants and followed their functioning over a test period of 6–8 weeks, during which we also took samples from

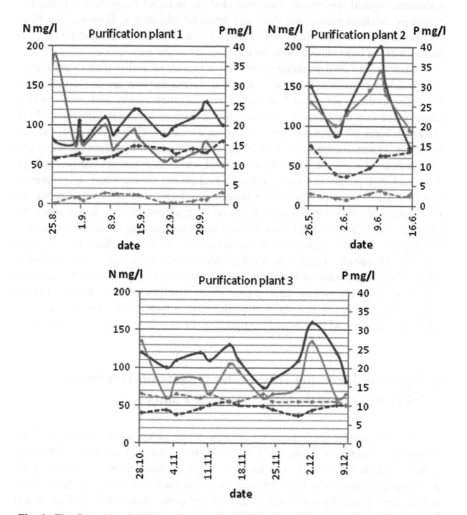

Fig. 1 The fluctuation in the quality of incoming (*solid lines*) and outgoing (*dashed lines*) wastewater in three different single-house purification plants in normal household use. Total nitrogen concentrations in *red* and total phosphorus concentrations in *green*

purified wastewater twice a week. According to our results, those purification plants that were correctly installed and regularly maintained met the purification requirements during the whole monitoring period, without significant variations in purification efficiency.

Flushing toxic chemicals, such as solvents or chlorine, is forbidden in the user manual of all package plants, and property owners generally are aware of this. However, commonly used cosmetics and household cleaning products often contain toxic chemicals that go down the drain. In order to find out how well the single-house purification plants could stand occasional loads of strong chemicals, we did two tests in five different purification plants. In the first one, we flushed two packets of hair coloring products into the drain and in the second one, we asked the owners of the purification plants to change their usual washing powder into a stronger one that contains phosphates and solvents. We monitored the effect of these chemicals with an YSI6600—series continuous turbidity/pH/oxygen/temperature sensor installed in the process tank of the purification plants and by taking the samples from the purified water after the chemical additions. Figures 2 and 3 show how the addition of hair coloring chemicals affected the process and purification performance in two different reactors. Although the pH and oxygen balance was disturbed during 3 days in the purification plant number 1, the perturbations do not significantly affect the purification results of nitrogen and phosphorus. In summary, the tested purification plants seemed to tolerate well the addition of strong household chemicals.

5 Servicing and Maintenance Issues

The servicing and maintenance emerged as one of the major issues in the functionality of different purification systems. Although the ease of maintenance is one of the major selling arguments of system providers, no purification plant can go on without regular maintenance. The required maintenance steps depend on the design of the treatment plant, but at least checking of pumps and air diffusers, adding of precipitation chemicals, and emptying of excess sludge are required for almost all plants. One common reason to low purification performance are problems with dosing of phosphorus precipitation chemical, which can be at too low level, if the settings are made in factory. The dosage amount should be adjusted based on number of users and their using habits to reach good phosphorus removal level.

Some property owners carry out these tasks regularly, and many have made a contract with a servicing company. However, many treatment plants in our study were found nonfunctional due to lack of maintenance or because of an installation fault. There is no data available in Finland on how many percentages of single-house purification plants are maintained properly. There is a large difference between the theoretical amount of sludge that should be produced in rural areas (based on the number of residents) and the actual amount of sludge received

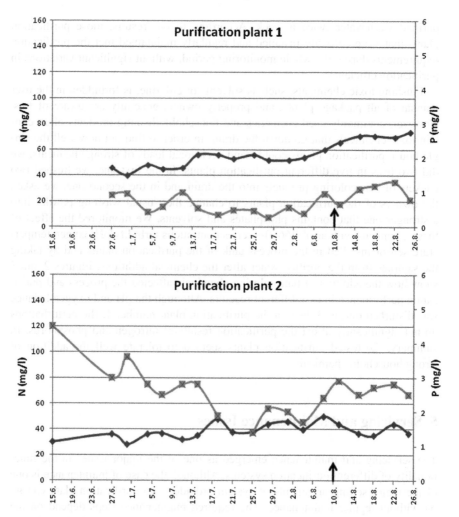

Fig. 2 The total phosphorus (*green*) and total nitrogen (*red*) concentrations in the purified waste-water of two different single-house purification plants. The *arrows* show the time of flushing of hair coloring products in the drain

by authorized treatment plants. This indicates that emptying of the sludge—the most basic step of maintenance—is not carried out properly in the majority of properties. In our study, we carried out a questionnaire where treatment plant owners were asked if they felt they had had sufficient information on wastewater legislation, purification systems, and maintenance issues. When the results of the questionnaire were compared to data on the maintenance level and purification performance of the same treatment plants (Table 2), it was clear that proper maintenance was the crucial factor in purification performance of the reactors. Interestingly, we found out that even though some property owners felt they had got sufficient information, they still did not take proper care of their treatment

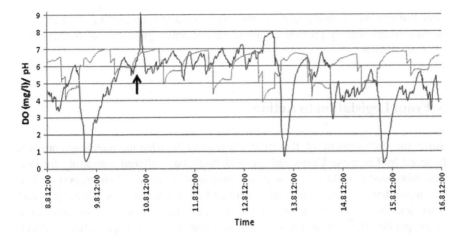

Fig. 3 The oxygen concentration (*blue line*) and pH (*green line*) in the reactor of a single-house purification plant during exposure to strong hair coloring products. The *arrow* shows the time of flushing of hair coloring products into the drain

Table 2 Links between information, action, and purification performance

Purification plant	1	2	3	4	5	6
Did you feel that you had got sufficient information on...						
...wastewater legislation?	No	Yes	Yes	Yes	Yes	Yes
...different purification systems?	Yes	Yes	Yes	Yes	Yes	No
...your own purification plant?	No	Yes	Yes	Yes	Yes	Yes
...maintenance of the purification plant?	No	Yes	Yes	Yes	No	No
Was the purification plant						
...properly installed and fixed?	No	No	Yes	Yes	Yes	No
...serviced regularly?	No	Yes	No	Yes	Yes	No
Did the purification plant meet the purification requirements over the whole monitoring period?	No	Yes	No	Yes	Yes	No

systems. A wastewater treatment system seems to be something that people rather forget, until it becomes for some reason unavoidable to do something. A study carried out in the Republic of Ireland found that many inhabitants of sparsely populated areas were unaware of what type of on-site wastewater system they had in their own property (Naughton and Hynds 2013).

It is of crucial importance that the package plant is correctly installed. Based on this study, it seems that different kinds of problems in installation are common. Property owners should make sure that they get competent contractors to install wastewater treatment systems. After installation, package plants must be monitored by their owners to ensure that treatment process has started to work properly. A sample from purified wastewater should be taken and analyzed few months after installation to ensure that the process has started functioning. Some of the manufacturers already provide sampling service and a guarantee of functionality after installation.

6 From Legislation to Action

Since the enforcement of the rural wastewater legislation, numerous projects financed by European Union and national funds have offered consulting for citizens who need to update the wastewater treatment systems in their properties. The consulting has been given through e-mail and telephone services, happenings and work demonstrations. In each district, there is an office responsible for consulting of the public. The Finnish Environment Institute hosts web pages where research information, environmental justifications for improved wastewater treatment, as well as clear and concise instructions for choosing and maintaining a purification system, are displayed. As a result, information on different wastewater systems is now available for those who are willing to modify their wastewater systems and are actively looking for information on different systems. Initial questions about the functionality of single-house purification plants have been addressed by independent studies, which have showed that the single-house purification plants generally work efficiently when they are properly maintained and installed. However, still it is estimated that more than half of the properties have not taken action to update their systems to meet the requirements of current legislation. Thus, it seems that either information is still not reaching concerned property owners, or then knowledge of the legislation and environmental reasons is not sufficient for making people to act. Rather, it seems like many people are expecting that the legislation will be changed again and that they may not need to do anything finally.

7 Conclusions

While working properly, package plants reached purification requirements easily. To achieve requirements, package plants need proper maintenance and regular observation from users. Regular observation helps to notice problems early and avoid expensive maintenance costs. Variation in purification results is typical for biochemically functioning package plants. If the treatment plant is well maintained, purification results meet the requirements despite the natural variation of biological process and the process recovers faster from occasional disturbance.

The purification results are not automatically similar in same kind of purification plants when they are installed in different households. The purification result depends significantly on how the package plant is used and maintained. Main reason for bad purification results are incorrect installation or wrong settings and lack of maintenance as this study demonstrates. Owners of package plants need more information about the maintenance procedures, and they need to be encouraged to look after their treatment plants by emphasizing advantages they gain by doing so.

Based on experiences of Finnish wastewater legislation, we conclude that it is more important to set requirements for purification plant manufacturers to test and develop their purification systems and for property owners in rural area to correct installation and maintenance than to set strict quantitative purification requirements. This is because a well designed purification system, when correctly installed and properly maintained, is likely to significantly reduce loading of nutrients and organic matter, whereas numeric values for purification requirements may cause confusion in general public and the fact that they are not monitored can give a misleading idea of the legislations' obligations.

Acknowledgments The paper is financed by EU's Central Baltic Interreg IV programme (The Minimization of Wastewater Loads at Sparsely Populated Areas project, MINWA), Maa-ja vesitekniikan tuki ry and Turku University of Applied Sciences. We like to thank Ilpo Penttinen, Hannamaria Yliruusi, Maiju Hannuksela, and Sirpa Lehti-Koivunen for their help during the project life cycle.

References

EU (2000) Water framework directive. Directive 2000/60/EC of the European parliament and of the council of 23 Oct 2000

European Environmental Agency (2001) Eutrophication in Europe's coastal waters. Topic report 7/2001

Garcia SN, Clubbs RL, Stanley JK, Scheffe B, Yelderman JC Jr, Brooks BW (2013) Comparative analysis of effluent water quality from a municipal treatment plant and two on-site wastewater treatment systems. Chemosphere 92:38–44

Government Decree on Treating Domestic Wastewater in Areas Outside Sewer Networks (542/2003). Available in www.finlex.fi

Government Decree on Treating Domestic Wastewater in Areas Outside Sewer Networks (209/2011). Available in www.finlex.fi

Hellström D, Jonsson L (2003) Evaluation of small wastewater treatment systems. Water Sci Technol 48(11–12):61–68

Lehtoranta S, Vilpas R, Mattila TJ (2014) Comparison of carbon footprints and eutrophication impacts of rural on-site wastewater treatment plants in Finland. J Clean Prod 65:439–446

Lowe KS, Tucholke MB, Tomaras JMB, Conn K, Hoppe C, Drewes JE, McCray JE, Munakata-Marr J (2009) Influent constituent characteristics of the modern waste stream from single sources. Final report. Colorado School of mines, Environmental Science and Engineering Division. Water Environment Research Foundation WERF

Naughton O, Hynds PD (2013) Public awareness, behaviours and attitudes towards domestic wastewater treatment systems in the Republic of Ireland. J Hydrol (in press). http://dx.doi.org/10.1016/j.jhydrol.2013.08.049

Nieminen J, Kallio J, Vienonen S (2013) Kiinteistökohtaisen talousveden laatu ennen käsittelyä. Vesitalous 6/2013 (in Finnish)

Putkuri E, Lindholm M, Peltonen A (2013) The state of the environment in Finland 2013. Finnish Environment Institute (in Finnish) (Abstract in English)

Tarasti L (2009) Hajajätevesiselvitys. Ympäristöministeriön raportteja 25. Helsinki 2009 (in Finnish)

Vesihuoltolaitosyhdistys (2012) 29 Vesihuoltolaitosten tunnuslukujärjestelmän raportti 2010. Helsinki 2012 (in Finnish)

Vilpas R, Santala E (2007) Comparison of the nutrient removal efficiency of onsite wastewater treatments systems: applications of conventional sand filters and sequencing batch reactors (SBR). Water Sci Technol 55(7):109–117

Authors Biography

Dr. Piia Leskinen acquired a Masters' degree in Environmental Engineering in the Tampere University of Technology in 2001 and a Ph.D. in Biochemistry in the University of Turku in 2006. She worked in the University of Turku as a senior researcher for five years, before moving to the Turku University of Applied Sciences in 2011. Currently Dr. Leskinen is working as the leader of the Aquatic Systems and Water Management Research team in Turku University of Applied Sciences.

Dr. Juha Kääriä has studied in the University of Turku in Southwest Finland and graduated (Ph.D.) in 1999. His background is fisheries and water biologist. His more than 20 years' working career includes Planning Officer responsible for water and fisheries issues in the environmental office of the town of Turku and Research Manager in Turku Region Water Ltd. From 2004, he has been working in Turku University of Applied Sciences as Research and Development Manager in the Faculty of Technology, Environment and Business. His main duty is to lead a wide environmental expertise program financed mainly by different European Union financial programs.

The Education, Research, Society, and Policy Nexus of Sustainable Water Use in Semiarid Regions—A Case Study from Tunisia

Clemens Mader, Borhane Mahjoub, Karsten Breßler, Sihem Jebari, Klaus Kümmerer, Müfit Bahadir and Anna-Theresa Leitenberger

Abstract The present study analyzes the interrelations of the education, research, society, and policy nexus on sustainable water use and agriculture in semiarid regions of Tunisia. The selected region of Tunisia is one of the most water-stressed regions in northern Africa, strongly exporting fruits and vegetables to European mainland whereas at the same time strongly lacking water resources and reducing production of food for

---

C. Mader (✉) · K. Kümmerer · A.-T. Leitenberger
Leuphana University of Lüneburg, Scharnhorststrasse 1, 21335 Lüneburg, Germany
e-mail: mader@leuphana.de

K. Kümmerer
e-mail: klaus.kuemmerer@uni.leuphana.de

A.-T Leitenberger
e-mail: leitenbe@inkubator.leuphana.de

B. Mahjoub
Institut Supérieur Agronomique de Chott-Meriem, BP 47,
4042 Chott-Meriem, Sousse, Tunisia
e-mail: Mahjoub.borhane@gmail.com

K. Breßler
Institute of Social Sciences, TU Braunschweig, Bienroder Weg 97,
38106 Braunschweig, Germany
e-mail: k.bressler@tu-bs.de

S. Jebari
National Research Institute for Rural Engineering, Waters, and Forestry (INRGREF),
BP 10, 2080 Ariana, Tunisia
e-mail: sihem.jebari@iresa.agrinet

M. Bahadir
Institute of Environmental and Sustainable Chemistry, TU Braunschweig,
Hagenring 30, 28106 Braunschweig, Germany
e-mail: m.bahadir@tu-braunschweig.de

C. Mader
Sustainability Team, University of Zurich, Binzmühlestrasse 14, 8050 Zurich, Switzerland

© Springer International Publishing Switzerland 2015
W. Leal Filho and V. Sümer (eds.), *Sustainable Water Use and Management*,
Green Energy and Technology, DOI 10.1007/978-3-319-12394-3_15

its own growing population. Water scarcity is the major problem in the agriculture of semiarid regions. Along with the population growth, water resources (qualitatively and quantitatively) for food production is exposed to severe strains and has become an important topic for science and politics as well as for the general public in these countries as well as globally. Natural water resources in Tunisia are faced with serious problems related to their quantity and quality (Mekki et al. 2013). Only 8.4 % of the total shallow groundwater has salinity levels that do not exceed 1.5 g/L (Benjemaa et al. 1999). Thus, there is also a lack of fresh drinking water for the population, caused by the extensive use of deep and fossil ground water by agriculture. Due to the lack of conventional water resources, water of marginal quality is used for agricultural irrigation.

Keywords Education · Tunisia · Society · Policy · Semiarid regions

1 Introduction and Methodology of System Analysis

Sustainable development is named as a core principle when it comes to planning of future scenarios for the society, environment, and economy, locally and globally (United Nations General Assembly 2012). The challenge behind is, that working with sustainability contexts is complex, as a huge variety of impact variables, perspectives, values, the present as well as future states need to be reflected and taken into consideration when it comes to find solutions and innovations that guide the way for a sustainable future (Mader 2013).

Water scarcity is a cause for major challenges in the agriculture and life of arid and semiarid regions (El Kharraz et al. 2012; Zeng et al. 2013). The United Nations Environmental Programme names the following major water challenges for Arica that is most relevant for semiarid regions in Tunisia (UNEP 2010: 124):

- Provide safe drinking water
- Provide water for food security
- Meet growing water demand
- Prevent land degradation and water pollution
- Manage water under global climate change
- Enhance capacity to address water challenges

Along with the population growth, water resource management—qualitatively and quantitatively—is exposed to severe strains and has become an important topic for science and politics as well as for the general public. As such, 75.9 % of all water withdrawals are being used by the agricultural sector, 3.9 % by the industry, and 12.8 % by municipalities (UNEP 2010).

Of special relevance to this paper, is the water situation in Tunisia. Tunisia is a northern African country, with a population of 10,778 million and a nominal GDP of 45.611 billion US$ (as of 2012, IMF 2014). According to the International Monetary fund staff estimates, the population will grow within the 7 years of 2011–2018 by app. 1 million (2011: 10.647 millions; 2018: 11.646 millions). The combination of

Fig. 1 Tunisian map, drainage area, and rainfall characteristics. Neighboring countries are Libya in the east and Algeria in the west (after DGRE 1983)

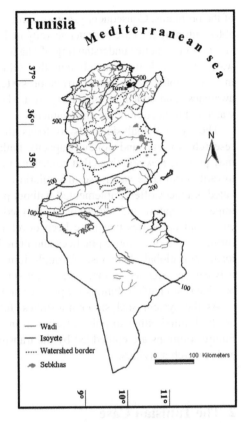

strong population growth and at the same time shrinking agricultural land resources due to salivation and domestication causes strong challenges for the present and future. A schematic map of the country is seen in Fig. 1.

Natural water resources in Tunisia are faced with serious problems related to their quantity and quality. The water availability per capita is about 450 m^3/year (2008), far from the 1,000 m^3 per capita set as threshold value for water scarce countries. Due to the lack of conventional water resources, water of low quality is used for agricultural irrigation.

Following those given challenges in Tunisia and as in semiarid regions, the focus of this paper should be the role of interactions to take place in the nexus of education, research, society, and policy to achieve sustainable solutions. *The hypothesis put forward here is that strengthening exchange and capacity building for sustainable development in general in the nexus of education, research, society, and policy supports the development of essential innovations and transformation for sustainable water use in the whole system of water use in semiarid regions for the present and the future.*

This paper is structured by the following scheme:

First, the problem and today's most conventional ways of solving the problem are briefly outlined. Immediately, the reader might recognize that those solutions might not provide strong future perspectives, but only cure the immediate symptoms of the problems. Consequently, authors provide a system analysis of the impact variables outlining the case of water scarcity in Tunisian agriculture. This system analysis supports a better understanding of challenges that are hidden by the problem of water scarcity. It becomes obvious that solutions do not only lie in the supply of more technologies and better irrigation systems, but are strongly connected to the awareness and education of society as a whole in regard to the sustainable use of water and protection of the environment.

The system analysis leads to focus the paper on the nexus of education, research, society, and policy. Capacity building in the nexus of those variables may provide new solutions that contribute to a holistic approach of transformation toward sustainable water use in semiarid regions. Modern or economically motivated practices often implemented without previous reflection on direct or indirect impacts on nature and society have often led to forget traditional practices, knowledge, and capacities that have been developed over generations. Through transformative research, those practices and knowledge are put in the context of today's local and global demands. Through joint research and knowledge exchange between researchers, farmers, and local stakeholders capacities are being built to provide sustainable solutions for practice and policy.

As the system analysis demonstrates, future solutions of water scarcity cannot be limited either to policy, technical innovations, or costly imports of goods. Future solutions are embedded by the transdisciplinary application of educational and research activities.

2 The Tunisian Case

In Tunisia, many challenges remain for the effective mainstreaming of water management and sustainable agriculture policies within the context of larger social and economic development policies. These include the following: (a) societal solutions, including economic incentives that are not always considered; (b) scientists do not play a sufficient strong role in defining public policies; (c) existing policies mesh poorly with economic development policies, which are further exacerbated by adverse subsidies and inappropriate incentives; and (d) international influences on national "mainstreaming" particularly in the form of development cooperation are typically not molded to the needs of the dry land peoples (Anderson et al. 2013).

Primary policy questions are usually underestimated or overlooked by decision makers. It is essential to find solutions to the current political institutions, and enable them to reverse their tendency of unsustainable behavior. For example, aridity is considered in Tunisia as a "fait accompli" and not an opportunity (Zafar et al. 2007). For that reason, it is difficult to develop strategies that promote investment in arid regions and to convince the government and other stakeholders to do so. Thus, a policy for sustainable agriculture in arid zones must be developed. This would aim, for instance, to optimize the use of water availability and to build on the comparative advantages of agricultural activities in arid lands. On the other hand, an effort should be made to raise environmental awareness and behavior among all citizens, including the sustainable use of water. Thus, the Tunisian authorities must overcome the notion that aridity and water scarcity linked to human consumption are inevitable.

There is often a differential prioritization of environment and development issues in national agendas. Environmental and developmental priorities should be properly harmonized at the national level in order not to become an obstacle to success. Furthermore, environment and development approaches should operate more transdisciplinary with sufficient social analysis and social exchange, and should not be only vertically and sector specifically focused (Zafar et al. 2007).

In Tunisia, national policy remains weakly connected to science. Scientific research activities do not have an appropriate focus on emerging challenges. Relevant scientific knowledge should be disseminated and used in order to plan and achieve national and local policies, laws, regulations, and action programs closely in line sustainable agriculture and water use.

The Tunisian government can incite a reorientation of the existing institutions through sustainable land management. Decision makers should also encourage paying environmental services, particularly in rural areas, for improving sustainable agricultural activities and preventing unsustainable water use. This reorientation could be made possible through a better transparency and accountability of Tunisian governments, the participation of multiple stakeholders, quantifiable results, and follow-up systems (FAO 2013).

Financial support may be required for the implementation of new technologies in semiarid environments because of the generally unpredictable profits and risks of such technologies, and because of institutional constraints such as land property rights issues. Focused financial inducements and deterrents, as well as awareness awakening, can be used to enlighten and educate landowners and land users, and hence let them become more directly involved. Such a commitment can lead to the conception and spreading of interventions that could be understood and streamlined by the local population.

Enhancing knowledge and understanding for people of how national factors can impact locally—and vice versa—is important. Issues related to lack of awareness of the fragility of the natural resource base should be examined more intensely at the national level, in order to reduce obstacles to the development of the core program and strategy in Tunisia.

Education and capacity building of local populations and policymakers should receive high priority. Mainstreaming sustainable water use and agriculture requires the capacity building, education, and better communication among local populations but also of policy makers (Scoullos 1998). For example, in Tunisia, the large deforested areas had resulted in serious land erosion. One lesson learned from the subsequent reforestation intervention demonstrated the value of incorporating activities to address the economic needs of the local population; this resulted in successful and sustainable programs.

Education for Sustainable Development (ESD) aims to balance human and economic welfare and nature for present and future generations with cultural values and respect for the environment. Besides, ESD empowers people to develop the appropriate knowledge and skills; to adopt attitudes and values, and shape behaviors in order to assume responsibilities for a sustainable future (Scoullos and Malotidi 2005). Higher education institutions (HEI) have a special responsibility to provide leadership on ESD. Indeed, HEI as facilities of interlinked education and research have the mission to promote development through research and teaching, disseminating new knowledge and insight, and building capacities of their students. As HEI educate and train decision makers, they play a key role in building sustainable societies (Mahjoub 2012). Some of the main key questions for Tunisian HEI are as follows:

- How the graduate will contribute to achieve sustainable water use, not as SD "specialist" but as a doctor, lawyer, teacher, journalist, chemist, etc.?
- How educational programs will impact the society of the 2050/consequences of the wrong short-sighted approaches and decision of yesterday and today?
- What is the quality and value of educational programs and skills of educators/ how it should be /how to improve it?
- How to shift from theory to specific actions (implementation)?

Furthermore, solid knowledge on science, technology, and economics is needed, but it is not enough. Understanding human behavior, social structures, culture, and cultural differences is critical when it is aimed to reach sustainable development (Scoullos and Malotidi 2005). Tunisia needs to pay attention to social and cultural sustainability. Without doing it, the global investment on environmental and economical sustainability will be lost. The recognition of practices, identity, and values plays a considerable role in setting directions and building commitments. It is important to investigate Tunisian's perceptions toward the environment and to meet with the current perceptions and values before setting up environmental education programs. HEI should contribute to social and cultural sustainability, and from that view point provide awareness, skills, and knowledge to solve the problems of environmental challenges (Mahjoub 2012).

3 Water Strategies and Participatory Approach

Throughout thousands of years, farmers have developed practices that quantitatively serve the agricultural production needs of the local population and at the same time do not harm the environment. These modes of agricultural production and water management are based on traditional knowledge. The latter fits perfectly into the geographic and social context of arid and semiarid areas. Population growth, internationalization of the food market, the use of chemical fertilization as well as the strong environmental pollution by industry and the population have caused tremendous challenges in qualitatively and quantitative water resource management. Consequently, the inherited traditional hydraulic systems, which were originally managed by farmers and the rural society, have obvious problems with keeping their traditions. They have often been affected by limited financial measures, the lack of technical references, the absence of attendance and up keeping, the disorganization of production units, and the insufficient profitability of the proposed technologies and of the socioeconomic neglects.

The Tunisian water development model that has been carried out since the 1960s did not consider and involve farmers. The state had acted during the last five decades regardless of participation and aspirations of the beneficiaries. The latter left their lands and their traditional ways, hydraulic structures, and know-how management heritage for other sectors (like industry and tourism) (El Amami 1984; Ennabli 1993). This context eventually reproduced many dependencies and has nourished more "spirit of assistance," and introduced some regulation of unemployment and labor market employment in the regions (PNUD-FAO 1991). The state has continued to treat the rural world with a spirit of support aggravating the context of their marginalization and its depletion.

During the past decades, the proliferation of state institutions has prolonged the process that deconstructs, marginalizes, or gets rid of the traditional social institutions, which amounted to the task of organizing, coding, decision making, and participation of the community in the preservation and conservation of natural resources. Consequently, the problem that has arisen today is how to reinitiate an association and implementation of agricultural and rural populations to the imperatives of appropriate water management measures. Especially, agricultural and rural experiences with participatory development approaches are rare. The ones that are available had rather limited success (Jebari and Berndtsson 2013).

The university, society, and policy nexus of sustainable water use and agriculture in semiarid regions.

As outlined above, the socio-environmental system of sustainable water use in Tunisia consists of a diversity of variables that are relevant and of which the interrelations need to be considered by stakeholders, be it on local scale of villages, the national policy scale, or either international development cooperation or UN and global policy scale.

In Fig. 2, 19 core impact variables are presented within a system grid. This system grid is the result of an activity (x-axis) and sensitivity/passivity (y-axis) analysis through which the impact variables have been assessed in the course of an impact matrix. The impact of each of the variables on each other and subsequently their sensitivity of being impacted by one another has been assessed.[1]

In this system grid, 19 indicators provide a holistic picture for the analysis of the sustainable water use in semiarid regions of Tunisia. After assessing the active and passive impact, system variables have on each other (0... no impact, 1 light or indirect impact, 2 strong and direct impacts), results are being transformed from an impact matrix into the shown system grid to demonstrate the variables' role in the system according to their activity and passivity. The activity score results

---

[1] In the course of a fact finding mission supported by the DAAD, the German academic exchange service, researchers from Germany and Tunisia met in Tunis in December 2013 to analyze the current situation of sustainable water use in agriculture of Tunisia.

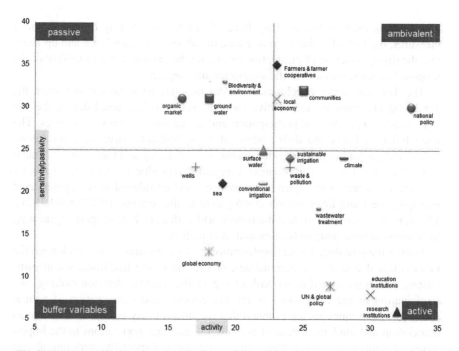

Fig. 2 System grid for sustainable water use in semiarid regions

from the sum of impact points the variable has on other variables. The passivity score results from the sum of impact points other variables have on the variable. According to the system grid, we can define four different groups of variables as they are separated by lines representing the mean activity and sensitivity/passivity scores (activity mean = 23; passivity mean = 25) (Scholz and Tietje 2002).

Ambivalent variables: The variables "national policy," "communities," "farmers," and "local economy" are considered above average in both sensitivity and activity, which places them in the ambivalent quadrant. This analysis demonstrates the strong relevance of those variables. It shows that for any upcoming system relevant actions the role of national policy, the needs, values and experience of communities, farmers as well as the local economy need to be considered, reflected, and involved.

Active variables: The five variables "UN/global policy," "education," "research," "wastewater treatment," "waste and pollution," "climate," and "sustainable irrigation" shown in Fig. 1 are considered above averages in activity and below averages in passivity; they are located in the active quadrant. As a consequence, it needs to be recognized that education, research, and global policy have a strong impact on either ambivalent or passive and buffer variables. Future strategies for sustainable development need to consider their central role for development. Aspects of waste and pollution, wastewater treatment, climate as well as sustainable irrigation have shown a strong impact on the system. Subsequently, their effects need to be considered for future strategies.

Passive variables: In contrast to this, four variables stand in the passive quadrant: "Organic market," "ground water," "surface water," and "biodiversity and environment." If one wants to change the future conditions of ground water, biodiversity, the environment as well as organic markets that contribute to healthy society and living local economy, the ambivalent and active nexus variables of society, education, research, and policy need to work together to develop sustainable innovations.

Buffer variables: Finally, five variables as "global economy," "conventional irrigation," "sea," and "wells" are called buffer variables because they are below average in both, activity and passivity/sensitivity.

Pointing out the impact variables representing the nexus, one experiences from this system grid, that education, research, and UN/global policy have a strong active role in the system and national policy, farmers, local community, and local economy have an ambivalent role which means they are either very active in impacting the whole system or and at the same time have a strong sensitivity by being affected by other variables.

The strong system activity of the nexus variables are education, research, society (farmers and local community) as well as policy (national and global policy/UN), consequently representing leverage points for the whole system of sustainable water use in semiarid regions, taken the representative case of sustainable agriculture and water use in Tunisia.

In the following system graph (Fig. 3), the role of the nexus system variables as leverage points becomes even more obvious:

The system graph (Fig. 3) shows that national policy through, e.g., introducing groundwater injection of treated wastewater has an immediate impact on the ground water (C4) and farmers, but would need to take place in combination with community involvement, research, and educational activities so to have a transformative effect on the whole system. Community needs to be aware on the positive and negative effects this technology accompanies. Surely, groundwater injection on the one hand may stabilize the ground water level on a very local scale, but on the other hand likely contaminates the aquifer with micro-pollutants as well as with high level of nitrate and microorganisms. So its consequences on the environment, soil fertility, quality of food, the biodiversity, and the community are hardly predictable. Through involvement of all nexus "parties," the challenge can be analyzed systematically and alternative solutions can be developed that tackle and transform the whole system toward sustainability.

In the nexus of education, research, policy, and society, the role of education and research is to reflect societal (in this case: community and farmers) needs, to take up experiences and knowledge that exist in the society, and that might almost have been forgotten (traditional knowledge) in research. This cocreative and transformative approach is called transdisciplinary research and education (Pohl 2008). Together with policy and society, transdisciplinary education and research build the sustainability nexus and contributes to the development of transformative solutions. Those solutions have long-term perspectives and are developed through

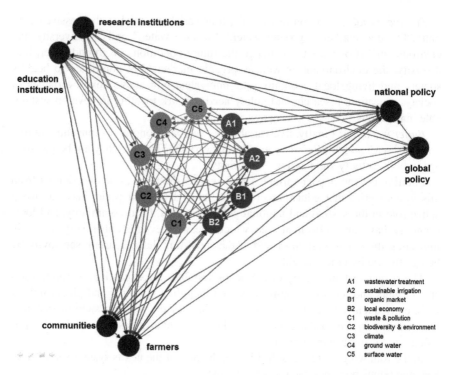

Fig. 3 System graph of education, research, society, and policy nexus in sustainable water use and agriculture in semiarid regions (own figure)

shared visions, responsibility, and agency. Society, policy, education, and research as well as economy need to change behavior and to become aware of the impacts their habits have.

Transformation toward sustainable development implies a shared system understanding, vision for the future, and agency for ones own responsibility, being part of the system (Mader et al. 2013). It is the policy responsibility to establish the necessary frameworks, so capabilities for education and research institutions as well as the society are available to take and to support agency of each individual and collective. Those necessary frameworks might be for example to implement global policy recommendations and programs like the UN Decade on ESD. It is not a coincidence that the UN has promoted the years of 2005–2014 as a decade on ESD and will follow this strategy in the course of the global action program on ESD with the aim of embedding sustainability competences into the scopes of all formal and nonformal educational institutions (UNESCO 2013). Through ESD, embedded from kindergarten up to universities and lifelong learning facilities, learners from all ages may acquire the necessary competences to take agency for sustainable development. This again implies system understanding as well as future envisioning, and reflective agency in multi-stakeholder environments (UNECE 2012).

4 Case Study: Combining Traditional Farmers Knowledge, Research, and Policy for Sustainable Development

Traditional knowledge is nothing else than techniques and practices that passed on through generations. It considers appropriate use of natural resources, allows protecting ecosystems, and plans sustainable agriculture. It was shown that innovative solutions can be driven from this indigenous know how. It is a dynamic knowledge that has always been able to renew and to adapt itself allowing societies to produce for the long-term benefit of the community while managing resources and environment in balance (UN 1992; in Agenda 21).

The World Bank promotes traditional knowledge as advanced innovative techniques appropriate to enhance local resources, to support diversity, and to promote human creativity. Based on research, several international organizations (e.g., UN, FAO, UNCCD, UNESCO, UNEP, and OECD) have confirmed the validity of the traditional knowledge and recognized its contribution to science and technology (TKWB 2007). Nowadays, most NGO's promote traditional knowledge as a new approach to international development and cooperation.

During hundreds of years, indigenous Tunisian people have traditionally harvested water and grown crops on sloping mountain valleys and harsh dry lands. They acquired and continuously developed innovative techniques and systems for efficient small-scale water management. These systems that are called nowadays water harvesting techniques and traditional hydraulic systems are scattered throughout Tunisia and have different shapes and specific characteristics according to the bioclimatic prerequisites (El Amami 1984).

Research efforts on better use of traditional hydraulic systems started from the 1960s at the CRGR (Research Center for Rural Engineering, currently the National Research Institute for Rural Engineering, Water and Forestry: INRGREF). Their impact on runoff, infiltration, and soil loss characterized the 1970s period through an experimental program investigating specific techniques. Recent and ongoing modeling work aims at defining the role of different hydraulic systems in providing blue and green water[2] for improved agricultural productivity to ensure sustainable rural development. The output of the latter research is crucial for setting suitable future water resources strategies (Jebari et al. 2014). In fact, the Tunisian authority that has based water development sector on mobilization policy and projects through large reservoirs is nowadays facing serious problems in managing limited water resources at regional scale. All the observed difficulties seem to be the consequences related to the absence of balanced hydraulic schemes at catchment level during the last five decades. In fact,

---

[2] Blue water: the fraction of water that reaches rivers directly as runoff or, indirectly, through deep drainage to groundwater and stream base flow.
Green water: is that fraction of rainfall that infiltrates into the soil and is available to plants.

considering simultaneously the large hydraulic projects managed by the state and the water harvesting techniques, which are in the responsibility of the farmers, would have certainly created a more sustainable agricultural context, less vulnerable rural society, and better availability of the water resources. Finally, combining modern and traditional hydraulic knowledge is becoming crucial to ensure a sound development (Berndtsson et al. 2014).

In Tunisia as well as in other semiarid regions, tackling the challenge of increasing water use in growing agriculture and increasingly polluted environments, people from all backgrounds need to learn to take responsibility while responding on others' values and perspectives.

5 Conclusions

Facing the challenges of population growth, environmental pollution, salinization of agricultural lands, increasing water scarcity, and climate change, Tunisia has to tackle huge challenges in the future. Immediate actions are required to prevent further noninverting damages to the natural resources of Tunisia. The paper has shown that such as actions need to be considered through cocreation of the whole nexus of education, research, society, and policy. And, only if this nexus works together on sustainable solutions, the growing problem of water scarcity can be tackled.

Future prospects:

Implementation of ESD in all formal and nonformal learning environments supports the establishment of a holistic system understanding of the individual and collective impact on the quality and quantity of water resources. ESD transforms the behavior of people toward a more conscious interaction with the natural resource of water and the environment. The consequences would be less environmental pollution and a reduction of water consumption through the use of, e.g., sustainable irrigation systems in agriculture and reduction of pollution through industry.

Research needs to adopt *transdisciplinary methods* to work together with society in the development of solutions and innovations for sustainable agriculture and sustainable water use.

Society including farmers, local economy, and communities need to *strengthen the market of sustainable agriculture*. Transparency in production, communicating the negative impacts of fertilizers on ground water as well as promoting the production and consumption of organic food, irrigated through sustainable systems could cause mind shifts among the community toward more conscious consumption as well as open up new business opportunities for farmers and the local economy competing the global market.

Policy needs to provide the adequate legal framework to enable education institutions, research, farmers, and community to make use of their capabilities in becoming agents for change toward sustainable development. Those requirements include the following:

- The development of a national ESD strategy reflecting the national sustainability challenges as mentioned above.
- Adopting quality criteria of research and incentivize transdisciplinary water research and higher education.
- Establishing legal frameworks to limit pollution by industry, farmers as well as citizen and incentivize organic production and sustainable irrigation modes, so society and farmers get easier access to sustainable products.

Finally, it is the responsibility of each individual to change the perspective from short term to long term and from problem orientation to system orientation to balance the human, economic, and environmental development of the country.

Acknowledgments The authors thank the German Academic Exchange Service for the funding of the first German-Tunisian fact finding mission on "Sustainable Agriculture in Semi-Arid Regions" SASAR held in December 2013 in Tunis, Tunisia. S. Jebari acknowledges helpful funding from the European project BeWater (Making society an active participant in water adaptation to global change) BEWATER project is funded by the European Commission, 7th Framework programme, Science in Society, Grant agreement Nr.: 612385.

References

Anderson J, Roseboom J, Weidemann Associates Inc (2013) Towards re-engaging in supporting national agricultural research systems in the developing world. USAID, Washington, DC

Benjemaa F, Houcine I, Chahbani MH (1999) Potential of renewable energy development for water desalination in Tunisia. Renew Energy 18:331–347

Berndtsso R, Jebari S, Hashemi H, Wessels J (2014) Traditional water management techniques—do they have a role in post Arab Spring Middle East? Hydrol Sci J forthcoming

DGRE (Direction Générale des Ressources en Eau) (1983) Carte du réseau hydrographique Tunisien, échelle: 1/1 000 000. Reproduction de l'Office de la Topographie et de la Cartographie, Tunis

El Amami S (1984) Les aménagements hydrauliques traditionnels de Tunisie. Centre de Recherche du Génie Rural. Ministère de l'Agriculture, Publication of the Research Center of Rural Engineering, Tunis, Tunisia

El Kharraz J, El-Sadek A, Ghaffour N, Mino E (2012) Water scarcity and drought in WANA countries. Procedia Eng 33:14–29

Ennabli N (1993) Les aménagements hydrauliques et hydro-agricoles en Tunisie. Institut National Agronomique de Tunis. Ministère de l'Agriculture et des Ressources Hydrauliques, Tunis

FAO (2013) SAFA—Sustainability assessment of food and agriculture systems. FAO, Rome

IMF (2014) Report for selected countries and subjects: Tunisia. Retrieved on 3 Mar 2014 from International Monetary Fund: http://www.imf.org/external/pubs/ft/weo/2013/01/weodata/we

orept.aspx?sy=2011&ey=2018&scsm=1&ssd=1&sort=country&ds=.&br=1&c=744&s=
NGDPD%2CNGDPDPC%2CPPPGDP%2CPPPPC%2CPPPSH%2CTM_RPCH%2CTX_RP
CH%2CLUR%2CLP&grp=0&a=&pr1.x=54&pr1.y=13

Jebari S, Berndtsson R (2013) Tunisian water resources policy at a cross road. Study presented at
the ISA Panel: water justice and stakeholder participation. 54th ISA annual convention. San
Francisco, California 3–6 April

Jebari S, Berndtsson R, Bahri A (2014) Traditional water collecting systems in Tunisia, challenges and way forward. Middle East Critique (forthcoming)

Kanzari S, Hachicha M, Bouhilila R, Battle-Sales J (2011) Characterization and modeling of
water movement and salt transfer in a semi-arid region of Tunisia (Bou Hajla, Kairouan)—
salinization risk of soils and aquifers. Comput Electron Agric 86:34–42

Mader C (2013) Sustainability process assessment on transformative potentials: the Graz model
for integrative development. J Clean Prod 49:54–63

Mader C, Scott G, Razak D (2013) Effective change management, governance and policy for sustainability transformation in higher education. Sustain Acc, Manage Policy J 4(3):264–284

Mahjoub B (2012) Environmental education for sustainable development in Tunisian
Universities: raising awareness and improving skills. In: International conference on emerg-
ing pollutants in the mediterranean basin—setting the bridges. EMPOWER project, DAAD
Hammamet, Tunisia, 10–16 Sept 2012

Mekki I, Jacob F, Marlet S, Ghazouani W (2013) Management of groundwater resources in relation
to oasis sustainability: the case of Nefzawa region in Tunisia. J Environ Manage 121:142–151

PNUD-FAO (1991) TUN. 86-020- Rapport d'évaluation sur les techniques de CES en Tunisie.
Direction de Conservation des Eaux et des Sols. Ministère de l'Agriculture et des Ressources
Hydrauliques, République Tunisienne

Pohl C (2008) From science to policy through transdisciplinary research. Environ Sci Policy
11(1):46–53

Scholz R, Tietje O (2002) Embedded case study methods—integrating quantitative and qualitative knowledge. Sage Publications, Thousand Oaks

Scoullos MJ (1998) Environment and society: education and public awareness for sustainability.
In: Proceedings of the Thessaloniki international conference. UNESCO and Government of
Greece, 8–12 Dec 1997

Scoullos M, Malotidi V (2005) Manuel sur les méthodes utilisées pour l'Education a
l'Environment et l'Education pour le Dévelopement Durable. MIO-ECSDE, Athens, Greece

TKWB (2007) International centre for traditional knowledge against desertification and for a
sustainable future in the Euro-Mediterranean. Retrieved on 7 March 2014 from Traditional
Knowledge World Bank: http://www.tkwb.org/web/?page_id=4&language=it

United Nations (1992) Agenda 21, United Nations Conference on Environment & Development,
http://sustainabledevelopment.un.org/content/documents/Agenda21.pdf, accessed 17.10.2014

UNECE (2012) Learning for the future—competences in education for sustainable development.
United Nations Economic Council for Europe, Geneva

UNEP (2010) Africa water atlas, division of early warning and assessment (DEWA). United
Nations Environmental Programme, Nairobi

UNESCO (2013) 37 C/57 proposal for a global action programme on education for sustainable
development as follow-up to the united nations decade of education for sustainable development
(DESD) after 2014. United Nations Educational Scientific and Cultural Organization, Paris

United Nations General Assembly (2012) A/res/66/288 the future we want. UN, New York

Zafar A, Borgardi J, Braeulel Ch, Chasek P, Niamir-Fuller M, Gabriels D, King C, Knabe F,
Kowsar A, Salem B, Schaaf T, Shepherd G, Thomas R (2007) Overcoming one of the greatest environmental challenges of our times: re-thinking policies to cope with desertification. A
policy brief based on the joint international conference: desertification and the international
policy imperative, Algiers, Algeria. United Nations University International Network on
Water, Environment and Health, Hamilton, Canada, 17–19 Dec 2006

Zeng Z, Liou J, Savenije H (2013) A simple approach to assess water scarcity integrating water
quantity and quality. Ecol Ind 34:441–449

Authors Biography

Clemens Mader Post-doctoral research associate, UNESCO Chair in Higher Education for Sustainable Development, Leuphana University of Lüneburg, Germany; Sustainability Team at University of Zurich (UZH), Switzerland

Borhane Mahjoub Assistant Professor of Environmental Chemistry, Higher Institute of Agronomy of Chott-Meriem, University of Sousse, Tunisia.

Karsten Breßler MA, Institute of Social Sciences, Technische Universität Braunschweig, Germany.

Sihem Jebari Researcher, National Research Institute for Rural Engineering, Water and Forestry, Tunisia.

Klaus Kümmerer Full Professor, Chair for Sustainable Chemistry and Material Resources, Institute of Sustainable and Environmental Chemistry, Leuphana University Lüneburg, Germany.

Müfit Bahadir Full Professor, Institute of Environmental and Sustainable Chemistry, Technische Universität Braunschweig, Germany.

Anna-Theresa Leitenberger Research Associate, Leuphana University Lüneburg, Germany.

Planning Under Uncertainty: Climate Change, Water Scarcity and Health Issues in Leh Town, Ladakh, India

Daphne Gondhalekar, Sven Nussbaum, Adris Akhtar and Jenny Kebschull

Abstract Access to safe drinking water is already a very serious issue for large urban populations in fast-growing economies such as India. This is further being impacted by climate change, leading to increase in water-related diseases. In regions where water is already scarce, integrated urban planning especially of water resources in conjunction with other sectors such as energy and taking health into consideration is urgently needed. The case study Leh Town, the capital of the Ladakh Region, is located in an ecologically vulnerable semi-arid region of the Himalayas and is undergoing very rapid transformation due to tourism and economic growth. Huge increase in water demand coupled with inadequate water supply and wastewater management are augmenting already serious environmental issues. In 2012–2013, we mapped point sources of water pollution using geographic information systems (GIS), analysed medical data and conducted questionnaire surveys of 200 households and ca. 300 hotels and guesthouses. Our study finds that occurrences of diarrhoea in Leh seem to have increased in the past decade, which may be related to groundwater pollution. Further, over 80 % of the water demand is currently being supplied from groundwater resources without regulation, so that these may be being depleted faster than their rate of recharge. This study discusses using GIS to support urban planning decision-making and advocates a partially decentralized sewage system for water resources conservation in Leh.

D. Gondhalekar (✉) · A. Akhtar
Centre for Urban Ecology and Climate Adaptation (ZSK), Technische Universität München, Arcisstr. 21, 80333 Munich, Germany
e-mail: d.gondhalekar@tum.de
URL: http://www.zsk.tum.de/.

D. Gondhalekar
Center for Development Research (ZEF), University of Bonn, Bonn, Germany

S. Nussbaum · A. Akhtar · J. Kebschull
Center for Remote Sensing of Land Surfaces (ZFL), University of Bonn, Bonn, Germany

© Springer International Publishing Switzerland 2015
W. Leal Filho and V. Sümer (eds.), *Sustainable Water Use and Management*,
Green Energy and Technology, DOI 10.1007/978-3-319-12394-3_16

Keywords Urban planning · Water resources management · Health–climate change · Geographic information systems (GIS) · India

1 Introduction

Rapid urbanization in developing economies such as India is inducing water-related environmental challenges (Marcotullio 2007) as urban water infrastructure planning is often unable to keep up with the pace of development. Resulting lack of access to safe drinking water and adequate sanitation is increasing water-related health risks (Galea and Vlahov 2005), which are further exacerbated by climate change (Vörösmarty et al. 2000). In particular, in regions where water is already scarce, integrated urban planning especially of water resources in conjunction with other sectors such as energy and taking health into consideration is urgently needed.

Health issues do not directly drive urban design, but they did provide the original impetus for the urban planning profession: the discovery in nineteenth-century London that cholera is a waterborne disease, for example, and that it was spreading from one particular contaminated water pump had huge implications for urban planning. Thus, urban design is considered a powerful tool for addressing new public health concerns (Jackson 2003a, b), but new frameworks linking public health and urban planning are needed (Naess 2006) to address contemporary challenges. Studies on the relation between the built environment and health are often confined to certain academic fields, making results difficult to share (Dannenberg et al. 2003). Further, such studies tend to focus on developed country contexts rather than developing countries like India. More cross-disciplinary (Jackson 2003a, b) and international research collaboration (Bork et al. 2009) as well as new approaches (Butsch et al. 2012) are needed to tackle complex water and health issues more effectively.

In India, although one of the earliest examples of public sewerage was found in the ancient Indus Valley (Jha 2010), only 16 % of the urban population today have access to adequate sanitation resulting in large-scale open defecation and thus ground and surface water pollution (WHO and UNICEF 2006). Under similar conditions in nineteenth-century Europe, centralized drinking water supply and sewerage systems proved very effective in curbing water-related diseases and improving public health. However, centralized sewage systems are very water intensive, and expensive to construct and maintain, and energy intensive to operate. Therefore, in regions where water is scarce and where urban areas are facing large-scale development pressures, centralized sewage systems may not be the most appropriate option in terms of water resources conservation. Thus, decentralized sewage systems are increasingly being recognized as a way to help conserve water resources (Lüthi et al. 2011). Although these have various inherent advantages such as the opportunity for nutrient recovery and lower maintenance cost (Tilley et al. 2008), they have rarely been implemented successfully (Sanimap 2009). Instead, the flush toilet and centralized sewage system, which has been termed "ecologically mindless", remains a preferred option (Narain 2002) and a symbol of "modernity".

In order to illustrate the challenges and opportunities in a development context such as India to implement a decentralized sewage system, we chose a case study where large-scale urban transformation is taking place in a water-scarce region, which could be a lighthouse example for alternative, innovative and more sustainable future development choices in terms of water and health.

2 A Case Study Town in the Desert: Potential Pilot for "Ideal" Sustainable Development

Our case study, Leh Town (hereafter Leh), is located in a remote semi-arid region in the Himalayas at an altitude of 3,500 m above sea level. Adjoining a dense historical town centre, Leh's urban area is spread throughout a green oasis, a valley of agricultural fields and groves of trees watered through a dense network of streams fed by glacial and snow melt water, surrounded by a desert landscape (Fig. 1). This intricate cultural landscape is the product of hundreds of years of very careful management of these limited and also often variable water resources,

Fig. 1 Geographical location and cultural landscape of Leh

on which many studies have been conducted (Angchok and Singh 2006;
Mankelow 2003; Laball 2000; Bhasin 1997; Tiwari and Gupta 2003). With such
a water management system enabling food and social security, Leh was a model
traditional agricultural irrigation society until only a few decades ago (Norberg-
Hodge 1991).

Today, traditional water management and agricultural practices are increasingly
making way to changing lifestyles and alternative sources of income especially
due to the rapidly growing tourism industry. Leh, the capital and cultural centre of
the Ladakh Region of Jammu and Kashmir State, is considered one of the fastest
growing small towns in India (Rieger-Jandl 2005: 124). Ladakh is a semi-autonomous region of India governed by the Ladakh Autonomous Hill Development
Council (LAHDC). Leh has a population of 30,870 (Census of India 2011). In
addition, more than 40,000 army personnel live in Leh (Skeldon 1985) and several
tens of thousands of migrant workers come to Leh every year.

Located in a remote region of India close to the borders of China and Pakistan,
Leh has only been open for tourism since 1974. Since then, the number of tourists
visiting Leh has risen exponentially, especially in the last decade: in 2012, there
were 179,000 tourists (Fig. 2), several times more persons than the local population. The vast majority of these tourists visit Leh in summer between April and
October because the winters are too harsh for most. In order to accommodate these
tourists, there has been a huge increase in hotels and guesthouses in Leh. Tourist
accommodations are increasingly building en suite bathrooms with flush toilets
and showers to enhance their attractiveness and thus income from tourism. Leh
does not have a sewage system, and hotels and guesthouses dispose of wastewater mainly through septic tanks and soak pits that are not being properly managed
according to our study. Therefore, we posit that the huge increase in tourist accommodations and the ensuing increase in wastewater may pose a human health risk
as the aquifer underlying Leh, which is fed by glacial and snow melt water, is used
for drinking water and may be polluted due to seepage.

In fact, Leh is almost wholly dependent on glacial and snow melt water. Rain in
the region is negligible and plays an insignificant role in the local water cycle due
to the high rate of evapotranspiration, a result of dry air and intensive solar radiation. Occasionally, cloudbursts occur in Ladakh which cause flash floods because
the landscape is so dry and without much vegetation, so that it cannot retain any

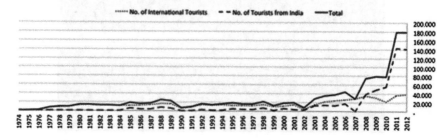

Fig. 2 Year-wise number of visitors to Leh (*Source* Leh Tourist Board)

water. Leh was hit by such a flash flood in 2010, which killed around 200 people and caused large-scale destruction. Surface water from glacial run-off also seems to be decreasing (Eichert 2009: 53) possibly due to climate change.

In our study, we set out to understand how urbanization processes have been affecting human and environmental health in Leh especially in terms of water resources management. We focused mainly on the last decade, when the largest increase in tourists took place. Our aim is to find out whether there are opportunities in Leh for integrating various sectors related to urban planning, like drinking water supply, wastewater and solid waste management, energy, and tourism infrastructure, in order to address human and environmental health issues in a comprehensive manner. We ask the question, could traditional wastewater management practices potentially hold a key to addressing water-related sustainability issues in Leh?

3 Rapid Urbanization and Human Health: Diarrhoea is a Common and Serious Health Risk

One of the leading water-related diseases, diarrhoea, is already a major public health concern in developing countries such as India, to which children are especially vulnerable: diarrhoea accounts for 16 % of deaths of children under 5 years of age globally, one-third of which occur in India (White Johansson and Wardlaw 2009: 5–7). Diarrhoea can have a number of causes, but the main transmission route is through drinking water (Howard and Bartram 2003; Sakdapolrak et al. 2011: 88), and nearly all deaths due to diarrhoea worldwide could be prevented through access to safe water, adequate sanitation and good hygiene (White Johansson and Wardlaw 2009: 10–13).

An increase in waterborne diseases such as hepatitis and diarrhoea was already reported in Leh over a decade ago (Bashin 1999). However, so far, no comprehensive study exists on incidences of water-related diseases and their potential causes in Leh. Our study tries to address this gap.

We were able to procure data on acute diarrhoea from the chief medical officer (CMO) for the whole of Leh District, which includes Leh Town and several surrounding villages, from 2001 to 2012. These data are hundreds of handwritten sheets in folders sorted by year, which we photographed at the CMO's office and then digitalized. There is a data gap between 2008 and 2010: apparently cows ate some of the folders during the renovation of the archive where they were being stored a few years ago. In addition, we conducted a socio-economic questionnaire survey of 200 households in Leh selected at random, representing 5 % of all households.

The CMO's data show that a significant portion of the population, namely over 10 %, seems to be affected by acute diarrhoea on a yearly basis. Further, when looking at the monthly occurrences of acute diarrhoea over the past decade, the

dashed trend line is following an upward course that seems to suggest an overall increase with peaks in the summer months (Fig. 3). However, these data also
include tourists, who are susceptible to diarrhoea in Leh due to unfamiliar bacteria and altitude sickness. Nonetheless, the data also show that over 10 % of the
under-5-year-olds in Leh seem to be affected by acute diarrhoea on a yearly basis.
Here, we assume that most small children are locals and not tourists because we
observed relatively few tourists in Leh accompanied by small children.

The figures seem high, but we cannot compare them because in a country like
Germany, for example, statistical data on a health issue like acute diarrhoea are not
collected unless there is an epidemic. Even if data were collected, definitions of
what constitutes acute diarrhoea may differ. Often, in developing countries, diarrhoea is not regarded as a health issue: because it is so common, it tends to be
regarded as a fact of life. The CMO's data only cover persons who visited government health institutions such as hospitals and clinics in Leh to see a doctor for
acute diarrhoea. Hence, we assume that the number of people suffering from diarrhoea in Leh but who do not consult the doctor is actually much higher. In addition
to being a serious health risk, incidence of diarrhoea also has economic implications as it can rend people unfit to work or reduce their working capacity.

Unfortunately, we found that it is difficult to gather information on diarrhoea
from the local population: culturally, it is a sensitive topic to talk about. Further,
measuring water quality to try to establish a causal connection between water pollution and diarrhoea incidences was beyond the scope of our project. Therefore,
we decided in our household survey to focus on people's perception of water
quality in Leh and its perceived impact on health and potential impact on water
consumption practices. Our hypothesis is that regardless of actual water quality, perception of it will influence how people consume it: for example, a rumour
about bad water quality of a particular well may stop people from using it without
knowing whether and why the water is polluted, and vice versa, the rumour may
well be based on experience values made by the local population.

We found that although 98 % of households thought that drinking water quality
is safe in Leh, 53 % of households thought drinking water quality today is worse

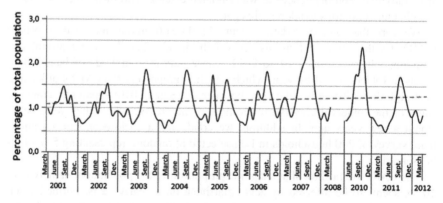

Fig. 3 Incidence of acute diarrhoea in Leh district

than 10 years ago. 34 % of households reported problems with their drinking water in terms of it having a strange smell, taste or colour. Lack of adequate wastewater management and treatment, i.e. septic tanks or soak pits, were thought by 26 % of households to be the main source for groundwater pollution. The local population also perceived increased use of chemical fertilizers over the past decades in agriculture as a water quality threat. 34 % of households thought diarrhoea is related to drinking water pollution. Thus, this study found drinking water pollution to be a serious concern of the local population.

4 Growth of the Tourism Industry: Implications for Food and Social Security

In order to accommodate the huge and increasing numbers of visitors, there has been a dramatic increase in the number of guesthouses and hotels in Leh in the past decades: in the 1980s, there were only 24 guesthouses and hotels in Leh, but by 1990 there were 62, by 2000 there were 117, by 2010 there were 282, and just from 2010 to 2012, the number had increased to ca. 360 guesthouses and hotels in business, with another ca. 60 not yet in business or under construction. Of 21 wards in Leh, 10 have agricultural land, while the others are predominantly desert like. We found that over 90 % of guesthouses and hotels in Leh are located in wards with agricultural land area (Fig. 4).

Only 40 % of Leh's population lives in the agricultural wards although these cover 1,358 ha compared to 535 ha for non-agricultural wards. In the agricultural wards, houses tend to be traditional Ladakhi multi-generation one-family clay-construction houses, many of which have been converted into guesthouses, and in non-agricultural wards, buildings are concrete slab constructions. 36 % of all households originate in Leh, of which 85 % live in agricultural wards, while for other households originally not from Leh, only 31 % do so.

In order to measure land use change in the two wards with the highest rate of urbanization over the past decade, we compared a high-resolution satellite image of 2011 with a Google Earth image of 2003. As Leh is located on a slope, individual fields are clearly visible in satellite imagery because each field is circumscribed by a stone wall which allows farmers to flood and thereby irrigate the individual fields. After digitizing each agricultural field, we compared the two images to see which areas that had formerly been agricultural land had been turned into built-up area.

We found that in these two wards, 14 % of agricultural land had been transformed into built-up land in the past decade. In addition, we found that in the same two wards at least 30 % of the land that was used for agriculture (32 of 97 ha) has fallen barren within only the last decade, between 2003 and 2011 (Fig. 5). Due to the property rights in Leh, fields are rarely divided up. Therefore, we hypothesize that when a household formerly active in agriculture constructs a guesthouse on an agricultural field or converts an existing house into a guesthouse, the income

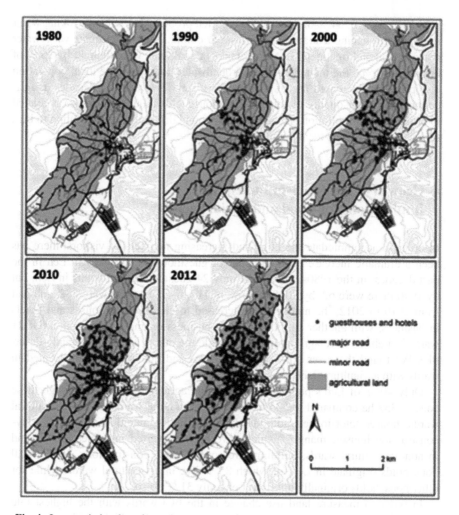

Fig. 4 Increase in hotels and guesthouses since 1974

from the guesthouse may make any income from agriculture redundant. Thus, the agricultural land area, even if only partially covered by the new guesthouse, may be left barren.

The trend of decreasing agricultural activity is also visible in the results of the household questionnaire survey. We found that average income in Leh has doubled in the last decade. Overall, agriculture was a source of income for 28 % of households 10 years ago, but is now only a source of income for 13 % of households. For those households still engaged in agriculture in Leh, the amount of land being farmed on average has decreased from 0.29 ha 10 years ago to 0.12 ha per household, a marked decrease of 59 %.

As recently as 40 years ago, Ladakh was a predominantly agricultural society (Norberg-Hodge 1991) that was to a large extent self-sufficient in terms of food

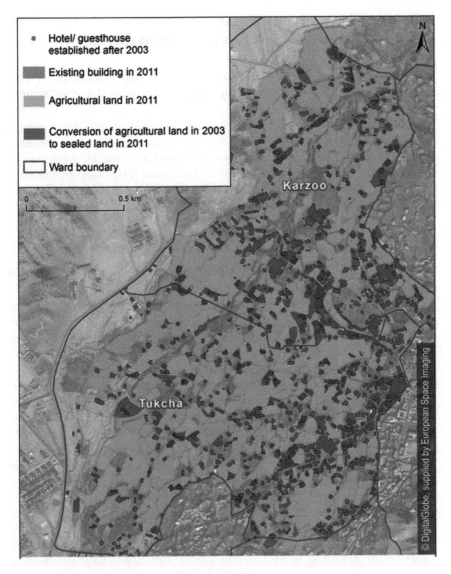

Fig. 5 Barren land in Leh's rapidly urbanizing wards

production. Today, Leh already has a food-grain import dependency ratio of 60 % (Pellicardi 2010: 89). The decrease in agricultural land measured in Leh is a significant amount, which needs to be addressed with a view to food security. Further, decrease in agricultural land also means decrease in irrigated land area, which in turn may be impacting the recharge rate of the groundwater aquifer. To address food security in Leh, LAHDC is planning to turn an expanse of desert area on the orographic left side of the Indus River into irrigated agricultural land. However, this may require huge additional amounts of groundwater extraction or diversion of Indus River waters.

5 Rapidly Rising Water Demand: Are Limited Water Resources Being Overexploited?

When we walk around Leh, ever accompanied by the sound of water that in some fields and marshy areas even bubbles directly from the ground, we tend to forget that Leh is situated in a desert.

The huge increase in tourists in Leh signifies a huge increase in water demand as guesthouses and hotels strive to provide flush toilets and showers as described above. In Leh, freshwater is supplied by a centralized and by a decentralized system. Currently, the public health engineering department (PHE) supplies following daily estimates during summer months (PHE 2013):

a. 1–2 million litres extracted via four tube wells from the Indus River aquifer;
b. 1.3 million litres extracted from various tube and borewells inside Leh;
c. 0.8 million litres channelled from various springs in the upper catchment area of Leh.

Thus, PHE is currently providing 3–4 million litres of water per day and most of it through groundwater extraction via bore and tube wells. The Indus River aquifer is used for PHE tube well extraction, while a deep aquifer underlying Leh, fed by glacial melt water, is used for private and PHE tube and borewell extraction. There is apparently another shallower aquifer underlying Leh. Water from the Indus River aquifer is being lifted about 300 m up to reservoirs distributed in Leh, which is very energy intensive, from where it is distributed by a gravity pipe system with several hundred public and private water taps and water tankers. For those without access to PHE water, public hand pumps are distributed throughout Leh that draw water from the shallower Leh aquifer at a maximum depth of about 10 m. According to our survey, Leh has 46 public hand pumps. 85 % of households use PHE taps, 18 % hand pumps and 8 % borewells as their primary drinking water source.

However, when we surveyed 318 guesthouses and hotels (90 % of total) in Leh, we found that 60 % of all guesthouses and hotels use a private borewell as a decentralized water supply source. One hotel owner interviewed of a hotel with 18 en suite rooms reported extracting up to 8,000 litres per day during the tourist season. Overall, guesthouses and hotels may be extracting up to about one-third the amount daily from the aquifer underlying Leh during the tourist season that PHE extracts daily. According to our interview survey with various local stakeholders, reasons for the increasing use of private borewells are water shortage in the centralized system; that is, PHE only provides water for 2–3 h in the mornings, which is considered insufficient to run a guesthouse or hotel with showers and flush toilets, and concern about PHE water quality and lacking water pressure.

When we look at a map, the high-density areas of groundwater extraction by private borewells in Leh (Fig. 6) predominantly overlap with the highest densities of guesthouses and hotels in direct proximity to the town centre. Outliers are large hotels that are removed from the town centre to profit from a quiet atmosphere. Interestingly, spatially, the location of the pipeline is also close to the highest densities of groundwater extraction, although water supply closer to the pipeline,

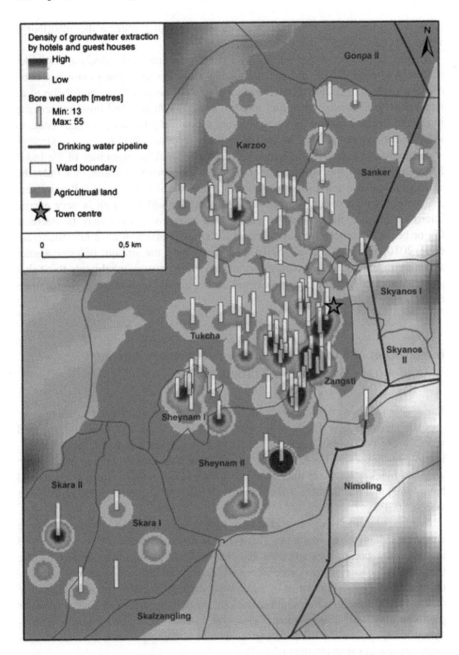

Fig. 6 Centralized and decentralized water supply systems

along which the service reservoirs are located from which water is then distributed via gravity pipe system, maybe better than far from it.

Hence, centralized water supply by PHE cannot match the demand for water in Leh. And even if PHE did provide for the water demand, guesthouse and hotel owners might still prefer to have private borewells to secure water for the tourism industry. However, we found that 99 % of guesthouse and hotel owners are interested in participating in a water-saving sanitation pilot study, which seems to indicate concern on the sustainable use of water resources in Leh.

Groundwater extraction in Leh is not regulated, and the capacities of the Indus River aquifer and the aquifer under Leh are not known. Further, not only the extraction but also pumping the water up hundreds of metres vertically and several kilometres horizontally from the Indus River aquifer to Leh is very energy intensive. Further, the amount of glacial and snow melt water may be decreasing or becoming uncertain due to climate change (Barnett et al. 2005; Bhutiyani et al. 2010; Immerzeel et al. 2010), hence affecting the amount of groundwater available in the aquifer. According to our interview survey, inhabitants think that some springs in Leh seem to have dried up because of high rates of groundwater extraction.

6 Pollution of Limited Freshwater Resources Through Inadequate Wastewater Management

Since about one-third of households, as described above, voiced concern over groundwater pollution due to lack of adequate wastewater treatment, we mapped where water pollution is occurring in terms of soak pits and septic tanks belonging to guesthouses and hotels, as we assume that these are producing much more grey and black wastewater than local households. Grey water is from kitchens and bathrooms, and black water is from toilets. To do this, we used global positioning systems (GPS).

We find that high-density guesthouse and hotel wastewater disposal sites are clustered around the town centre. This is not an obvious product of guesthouse and hotel density, because guesthouses and hotels closer to the town centre could be receiving more tourists, hence be richer and thus more likely to invest in wastewater treatment. Highest densities of wastewater disposal are also found in proximity to the PHE drinking water supply pipeline, so that seepage and thus freshwater pollution may have to be assumed. However, households in agricultural and non-agricultural wards alike think that groundwater is being polluted by lack of adequate wastewater management (Fig. 7).

According to the World Health Organization (WHO 1996), freshwater extraction locations should be a minimum of 30 m away from wastewater discharge locations. To estimate to which degree the quality of the groundwater in Leh is potentially at risk from sewage seepage, we spatially related areas of high wastewater production to areas of freshwater extraction such as borewells and hand

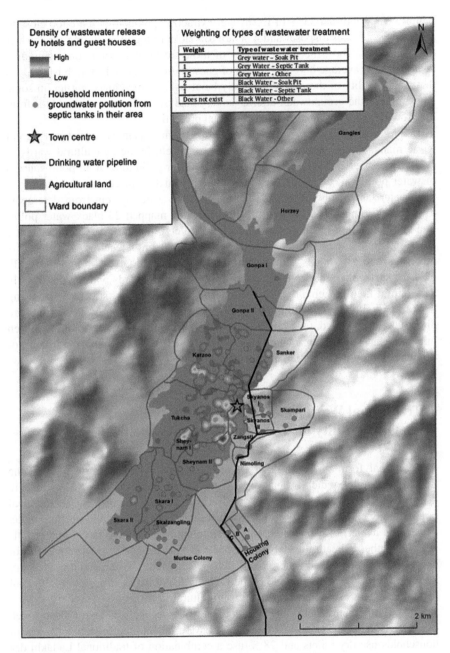

Fig. 7 Wastewater production and perception of groundwater pollution

pumps. We found that 33 % of freshwater extraction points in Leh are too close to areas of wastewater disposal and 4 % are too close to highly polluting wastewater disposal areas. Thus, the water quality of these freshwater extraction points may be at risk. The average bore well depth of guesthouses and hotels in Leh is 33 m, so that water quality of these may generally be at risk. As mentioned earlier, in some areas in Leh, the groundwater aquifer is very shallow and water just bubbles from the ground, and these areas may need special protection.

In addition, pollution of surface waters due to inadequate wastewater and solid waste management is also a significant issue in Leh. In the agricultural wards of Leh, we mapped 270 surface point sources of water pollution. Of all point sources of water pollution, 80 % are grey water inlets, which is of concern because with grey water, chemicals are being released into the water system from detergents used for cleaning and washing purposes. We also mapped 23 black water pollution sites including black water inlets (from toilets), public toilets without septic tanks and foul-smelling empty lots being used for open defecation, and soak pits other than those of hotels and guesthouses, and 18 open garbage dumps. Overall, 62 % of surface point sources of water pollution in the wards with predominantly agricultural land in Leh Town are within 100 m of rivers and streams. According to World Health Organization guidelines, this implies that open waters in Leh are directly being polluted. Overall, the type and distribution of water pollution in Leh indicates that strict environmental planning is needed as currently the quality of limited drinking water resources may be at risk.

7 Traditional Wastewater Management: An Opportunity for Development Innovation

A traditional form of decentralized wastewater management infrastructure, the Ladakhi dry toilet, which is very well adapted to the local conditions, has been in use in Ladakh for hundreds of years and is still used by the majority of the local population. The Ladakhi dry toilet is an elevated slab with a hole in the middle, sometimes as part of a house or as a separate outhouse, where faecal matter falls into a chamber beneath the slab and is covered after each visit by a shovel full of earth—hence "dry" as no water is used. The faecal matter is stored and used as dry agricultural fertilizer on the fields. The Ladakhi dry toilet is still used by 30 % of local households as a source of organic agricultural fertilizer.

Although it seems that richer households are more likely to own a flush toilet, overall, in Leh, 60 % of households do not have a flush toilet. In summer, 67 % of households use dry toilets and 28 % use a combination of traditional Ladakhi dry toilets and flush toilets. In contrast, only 1 % of tourists admit to using a Ladakhi dry toilet in Leh (Akhtar and Gondhalekar 2013: 31). In winter, 91 % of households use a Ladakhi dry toilet as the piping systems of flush toilets tend to freeze.

In order to deal with increasing amounts of wastewater, LAHDC will start this year to implement a centralized sewage system designed for the year 2040 through a private consulting and engineering company. This system is planned to comprise about 75 km of piping to be laid at a depth of 2 m below the surface to avoid freezing in winter. The collected wastewater is to be channelled to a central wastewater treatment plant below Leh on a barren land area, from where treated water is to be discharged into the Indus River (Tetra Tech 2009). However, such a centralized sewage system may require increased water supply just in order to flush long pipes, which will in turn require more energy for extraction. Such energy will need to be supplied, but energy provision is already a challenge in Leh, with the town facing regular power cuts. Further, a centralized sewage system may entail high operation and maintenance costs due to the harsh climate and rugged topography.

Despite these seemingly natural constraints to the implementation of a centralized sewage system, nonetheless, such a system represents an opportunity for the local government to invest in large-scale infrastructure. Further, the centralized sewage system may symbolize the "modernity" that a society facing the burdens of rapid transition and as recently still as traditional as Ladakh wishes to strive for. With its apparent record of success, the centralized sewage system still stands for the "business as usual" option to deal effectively with wastewater in an urban context.

In contrast, however, a decentralized sewage system in the parts of Leh that are less dense than the historic centre and have much agricultural land area may help address wastewater management challenges as well as to conserve groundwater resources by enabling the following:

- Length of overall piping system may be much less and may require less water for flushing
- Nutrient recovery in the form of organic as opposed to chemical fertilizer, enabling lower environmental impact of agriculture and continuation of traditional agricultural practices
- Reuse of treated wastewater in agricultural irrigation locally
- Less environmental pollution of soil and water resources and loss of water due to less seepage due to shorter pipes
- Lower energy consumption due to less water having to be lifted from groundwater resources and pumped uphill and pumping water within the pipe network to overcome topographic differences
- Renewable energy use potential through smaller pumps that can be powered by solar power
- Renewable energy production potential such as biogas from faecal sludge
- Lower costs of construction, operation and maintenance

In addition, wastewater could be treated and channelled back to replenish the aquifer underlying Leh proportionally to water demand and rate of extraction locally. One hotel in Leh has already implemented its own decentralized wastewater treatment plant out of environmental considerations. However, this is so far an exception.

8 Visioning Alternative Future Development in Leh, and Getting the Vision to the Ground

Leh in many ways is an ideal case study: until only a few decades ago a purely agriculturalist society, many inhabitants still practice agriculture for their own food production and also to cook for tourists in the guesthouses. Traditional wastewater management practices of collecting faecal sludge and using it as organic fertilizer on the fields of Leh are still widespread. Leh has been facing large-scale development pressures in a very fragile ecological environment in the very short space of only a few decades. However, the continuation of traditional agricultural and wastewater management practices may hold the key to enabling an alternative form of "modernity" in Leh that may be more suited to its locational characteristics and more appropriate in terms of water resources conservation than a centralized sewage system.

Yet devising such an alternative solution for Leh is a very complex affair that requires new tools as well as new ways of thinking. A geographic information system (GIS) can be used as a spatial decision support system (SDSS), generally described as a computer-based system to assist decision-makers while solving a spatial problem (Sprague and Carlson 1982). This can be a very useful tool to model alternative future development scenarios highlighting the potential water and energy savings and can thus support long-term decision-making on urban planning issues in Leh. For example, our mapping of the private borewells of guesthouses and hotels enabled us to tell where the highest densities for water demand are spatially located and how they relate to available water resources of the PHE pipeline. Or, a suitability site analysis determining the best places for establishment or construction of new guesthouses and hotels could be conducted. This analysis process is a typical example of a so-called multi-criteria approach where different actors with competitive interests and goals need to be considered. However, currently, LAHDC does not use GIS and faces severe constraints in terms of personnel and budget, and seemingly more pressing issues that need to be dealt with on a daily basis in order to supply water to the local population. Hence, for a comprehensive SDSS, models and tools need to be developed which enable political decision-makers to utilize geospatial analysis without too much capacity building.

In our interview survey, various key stakeholders agreed that it is a pressing issue to manage limited water resources in a more comprehensive manner in Leh. Our project is being supported by a local non-governmental organization, the Ladakh Ecological Development Group (LEDeG), and has been approved by LAHDC at a joint inception workshop. In parallel, another research project in collaboration with the Indian non-governmental organization ACWADAM is ongoing to test groundwater quality in Leh, results of which are pending. Therefore, the levels of water pollution are currently not known. So far, also there has been no systematic study to determine the volume of the aquifer underlying Leh, and assessing this is very costly. Therefore, it is currently not known whether groundwater resources are being overexploited, for example whether the rate of extraction of the aquifer underlying Leh is higher than the rate of recharge. Hence, also in light of projected continuation of guesthouse and hotel construction in Leh, LAHDC needs to plan under uncertainty. However, defining a point in time to act under such uncertainty is extremely difficult. Also, taking a political decision to implement an alternative solution such as a decentralized sewage system is difficult when the facts are not at hand to throw light on its potential benefits.

In this situation, and wanting to address the issue of inadequate wastewater management quickly, LAHDC is prey to companies who want to sell the "business as usual" option, namely a centralized sewage system in Leh. Further, being semi-autonomous, LAHDC is also attracted to a large-scale investment opportunity that such a centralized sewage system is. Despite evident need for wastewater management infrastructure, implementing a large-scale technological option like a centralized sewage system may also seem more feasible to the government than getting embroiled in a potentially time-consuming sociocultural process of trying to implement a decentralized alternative to it. But which role should the government play in order to enable sustainable development choices? It seems that a stronger role by the government is required as the only entity with the resources to implement long-term alternative development options. An SDSS could support LAHDC in its role as a "parent" to sustainable development in Leh.

If funds are short, there may be other ways to procure the finances to implement an alternative decentralized sewage system without an investor being needed. Leh currently does not levy a tourist tax. A tourist tax could be implemented, modelled on tourist destinations such as Bhutan, and collected either on arrival at the airport, or as many tourists arrive over land, in the guesthouses and hotels. At the moment, a tourist tax of significance is only needed to visit nature conservation areas in Ladakh, which, however, has been very effective in creating revenues to protect such areas. Further, although organizations like LEDeG have mounted many awareness-raising campaigns on this issue, strong government support is needed to systematically curb possible over-consumption of water in Leh mainly by tourists. Water conservation strategies could also present innovative opportunities for eco-tourism. In any case, this study advocates an independent evaluation of which type of wastewater management system could be most beneficial for water resources conservation in Leh.

9 Conclusion: "If Not Now, When?"

The case of Leh Town, as it is facing large-scale pressure to take decisions concerning future development under climate change uncertainty, highlights the question of, as Primo Levi put it, "If not now, when?" It is human to tend to think that innovation is coupled with risk. But in a world of climate change, the opposite may be true: implementing "business as usual" options under uncertainty may hold risk for us. Alternative and innovative approaches, which may be more flexible, may be more appropriate for dealing with uncertainty-related challenges. In particular, decentralized wastewater management seems to have much potential to address various aspects of water-related uncertainty. To foster innovation, courage by decision-makers is needed in order to lead the way on new sustainability pathways to be followed by others. With an appropriate vision, Leh has the full potential to become an international lighthouse example of an "ideal ecosociety".

Acknowledgements We thank our research partner, Ladakh Ecological Development Group (LEDeG), Leh, India, for supporting the project in terms of conducting field and interview surveys, organization of stakeholder workshops and other project-related work. This research is supported by a Marie Curie International Reintegration Grant within the 7th European Community Framework Programme (PIRG06-GA-2009-256555) and the German Research Foundation (DFG) (KE 1710/1-1).

References

Akhtar A (2010) Tourism and water resources in Leh Town (NW-India): analysis from a political ecology perspective. Master thesis in geography, Ruprecht-Karls-Universität Heidelberg, Jan 2010

Akhtar A, Gondhalekar D (2013) Impacts of tourism on water resources in Leh town. Int Assoc Ladakh Stud 30:25–38

Angchok D, Singh P (2006) Traditional irrigation and water distribution system in Ladakh. Indian J Tradit Knowl 5(2):397–402

Barnett TP, Adams JC, Lettenmaier DP (2005) Potential impacts of a warming climate on water availability in snow-dominated regions. Nature 438:303–309

Bashin V (1997) Water sharing and human solidarity in Ladakh. J Hum Ecol 8(4):279–286

Bashin V (1999) Leh—an endangered city? Anthropology 1(1):1–17

Bhutiyani MR, Kale VS, Pawar NJ (2010) Climate change and the precipitation variations in the northwestern Himalaya: 1866–2006. Int J Climatol 30:535–548

Bork T, Butsch C, Kraas F, Kroll M (2009) Megastaedte: Neue Risiken fuer die Gesundheit. Deutsches Aerzteblatt 39:1609 (in German)

Butsch C, Sakdapolrak P, Saravanan VS (2012) Urban health in India. Int Asienforum 43(1–2):13–32

Census of India (2011) http://www.censusindia.gov.in/pca/SearchDetails.aspx?Id=4048. Feb 25, 2014

Dannenberg AL, Jackson RJ, Frumkin H, Schieber RA, Pratt M, Kochtitzky C, Tilson HH (2003) The Impact of community design and land-use choices on public health: a scientific research agenda. Am J Public Health 93(9):1500–1508

Eichert D (2009) Die historische Altstadt von Leh: Wandel und Sanierung eines historischen Stadtquartiers im indischen Trans-Himalaya. Master thesis in geography, Ruprecht-Karls-Universität Heidelberg, Sept 2009 (in German)

Galea S, Vlahov D (2005) Urban health: evidence, challenges and directions. Annu Rev Public Health 26:341–365

Gondhalekar D, Akhtar A, Keilmann P, Kebschull J, Nussbaum S, Dawa S, Namgyal P, Tsultim L, Phuntsog T, Dorje S and Mutup T (2013) Drops and hot stones: towards integrated urban planning in terms of water scarcity and health issues in Leh Town, Ladakh, India. In: Gislason M (ed) Ecological health (advances in medical sociology, vol 15), Emerald Group Publishing Limited, Bingley, United Kingdom

Howard G, Bartram, J (2003) Domestic Water Quantity, Service Level and Health. Publication of World Health Organization, WHO Press, Geneva

Immerzeel WW, van Beek LPH, Bierkens MFP (2010) Climate change will affect the Asian water towers. Science 328:1382–1385

Jackson LE (2003a) The relationship of urban design to human health and condition. Landscape Urban Plan 64:191–200

Jackson RJ (2003b) The impact of built environment on health: an emerging field. Am J Public Health 93(9):1382–1384

Jha N (2010) Access of the poor to water supply and sanitation in India: salient concepts, issues and cases. International policy centre for inclusive growth working paper no. 62

Labbal V (2000) Traditional oases of Ladakh: a case study of equity in water management. In: Kreutzmann H (ed) Sharing water: irrigation and water management in the Hindukush —Karakoram—Himalaya. Oxford University Press, Oxford

Ladakh Autonomous Hill Development Council (LAHDC) (2005) Ladakh 2025 vision document. LAHDC, Leh, Ladakh

Lüthi C, Panesar A, Schütze T, Norström A, McConville J, Parkinson J, Saywell D, Ingle R (2011) Sustainable sanitation in cities—a framework for action. Sustainable sanitation alliance (SuSanA) & international forum on urbanism (IFoU), Papiroz Publishing House, The Netherlands

Mankelow S (2003) The implementation of the watershed development programme in Zangskar, Ladakh: irrigation development, politics and society. Master thesis, School of Oriental and African Studies (University of London), Sept 15 2003

Marcotullio PJ (2007) Urban water-related environmental transitions in Southeast Asia. Sustain Sci 2:27–54

Naess P (2006) Urban structure matters. Routlegde, Abingdon

Narain S (2002) The flush toilet is ecologically mindless. Down Earth 10(19):28

Norberg-Hodge H (1991) Ancient futures: learning from Ladakh. Oxford University Press, New Delhi

Pellicardi V (2010) Sustainability perspectives of development in Leh district (Ladakh, Indian Trans-Himalaya): an assessment. Doctoral thesis submitted to CIRPS (Interuniversity Research Centre for Sustainable Development) University of Rome "Sapienza"

Public Health Engineering Department (PHE) (2013) Leh Ladakh. http://leh.nic.in/dept.htm May 14 2013

Rieger-Jandl A (2005) Living culture in the Himalayas anthropological guidelines for building in developing countries. Facultas, Wien

Sakdapolrak P, Seyler T, Prasad S (2011) Measuring the local burden of diarrhoeal disease among slum dwellers in the megacity Chennai, South India. In: Krämer A, Khan MH, Kraas F (eds) Health in megacities and urban areas. Physica-Verlag, Springer, Berlin

Sanimap (2009) Puzhehei ecosan project, Xianrendong village, Yunnan province, China. http://www.sanimap.net/xoops2/modules/gnavi/index.php?lid=126. Jan 21 2013

Skeldon R (1985) Population pressure, mobility, and socioeconomic change in mountainous environments: regions of refuge in comparative perspective. Mt Res Dev 5:233–250

Sprague RH, Carlson ED (1982) Building effective decision support systems. Prentice-Hall Inc, Englewood Cliffs

Tetra Tech (2009) Plan for implementation of centralized sewage system in Leh Town, Ladakh. Public presentation Leh, Ladakh, 17 Dec 2009

Tilley E, Lüthi C, Morel A, Zurbrügg C, Schertenleib R (2008) Compendium of sanitation systems and technologies. Swiss Federal Institute of Aquatic Science and Technology (Eawag), Dübendorf

Tiwari S, Gupta R (2003) An ethnography of the traditional irrigation practices of Leh town: changing currents. Paper presented at the 11th colloquium of the international association for Ladakh Studies, Leh 21–25 July 2003

Vörösmarty CJ, Green P, Lammers RP (2000) Global water resources: vulnerability from climate change and population growth. Science 289:284–288

White Johansson E, Wardlaw T (2009) Diarrhoea: why children are still dying and what can be done. WHO Press, Geneva

World Health Organisation (WHO) (1996) Fact sheet 3.9: septic tanks. Factsheets on environmental sanitation series, WHO Press, Geneva, pp 328. http://helid.digicollection.org/en/d/Js13461e/3.9.html. Aug 19 2013

World Health Organisation (WHO) and United Nations Children's Fund (UNICEF) (2006) Meeting the MDG drinking water and sanitation target: the urban and rural challenge of the decade, WHO Press, Geneva

Authors Biography

Dr. Daphne Gondhalekar is an urban planner and Scientific Director of the Centre for Urban Ecology and Climate Adaptation (http://www.zsk.tum.de), Technical University Munich, Germany. Dr. Gondhalekar specializes in integrated urban planning, with focus on water and health in Germany, China and India. Previously, Dr. Gondhalekar worked as Senior Researcher at the Center for Development Research (ZEF), University of Bonn (2009–2013), at Environmental Planning Collaborative, an NGO in Ahmedabad, India (2008), and at the Department of Urban Studies and Planning at Massachusetts Institute of Technology (MIT) in Cambridge, MA, USA (2007–2008). She holds a PhD in Urban Planning from the University of Tokyo, Japan.

Dr. Sven Nussbaum works as a Project Manager at the German Aerospace Center (DLR) in the area of Technical Innovation in Business. He has over 10 years of working experience in Remote Sensing, Geodata and Information Systems. From 2011 to 2014, he worked as a Scientific Coordinator at the Center for Remote Sensing of Land Surfaces (ZFL) at the University of Bonn. Before that he was for 4 years GIS Team Lead at the International Atomic Energy Agency (IAEA) charged with task in the field of International Treaty Verification. His PhD dealt with the topic of object-based image analysis applied to critical security infrastructure.

Mr. Adris Akhtar is a Geographer with expertise in hydrology and GIS. He has worked with local as well as international NGOs in South Asia, in research and development practice, for the last 4 years. His research is based on extensive field work and has focused on urban human–nature interactions such as impacts of tourism on and health impacts of water resources.

Ms. Jenny Kebschull studied Geoecology at the TU Bergakademie Freiberg from 2007 until 2010 and continued her studies at the Department of Geography University of Bonn with focus on ecology, water, soil and geoinformation systems. Since 2010, she has been working for the Working Group for Agriculture, Water and Soil (http://www.alwb.de), the Center for Remote Sensing of Land Surfaces (http://www.zfl.uni-bonn.de) and Center for Development Research (http://www.zef.de) in the field of geoinformation systems, spatial data analysis and map design.

Rainwater Harvesting—A Supply-Side Management Tool for Sustaining Groundwater in India

Claire J. Glendenning and R. Willem Vervoort

Abstract Much of India's agricultural production is reliant on groundwater for irrigation, which has led to declining water tables. Rainwater harvesting (RWH), the small-scale collection and storage of run-off to augment groundwater stores through recharge, is an important supply-side management tool to sustain this precious resource. Understanding the impact of RWH is crucial to ensure that the net effect on groundwater and the watershed water balance is positive both locally and within a watershed. Using a case study of a watershed in rural Rajasthan, the Arvari River, this chapter describes the hydrological impacts of RWH for groundwater recharge carried out by the local community and a non-government organisation (NGO). The chapter first defines RWH and its potential to change the water balance. It then describes the field- and watershed-scale impacts of RWH in the Arvari River watershed. Finally, the chapter explores the operation of the local community watershed organisation that supports demand-side water management. This study shows that for sustainable management of groundwater, RWH construction must be balanced with groundwater demand management.

Keywords Water storage · Groundwater · Managed aquifer recharge · Water balance · India · Watershed development

1 Introduction

Eighty percentage of the global groundwater use occurs in Bangladesh, China, India, Iran, Pakistan and the USA (Shah et al. 2007), with India being the largest groundwater irrigator in the world (Shah et al. 2006). In India, groundwater

C.J. Glendenning (✉)
International Food Policy Research Institute, New Delhi, India
e-mail: claireglendenning@gmail.com

R.W. Vervoort
University of Sidney, Sidney, Australia

© Springer International Publishing Switzerland 2015
W. Leal Filho and V. Sümer (eds.), *Sustainable Water Use and Management*,
Green Energy and Technology, DOI 10.1007/978-3-319-12394-3_17

313

accounts for more than 45 % of the total irrigation supply (Kumar et al. 2005) and accounts for about 9 % of India's Gross Domestic Product (Mudrakartha 2007). This has not always been the case; over the last 50 years India has seen a huge boom in the use of groundwater, resulting in an exponential increase in the number of tube wells—in 2000, there were about 19 million (Shah et al. 2003).

Development of groundwater irrigation has been extremely important for rural poverty alleviation with dramatic improvements to small-holder farmers' livelihoods. This is because groundwater requires little transport, can be accessed relatively easily and cheaply, is produced where it is needed and provides a relatively reliable source of water. Also, groundwater irrigation tends to be less biased against the poor compared with large-scale surface water irrigation projects, partly because it can be developed quickly by individuals or small groups. However, groundwater development in India has contributed to serious groundwater depletion, with the water table declining at a rate of 1–2 m/year in many parts of the country (Rodell et al. 2009).

The main replenishment of groundwater is through recharge from rainfall. Recharge is the movement of water beyond the root zone that reaches the underlying aquifer and can be highly variable. Total recharge volumes are also difficult to predict. In India, because rainfall patterns are monsoonal with approximately 75–90 % of rainfall concentrated in the summer months (June to September), there is very little time for natural recharge to the aquifer, due to rapid run-off. This is exacerbated by changing land use, including deforestation which increases run-off potential. Nevertheless, as a result of this rainfall pattern, India has a long history of rainwater harvesting (RWH). In many rural areas of India, a specific purpose of RWH is to catch and store monsoonal run-off, which then percolates to groundwater tables. RWH changes the water balance of a watershed where water is stored and delayed with a transfer of surface run-off into groundwater through recharge, as well as evaporation and transpiration. But because the potential increase in available groundwater may encourage increased groundwater abstraction for irrigation or other uses resulting in socio-economic impacts, the impact on the water balance may be zero or negative.

In order to assess whether RWH supports groundwater sustainability, it is important to quantify the hydrological impact of RWH structures and the related downstream trade-offs for a given level of watershed development. This subject is explored in this chapter. The chapter first reviews RWH and how it changes the water balance, and then describes field- and watershed-scale impacts of RWH in a case study watershed, the Arvari River. Finally, the chapter examines the local community watershed organisation that supports demand-side management of water, and then concludes with some remarks.

2 Background—RWH Defined and Potential Impacts

In India, the technological advance from shallow wells, with animal pulling and human labour, to diesel and electric pumps has greatly impacted the amount of groundwater extraction. While the growth in groundwater use over the past few decades has improved rural livelihoods, there are increasingly serious issues with aquifer depletion. Currently, the response to this depletion has focussed on supply-side groundwater management. In India, a massive integrated watershed development programme provides public resources to local communities including for constructing RWH structures. Methods to recharge aquifers, including RWH, have become so widespread in India that it is sometimes referred to as 'a groundwater movement' or 'artificial recharge movement'. However, one of the difficulties of assessing the hydrological impact of RWH, and any improvement in groundwater sustainability, is the lack of evaluation of the project impacts on groundwater, as well as any upstream–downstream trade-offs.

In general, RWH encompasses methods to induce, collect, conserve and store run-off from various sources and purposes, by linking a run-off-producing area with a separate run-off-receiving area (Fig. 1). Small-scale structures collect run-off for either domestic use or supplemental irrigation, as is most common in parts of Africa, or for groundwater recharge, as is typical in many regions of India. Methods of RWH have three common characteristics (Boers and Benasher 1982):

- depends upon small-scale capture of local rainfall and/or run-off (does not include storing river water in large reservoirs or mining groundwater);
- can be applied in arid and semi-arid regions, where run-off has an intermittent character and rainfall is highly variable, so drought and flood hazards to agriculture are significant;
- is a relatively small-scale operation in terms of watershed area, volume of storage and capital investment, ranging from household, to field or small watershed.

From as early as 4500 BC, RWH has been practised in various parts of the world and is most commonly found in developing countries due to its decentralised, low-cost and local-scale aspects. In India, RWH has been practised for at least 1,000 years (Agarwal and Narain 1997). Despite the long RWH tradition, it was neglected from the time of British rule. But in the last few decades, RWH has seen a strong revival, involving the participation of communities, government and non-government organisations (NGOs).

One of the purposes of RWH in India is to store run-off which then recharges shallow groundwater aquifers (Fig. 1). Due to the monsoon rainfall pattern, RWH

Fig. 1 Schematic representation of RWH functioning for groundwater recharge

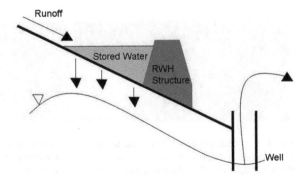

stores run-off that might otherwise continue downstream. Depending on the geology, stored water can percolate into the underlying groundwater table (Fig. 2), which is subsequently used for irrigation and domestic purposes via dug wells or tube wells. However, RWH consists of open storages which can be subject to high evaporation losses, due to high surface area to volume ratios. Thus, if the infiltration rate of the stored water is low, most of the water will be lost via evaporation. Conversely, once the water is stored in aquifers, evaporation is essentially zero. RWH often leads to increased crop production intensities and greater crop yield, because rises in the water table mean better accessibility and yields of groundwater for irrigation. This feedback between RWH development and increased irrigation area is an important to consider due to the impacts on groundwater sustainability and watershed water balances.

Definitions for groundwater sustainability are still argued, and it is often defined as safe yield, or the maintenance of a long-term balance between the annual groundwater withdrawals relative to recharge (Sanford 2002). This may be considered too simplistic as it does not take into account 'capture', which is the reduction in groundwater discharge or increase in recharge (Kalf and Woolley 2005). Hence, understanding the impacts of demand and supply-side groundwater management, particularly extraction for irrigation and recharge from RWH, is important to understand in order to enhance groundwater sustainability.

Existing studies of RWH impacts on groundwater in India highlight the variability in the aquifer response to RWH and the complexities of taking a full range of physical measurements to quantify changes in the water balance. Recharge is one of the most difficult components of the water balance to measure, because it needs to be measured below the visible surface and is highly variable; in arid environments, it can be the smallest component of the water balance. Nevertheless, previous studies have reported that groundwater levels have risen 2–8 m, and about 3–8 % of rainfall is recharged through RWH structures. Considering the difficulty and time required for physical measurements of recharge, modelling provides a cheap and fast way to consider larger-scale watershed effects of RWH. In fact, a number of modelling studies have looked at the amount of run-off that can be captured by RWH to prioritise watersheds for RWH development (see Glendenning et al. 2012 for a detailed literature review on RWH impacts). The

Fig. 2 a, b Example of a RWH structure known as an Anicut in Rajasthan, India. At the end of the monsoon in September, the structure is full. Three months later, the storage is almost empty, through evaporative loss, lateral sub-surface flow and recharge

next section explores in detail the change on the water balance due to RWH in the Arvari River watershed, firstly at the field level and then at the watershed level using a hydrological model.

3 RWH Impacts—Case Study of the Arvari River Watershed

3.1 Field-Scale Impacts

The semi-arid ephemeral Arvari River is located in the state of Rajasthan, India's largest state, which has a predominantly agrarian population (Fig. 3). Although the state covers 10.5 % of India's geographical area, it shares only 1.2 % of its water resources.

The Arvari River watershed (476 km²) is in the eastern part of the state. The watershed drains into Sainthal Sagar dam, a medium-size irrigation project built in 1898. In the watershed, there has been significant reinvestment in RWH over the last 25 years, which is the result of the work of the community and a local NGO,

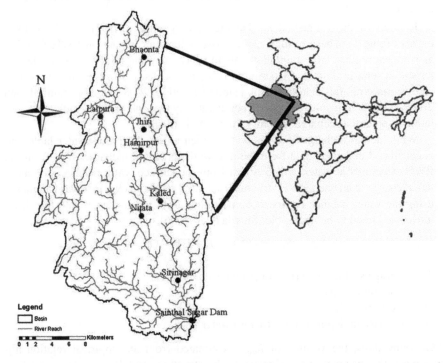

Fig. 3 Position of the case study catchment, the Arvari River watershed in the eastern part of Rajasthan. The villages, which were the focus of the data collection study, are highlighted

Tarun Bharat Sangh (TBS). However, the watershed is ungauged with no climate station and available geophysical information does not capture spatial and temporal variability across the watershed.

Since 1985, the community, with TBS support, has built around 366 RWH structures throughout the Arvari River watershed. The different structure types have different physical specifications and include the following:

- Anicuts dam the main reach of the river and are generally made of cement and stone or concrete. These structures are locally reported to have a very large impact on local groundwater tables (Fig. 4a);
- Bandhs are similar to Anicuts, but dam tributaries to the main river reach and are made of concrete in the middle with have earthen outer edges, while some are entirely made of earth (Fig. 4b);
- Johads are small earthen dams shaped like a crescent moon. They are found at the foothills of slopes, collecting water from a small hilly watershed area. The main purpose of Johads is for livestock watering (Fig. 4c).

Data were collected from wells and RWH structures in six villages in the middle and upper reaches of the Arvari watershed in 2007 and 2008. In each village, a RWH structure was monitored (in total two Bandhs, two Anicuts and two Johads), along with dug wells in the vicinity of the structure (in total 29). Three rain gauges were set up in the villages of Bhaonta, Hamirpur and Sirinagar to collect daily rainfall.

Rainfall in the watershed is characterised by high variability and localised events (Table 1). These rainfall characteristics have an impact on RWH function as storages require large rainfall events to fill, a common rainfall-run-off characteristic in semi-arid areas. Also rainfall is highly localised, so some structures may receive more rainfall than others in a similar location. 2008 had more rainfall than 2007, so the field results compare a comparatively wet and dry year indicating variation in potential recharge (R_{ep} m^3) from RWH structures.

While actual recharge (R_{gw}) is water which enters the groundwater table and is calculated using the data from the dug wells, potential recharge (R_{ep}) from the RWH structures is water, which is lost from the surface water balance. To calculate recharge from each RWH structure, potential recharge (R_{ep} m^3) was calculated using the water balance approach on non-rainy days when run-off into the structures is assumed to be negligible (Sharda et al. 2006):

$$R_{ep} = -As \cdot \Delta h - ET - Ot$$

As average surface area (m^2) of stored water;
ET evaporation (m);
Ot overflow (m^3);
Δh decrease in depth (m) in structure water level.

On rainy days, the volume of R_{ep} is estimated from an empirical relationship between the average depth of water in the structures and R_{ep} on non-rainy days (Sharda et al. 2006);

Fig. 4 **a–c** Example
rainwater harvesting (RWH)
structures present in the
Arvari River watershed, **a**
Bandh, **b** Johad, **c** Anicut

Table 1 Annual rainfall (mm) and number of rainy days in the Arvari River catchment

Focus area	Total rainfall (mm) 2007	Total rainfall (mm) 2008	Rainy days 2007	Rainy days 2008
Bhaonta	361	751	40	32
Hamirpur	449	897	27	56
Srinagar	499	494	18	38
Average	436	714	28	42

$$R_{ep} = ah_{av}b$$

a curve-fitting parameter,

b curve-fitting parameter

h_{av} average depth $(h_t + h_{t-1}/2)$

From the monitored RWH structures, total R_{ep} volume was greatest at Sankara Bandh (Fig. 5a). Beruji Bandh was directly below Sankara Bandh, so recharge volume here is less, as less run-off reached this structure and average storage depth was lower (Fig. 5b). Bandhs had a shorter storage time than Johads (Table 2). Beruji Bandh emptied the fastest of all structures, probably mainly due to the smaller run-on volume, but also possibly due to the underlying geology, which appeared to have higher infiltration rates. The structure type reflects a range of R_{ep} values. Because structure types were purposely built to fulfil certain objectives, each structure type has similar properties across the watershed.

The heavier rainfall in 2008, which meant more stored water in the RWH storage areas, is reflected in the R_{ep} values with a longer storage time and greater volume compared to 2007 (Fig. 5). However, the rate of R_{ep} also depends on the shape and size of the RWH structure storage area, run-off characteristics and infiltration characteristics in the storage area. These R_{ep} rates may change over time due to siltation, which occurs when particles brought by run-off are deposited in the structures and build over time to form a layer with lower hydraulic conductivity. In the Arvari watershed, regular maintenance of RWH structures, including desilting, is encouraged by the NGO through 'Gram Sabhas' or village councils.

In all structures, R_{ep} reaches a maximum daily depth limit with increasing cumulative rainfall as described in Sharda et al. (2006). This reflects the engineering design of the structures, which can only store a certain amount of run-off and so can only induce a maximum depth of R_{ep}.

In 2007, around 6.6 % of rainfall became R_{ep} and in 2008 about 7 %. The fraction of rainfall that becomes recharge is very similar between a dry and a wet year, which probably reflects the recharge efficiency of the structures.

Using the Water Table Fluctuation method to calculate actual recharge using water table levels in the dug wells (R_{gw} mm/day), well response also reflects the increase in rainfall in 2008 (Fig. 6). Rainfall in 2008 was mostly sufficient for meeting crop needs, resulting in less pumping from wells than in 2007, so the water table rise is seen more clearly in 2008 and recharge more effectively measured. The rate of

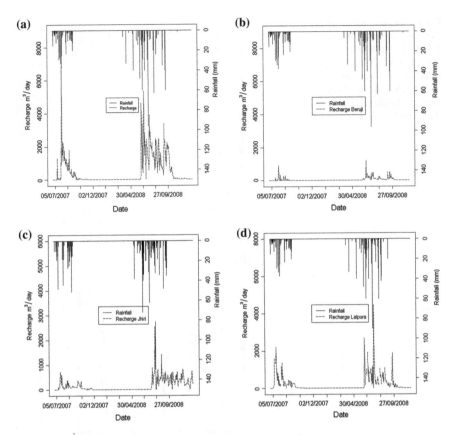

Fig. 5 **a–c** Estimated volume of potential recharge (R_{ep} m³/day) with daily rainfall (mm) from Bhaonta gauge for **a** Sankara *Bandh*, **b** Beruji *Bandh*, and with daily rainfall (mm) from Hamirpur gauge for **c** Jhiri *Johad*, and **d** Lalpura *Johad*

Table 2 Average daily recharge R_{ep} (mm/day) and number of days water was stored in monitored RWH structures in 2007 and 2008

RWH structure monitored	Year	Total days of storage	Average daily Rep (mm/day)	Standard deviation of Rep (mm/day)
Sankara *Bandh*	2007	169	45.6	0.07
	2008	206	55.6	0.15
Beruji *Bandh*	2007	138	20.9	0.07
	2008	201	27.8	0.03
Jhiri *Johad*	2007	273	12.3	0.01
	2008	308	19.5	0.01
Lalpura *Johad*	2007	240	15.7	0.02
	2008	382	23.5	0.04

Fig. 6 The six monitored
wells at focus area Bhaonta,
a ASL (m) water level,
b Relative water level height,
and **c** Relative well height
at Well BK5 and depths at
Sankara Bandh and Beruji
Bandh (m)

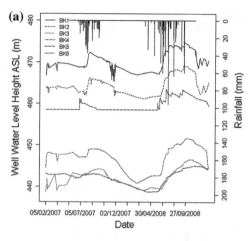

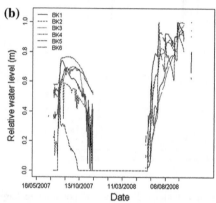

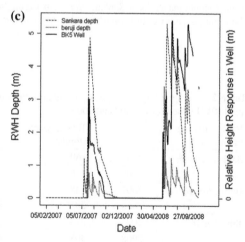

groundwater-level increase is different for each monitored well, which could be due to a range of factors, including aquifer properties, number of RWH structures nearby, pumping from other wells in the vicinity and the amount and intensity of rainfall in the local area. The variability highlights local watershed variability, a common feature of semi-arid areas and also the local recharge impacts of RWH.

Estimates of R_{gw} ranged from 7.2 to 11.3 mm/day. The recharge values derived from the well data were much lower than the estimates of R_{ep} from the RWH structures, suggesting that the aquifer has a large lateral transmissivity. As a result R_{ep} from the RWH structures would initially cause local mounding, which then dissipates across the aquifer. Quantification of such lateral flow would require greater knowledge of aquifer properties, which is limited in this watershed.

Based on this field analysis, RWH structures clearly have a large impact on the amount of recharge occurring in a local area and RWH function is dependent on rainfall, with larger R_{ep} volumes in the higher rainfall year. However, the size of the storage area in the structure influences the maximum daily depth of recharge and this reaches a maximum limit (see Glendenning and Vervoort (2010) for more details on this field study). Using these field observations, a conceptual model simulates watershed trade-offs with and without RWH next.

4 Watershed-Scale Impacts

A conceptual water balance model, based on field data from the Arvari River watershed, describes watershed-scale trade-offs with the presence of RWH. Sustainability indices are used to compare different scenarios. Due to collection and storage of run-off, RWH may change the watershed water balance and therefore increase the sustainability of irrigated agriculture as a result of more reliable groundwater stores. To examine this impact, sustainability indices are a useful way to measure changes in a water resource system. Quantification of the sustainability of water resource is achieved by using indices of reliability, resilience and vulnerability, which are based on whether a specified demand threshold is met by a defined water resource system, in this case whether enough groundwater is available for irrigated agriculture (Hashimoto et al. 1982).

Reliability (RE) is the frequency or probability that a system is in a satisfactory state:

$$RE = prob[X_t \in S] \quad \text{where} \quad RE = \sum_{t=1}^{T} Z_t$$

X_t is the output state or status of the system at time t, and Z_t is a binary measure where X_t is either an element of S values (output or performance of the water resource system in a satisfactory state) or an element of F values (output or performance of the water resource system in a failure state). The RE index ranges from 0 to 1, where 1 reflects 100 % reliability.

Resiliency (RS) is the probability that when a system is in a failure state, the next time step is a satisfactory state:

$$\text{RS} = \text{prob}\{X_{t+1} \in |X_t \in F\} \quad \text{where} \quad \text{RS} = \sum_{t=1}^{T} W_t \Big/ \left(T - \sum_{t=1}^{T} Z_t \right)$$

$W_t = 1$ if X_t is element of F and then X_{t+1} is an element of S; otherwise, $W_t = 0$. The resilience index ranges from 0 to 1, where 1 is a system with 100 % resilience. Vulnerability (V) is the maximum of the sum of the difference between the threshold (criteria C) and the actual level (X_t) for any failure periods (J_i). It thus reflects the severity of the failure. The vulnerability index can have a wide range depending on the difference between X_t and C, but higher values indicate higher vulnerability, lower values represent a less vulnerable system

$$V = \max \left\{ \sum_{t \in J_i} C - X_t, \quad i = 1, \dots, N \right\}$$

In order to quantify these indices, the status of the water resource system needs to be described as either satisfactory (S) or unsatisfactory (F), using defined thresholds, which in this case study relate to water demand for irrigation. As RWH increases groundwater recharge, irrigated agriculture tends to increase. Consequently, the sustainability indices provide a useful method to compare differences in modelled watershed water balances under different management scenarios.

5 The Model

A conceptual water balance model was purpose built to capture the relevant hydrological processes in the watershed that are influenced by RWH. This includes surface water—groundwater interactions and recharge volumes from RWH. The Arvari River watershed displays great variability in climatic and landscape conditions. To capture some of the variability, the modelled watershed was divided into three smaller units or sub-basins. Based on the local hydrogeological information, an upper shallow alluvial aquifer, which is hydraulically connected to a deeper aquifer was included in each sub-basin.

A stochastic model to simulate the rainfall was used to create a time series of daily realisations of rainfall for each sub-basin. The partitioning of rainfall into run-off uses the USDA-SCS curve number method. Within each sub-basin hydrological response units (HRU) are defined, which represent a land use with unique management factors and soil type and have no spatial interpretation. The water balance is calculated for each HRU in each sub-basin.

The dominant land uses in the Arvari River watershed were incorporated into the conceptual model. In the hills, with higher elevations and shallow rocky soils, land is usually used for grazing with thinly covered forest areas or open land known as commons. In the plains along the river, where the soils are deeper and richer in clay content, land use is predominantly agriculture, including irrigated agriculture. The main cropping seasons are the Kharif, the monsoon crop, and Rabi, the winter crop. Further from the river, agriculture still persists, but there is less irrigation and the area for Rabi is reduced. The main land uses therefore used in the conceptual model are agriculture, commons and RWH. Rainfall, irrigation and evapotranspiration (ET) influence the soil layer, which represents the layer where root activity takes place. Input into the root layer is rainfall (P)–runoff (RO). Recharge of groundwater occurs when the soil layer reaches field capacity (FC), and all excess input water becomes recharge. Irrigation water is taken from the shallow aquifer in each sub-basin and takes place when the soil profile is at critical available water level (AW_{cr}). Evaporation for RWH structures is set to potential evaporation (PotET), predicted using a similar approach used to predict rainfall. Actual ET was based on a piecewise linear function, which has been widely used in ecohydrological water balance models (Teuling and Troch 2005).

The total volume of RWH is considered as one large reservoir at the end of each sub-basin. There are different parameter values for Bandhs and Johads, which was based on the field observations for each structure type. Recharge depth (R_{ep} in mm) is calculated using Darcy's Law.

A qualitative adjustment of the model to realistic values was based on the 2 years of field data (Glendenning and Vervoort 2010). This adjustment brings the model in the realm of possible real watersheds rather than being purely theoretical. It was observed that local farmers increase the area of irrigated agriculture based on the groundwater availability, which they judge from the water levels in their dug wells. The actual groundwater use is a complex function, which relates to how the farmers adjust the irrigation area to groundwater availability. In the model, a simplified stepwise function was introduced to mimic this behaviour (more details on the model in Glendenning and Vervoort Glendenning and Vervoort 2011).

6 Simulation Analysis

To calculate the sustainability indices, the water resource system was based on the availability of water in the sowing months of the Kharif (July) and Rabi (November) crops. To meet crop water demands in the Arvari River, the farmers rely on a combination of groundwater and rainfall in Kharif and groundwater in Rabi. Kharif, the monsoon crop (mostly maize), is planted after the first rains in July and harvested in October. Rabi, the winter crop (mainly wheat), is planted in November and harvested around March. The daily available water in the system is defined as the groundwater storage combined with any rainfall. Because the initial available water at the start of each season strongly determines water availability for the rest of the season and the amount of irrigated area planted, any impact of RWH is best based on the first month of each season (July and November). To calculate daily values of the sustainability indices, the water available was compared with a defined threshold, based on FAO crop factors for maize and wheat (Allen et al. 1998).

Different scenarios were modelled based on the area of irrigation and the area of RWH. The percentage area of RWH land use varied from 0, 0.5, 1, 1.5 and 3 % and the irrigated land-use percentage area was varied at 0, 5, 10, 15 and 20 %.

In the Arvari River watershed, groundwater supply is not only affected by the irrigation demand (represented as the irrigation area) and recharge from RWH, but also by rainfall amount. While irrigation area and RWH area are determined by societal management actions, rainfall is an exogenous variable. To understand the influence of annual rainfall and drought conditions (defined as below-average rainfall) on irrigated agriculture and RWH, the model was run with different numbers of years of below-average rainfall.

7 Results

The system without RWH—All indices show that as the area of irrigation increases, sustainability decreases in the system. A greater area of irrigation represents a greater crop water demand, and therefore, more groundwater extraction would occur, so that water levels are below the required water availability threshold more frequently.

The system with RWH—Introducing RWH creates a more viable water resource system for irrigated agriculture, compared to the previous analysis without RWH. This is indicated by overall higher reliability and resilience, and lower vulnerability (Figs. 7, 8 and 9). Within each irrigation scenario, as RWH area doubles from 1.5–3 %, the benefit to the system is not large and in some cases the sustainability indices actually decrease. This suggests that there is a limit in the area of RWH that gives a maximum recharge benefit and beyond which the benefit is marginal. In addition, the fact that farmers respond to higher water levels and increase their irrigation area means that some levels of RWH area are less sustainable depending on management practices.

The influence of rainfall—To understand the influence of annual rainfall variations on irrigated agriculture and RWH, the indices were compared between below and above-average annual rainfall years. The analysis concentrated on the Rabi season, which has less daily rainfall, so the impact of recharge from RWH would be the primary influence on whether the threshold demand is met. Systems with and without RWH have higher reliability with above-average rainfall. However, when rainfall is below-average, the reliability of the system with RWH is greater than a system without RWH (Fig. 10). Without RWH, resilience is lower in both rainfall scenarios. But a system with RWH has better resilience than a system without RWH (Fig. 11). The vulnerability index shows that a system with RWH is less vulnerable than a system without RWH in both rainfall scenarios. When RWH is present, the sustainability indices are lower in dryer years than wet years,

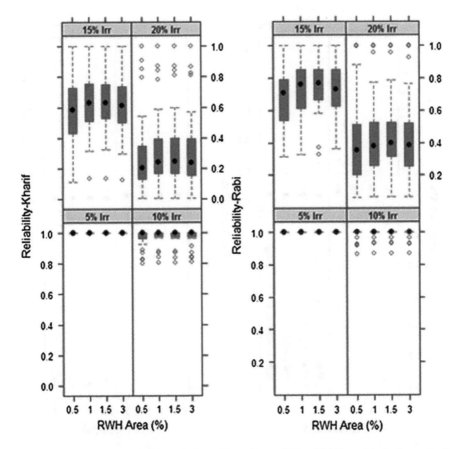

Fig. 7 Distribution of the reliability index, with median and 50 and 95 % confidence intervals of the distribution, for different areas of irrigation and RWH for *Kharif* and *Rabi* season

due to lower recharge from RWH. However, in below-average rainfall years, the indices are greater with RWH in the system than without RWH. This means RWH alleviates some of the deficit in below-average rainfall years, even though recharge would be less than in an above-average rainfall year.

Drought conditions and RWH impact—RWH provides within season buffering, but may not provide longer-term supply if drought occurs, because the local pumping practices for irrigation reduce the buffer between seasons and recharge from RWH would decrease. However, the overall sustainability indices were similar if the number of below rainfall years (the drought period) increases. This suggests that RWH is able to provide a limited inter-annual buffer in low rainfall periods. Overall, the resilience index was low, but decreases in resilience and reliability due to the lengthening drought are smaller with RWH in the system, so RWH interventions are valuable.

RWH stream flow impacts—The model simulations showed the obvious result that increasing rainwater harvesting decreases catchment stream flow, as water

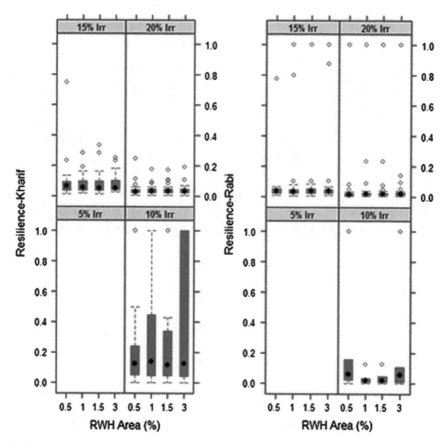

Fig. 8 Distribution of the resilience index, with median and 50 and 95 % confidence intervals of the distribution, for different areas of irrigation and RWH for *Kharif* and *Rabi* season

is transferred from blue water to green water. In addition, due to the increased buffering of RWH, larger rainfall events would be needed to generate significant stream flow.

8 Discussion of Model Results

Under the RWH scenarios in the model, irrigated agriculture is more viable than in a system without RWH. Continuously increasing RWH area does not bring additional benefits though, because as RWH area increases, the system reaches a limiting point, where the sustainability indices do not increase further, and in some cases decrease. This is firstly because the run-off from a watershed area is finite and more RWH means that water is spread across a larger storage area. As a result,

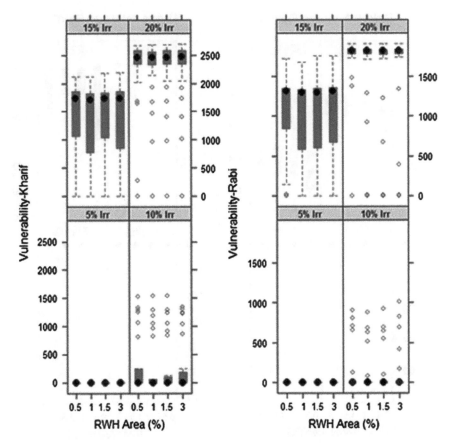

Fig. 9 Distribution of the vulnerability index, with median and 50 and 95 % confidence intervals of the distribution, for different areas of irrigation and RWH for *Kharif* and *Rabi* season

the depth in each of the structures is lower and recharge per structure decreases. Recharge cannot increase beyond a certain limit for each structure. Secondly, the aquifer storage capacity is limited. If the area of irrigation is large, RWH many not be able to reduce the stress on groundwater, because demand is too large. Finally, the modelled farmer response is to increase irrigation area with increased water availability.

At high levels of RWH development in a watershed, the benefit to irrigated agriculture may not be worth the cost. This reinforces theoretical speculations about watershed-scale impacts of RWH by Kumar et al. (2006). Their study concluded that a greater degree of RWH development would decrease the social, economic and environmental marginal benefits of building additional RWH structures. If there are already many RWH structures in a watershed, the marginal benefit for every new RWH structure is smaller than for structures already built, and the marginal cost is greater due to the social and environmental costs of harvesting every

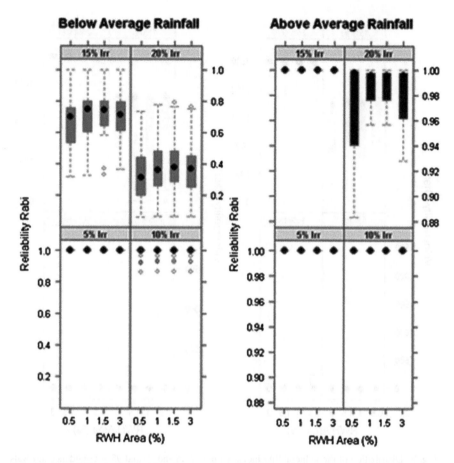

Fig. 10 Reliability in *Rabi* with below-average rainfall and above-average rainfall with various scenarios of RWH and irrigated agriculture, with median and 50 and 95 % confidence intervals of the distribution

unit of water and potential for over-appropriation. At higher RWH development, the marginal benefit may decrease because of low aquifer storage capacity and lower chances of finding appropriate sites for RWH development (Kumar et al. 2006). In the conceptual model presented here, the assumed small aquifer storage capacity could explain why increasing RWH area does not strongly increase the benefit. Before further recommendations, these results must be compared with other models that have been calibrated and validated to a specific watershed, based on more extensive field data than presented here.

The model confirms that RWH increases the viability of groundwater-irrigated agriculture if short-term mild drought conditions occur. However, in the model, longer-term drought does not seem to be alleviated by RWH. While the sustainability indices decrease when drought occurs, the modelled system functions better with RWH than without.

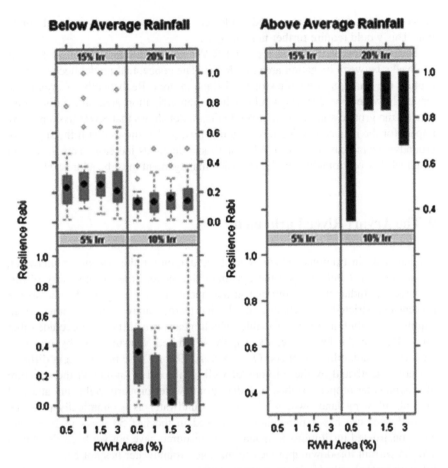

Fig. 11 Resilience in *Rabi* with below-average rainfall and above-average rainfall with various scenarios of RWH and irrigated agriculture, with median and 50 and 95 % confidence intervals of the distribution

When RWH is introduced, stream flow strongly decreases. Capturing local run-off upstream in RWH storages addresses problems of frequent drought and widespread poverty in upper watersheds. However, the 'blue' water investments, like irrigation canals, are generally located downstream and depend on large volumes of run-off. As RWH leads to an increase in 'green' water used for irrigated agriculture upstream, this will affect 'blue' water availability for downstream users, including irrigators and ecosystems, because flows decrease. However, RWH could also have positive environmental impacts as a result of reduced land degradation. For example in the Arvari River watershed, more area has been re-forested since RWH development as the community recognises the importance of erosion control. There could also be improvements in water quality as run-off is slowed through the watershed, so larger sediments would filter out and the erosive power of the flows is reduced. Finally, longer residence times increase possible feedbacks between water and vegetation and could lead to improved environmental condition. This would require further field analysis.

The scenarios of irrigation area and RWH area highlight the link and strong feedbacks between irrigation area and RWH. The conceptual model does not fully capture the changing dynamics of the irrigation area. Realistically, if groundwater levels were low, farmers would reduce their irrigation area, alleviating pressure on the groundwater resource. While the model shows that RWH has a positive impact for the irrigated agriculture sustainability, it also shows that if the supply of groundwater increases due to RWH, then local demands increase. This means for sustainability of groundwater, demand-side management must be addressed.

9 The Arvari River Parliament

Groundwater institutions, which regulate and monitor the use of groundwater, are an important influence on the dynamics that govern the sustainability of this resource. In India, the institutional arrangements related to groundwater do not prevent unrestricted use, which is partly due to the nature of the resource itself. Aquifers are often large and invisible, which means it is difficult to exclude other users. Despite the large size, the supply is limited; consumption by one user reduces its availability to others (Ostrom et al. 1994). These management difficulties are exacerbated by the millions of wells that exist in India and the structure of groundwater property rights. In India, groundwater property rights are attached to land, allowing land owners to extract groundwater as economically possible. Groundwater use has exponentially increased in the last 50 years in India, but well-constructed institutions to manage this resource have not yet been developed and the current top-down approach to manage groundwater has not been successful, with water levels declining in many states.

Giving resource property rights to communities, so that it is managed as a commonly owned or common property resource, is an alternative to effectively manage resources like groundwater, particularly in developing countries. But there are very few documented examples of successful community-based groundwater institutions, and relatively little is known about those institutions that do exist and how they govern groundwater use. This section examines one such institution in the Arvari River watershed, the Arvari River Parliament (ARP), initiated in 1998 with the support of TBS.

Since 1987, TBS has been building RWH structures with the local community. TBS works in villages after being approached by a village community for support. For each structure that is built, the village community covers a proportion of the construction costs; either monetarily or through voluntary labour. Where structures are built, TBS encourages the formation of a Gram Sabha (village council), to discuss where the structures ought to be built and how the structures would be maintained. The general results of RWH for communities in this area have been very positive. Moench et al. (2003) found that 85 % of RWH structures have benefited small and marginal farmers by increasing groundwater supplies. This has allowed farmers to increase the area under irrigation and decrease their dependency on the rain-fed Kharif crops.

Until 1996, there were no watershed-scale management plans in place. In 1996, a conflict between State Government officials from the State Fisheries Department and the village of Hamirpur led to the formation of the ARP. In 1995, a large Anicut was built on the river reach at Hamirpur, which held a substantial amount of water. In November 1996, the State Fisheries Department gave licenses to fish in the Anicut to a Jaipur contractor. The village people opposed the contractor. The community felt that because they had built the RWH structure that held the water, the government did not have the authority to decide how that water could be used.

> When there was no water, where was the government? We requested and appealed many times to the government. But they did not listen to us; they did not do anything to provide us with water. Now, we have water because of the efforts of TBS and ourselves. All of us worked hard without any government support. Our efforts were not supported by the government... ...therefore the government has no right to give fishing-contracts for our water (interview with Gram Sabha member, 2008).

With the support of TBS, the villagers held a satyagraha (non-violent protest) against the State Fisheries Department for 2 months. The result was that the fishing contracts were cancelled in March 1997. Due to this incident, the community and TBS realised that others outside the community could use the water provided through RWH, which were considered the common property of each village. Consequently in December 1998, TBS called representatives of all Gram Sabhas of the Arvari River watershed to a meeting, where the ARP was initiated. On 26 January 1999, the ARP was officially affirmed. The ARP has 110 representatives from the 72 villages of the Arvari River watershed. The rules of the ARP were first set up in 1998 and focus mainly on water conservation and utilisation and forest conservation. All members agree to enforce the following informal rules:

- Water-intensive crops such as sugarcane, rice and cotton are not to be planted
- No one shall draw water from the river or RWH structures. But those people, who gave their land for RWH structures or whose land is under water because of RWH, can take water from RWH structures and the river
- No commercial fishing is allowed in water stored in RWH structures
- Tube wells, which tap deeper aquifer, are not allowed
- Construction and maintenance of RWH is encouraged
- Land is not to be sold for mining/quarrying or any other industrial activity
- Protection and planting of forests in encouraged

These informal rules are discussed at biannual meetings to highlight practical problems in their implementation and to suggest new guidelines if needed. Suggestions, if any, are debated and discussed. Members also seek guidance for resolving conflicts and report any violations of the rules listed above. The informal rules are then supposed to be conveyed to individual villages through the elected Gram Sabha representatives. These are then discussed and implemented at the village level either through social or moral pressure, depending on the activeness of each village Gram Sabha.

Despite the presence of the informal rules of the ARP, unrestricted use of groundwater still occurs within the Arvari River watershed. This is due to a number of factors including information problems about the resource size and capacity. Monitoring use of a resource that is easily accessed, yet invisible, has high transaction costs because it would be difficult to monitor without understanding the capacity of the resource and the number of users. Scientific information about hydrogeological properties or aquifer boundaries in this area is not readily available. Depending on the size of the aquifer, farmers using groundwater may not have any appropriation or provision problems until irrigation is fully established and only then may they be motivated to work together.

Farmers in the ARP considered the amount of groundwater available a problem; and, to them, this was the result of the amount of rainfall, rather than the area of irrigation or amount of groundwater extracted. Those farmers who were members of the ARP believed more RWH structures ought to be built to alleviate any groundwater shortage. The solution to limit extraction of groundwater could be to limit number of wells built, which for people in this watershed, who are dependent on groundwater for their livelihood, might not be an attractive solution. Annual fluctuations are very strong due to the shallow aquifer system that most farmers in this area access. This reflects the amount of rainfall, so the aquifer may self-limit extraction.

Groundwater demand rules in the ARP apply mostly to crop choice. Crops are visible, and so it is easy to see whether other users are abiding by the informal rule to not plant certain crops. However, farmer interviews suggest that most people in the Arvari River watershed are not aware of the ARP or its rules. Those farmers who were not members of the ARP and owned wells near RWH structures had not heard of the ARP. And many of them were unaware of a Gram Sabha in their village. While the rules suggest no new tube wells to be built, there are no direct restrictions on dug well drilling or pumping. In 2007, several new tube wells were seen, and in one village, three new tube wells had been installed in the last 3 years. There is nothing to stop well owners from deepening existing wells, digging new wells, or increasing pump capacity. Also, pumping has occurred directly from the river and from the storage of the RWH structures.

Enforcement and monitoring of the rules depends on the strength of the moral sanctions within the community. This depends on the activity and strength of the Gram Sabha to convey the ARP informal rules to the village community. An evaluation of TBS's work found that most Gram Sabhas remained dormant after TBS had withdrawn from the village. However, in some villages, the Gram Sabha had taken up further activities (Kumar and Kandpal 2003). Where the Gram Sabha is most active, there has been significant impact on the management of common resources, such as forests, grazing lands and RWH construction and maintenance (Moench et al. 2003). The most active Gram Sabhas were found in villages at the upper end of the watershed. The percentage of land under RWH in these areas is higher than other areas on lower elevations. These landscape positions also mean that RWH has had the largest impact on groundwater tables, so the community is more likely to work collectively because of the individual benefit from working together and the direct reward of their collective action in well water levels.

The ARP does not have many of the factors that Ostrom et al. (1994) consider to favour collective action. While each user is heavily dependent on the resource, the boundaries are not clear. New users can only be excluded if they do not own land, as groundwater rights legally lie with the land. There is no structured monitoring of the resource or operational rules regarding the amount of groundwater extraction, so 'free riding' can occur, which could discourage collective action. Currently, the Arvari River watershed is not facing serious supply issues in groundwater. This is because the farming practices are such that the level of irrigation does not have large or strong negative impacts on other users in the watershed. Also the resource itself encourages self-limiting behaviour in pumping. Most farmers access the shallow aquifer through dug wells, which varies significantly within a year. Consequently, irrigation area fluctuates depending on the level of water at the end of each monsoon.

As population grows and more area is developed for irrigation, perhaps longer-term water table decline patterns could appear. In this instance, the ARP could provide a forum through which members can discuss such issues, More importantly, the institution could act as a lobby and empower users to exclude other larger users of groundwater from entering the watershed, for example industrial units.

The ARP encourages the implementation of RWH across the watershed. All farmers, when asked about the downstream impacts of RWH, said that any impact would be positive, because groundwater would move towards them in the aquifer. Strong lateral flow was also suggested in the field analysis, with almost 30 % of potential recharge from RWH structures moving away laterally in the unsaturated zone of the shallow aquifer. So while RWH works are carried out at the village scale, the water stock that the village would hope to access from the construction of RWH may not be exclusive to the group of users who initiated RWH. Instead that water may move away or could even be pumped away from the control of that group, if there is no physical boundary. By encouraging RWH across the watershed, any movement of groundwater away from some users could be balanced by more RWH construction upstream. The effects of RWH are largely local, and in the upper rockier areas of the watershed, wells almost 6 km away are influenced by recharge from RWH. However, the model simulations suggest that RWH impacts reach a maximum level of efficiency beyond which the benefit decreases, i.e., the cost–benefit ratio increases to a point beyond which the benefits do not warrant the cost. This consequence means that demand-side management of groundwater would be needed in tandem with management of groundwater supply.

As seen in the ARP, a community-based approach to manage groundwater faces challenges and constraints, partly because of the lack of information about the resource that is easily available for resource users. Often for community-based institutions to work, it must serve a private purpose important to the users; otherwise, they may not all participate or may even try to work against it. Currently in the Arvari River, watershed that private purpose across the watershed is not strong, because dug wells tap the shallow aquifer, which fluctuates annually and is largely dependent on rainfall and recharge from RWH. But if irrigation area increases, and more tube wells are sunk, then perhaps the need for collective action will arise, and an institution like the ARP will have more relevance. At this time, the need for reliable information about aquifer characteristics and user behaviour will be important for the ARP to function effectively. Ultimately though, if groundwater shortage becomes a serious problem in future, then change in livelihood patterns from the current heavy dependence on irrigated agriculture would be important to consider for sustainable groundwater management in addition to improving farming practices that are more water efficient with the support of extension services.

10 Conclusion

There is continued need for improved understanding of how RWH functions and what impacts RWH structures have on groundwater availability, as well as on the local and downstream environment. While local hydrological impacts of RWH for recharge are positive and increases groundwater supply, hence crop production, it is also clear that the cumulative hydrological impact of RWH on stream flow can be significant. In addition, land-use changes, such as the development of more irrigated agriculture can be the result of perceived or real groundwater increases as a result of RWH. This land-use change could result in a net decrease in the amount of locally available water in addition and decreased availability of water for downstream users.

The complexities associated with understanding and measuring groundwater recharge have made it very difficult to quantify the hydrological impacts of RWH at the local and watershed scales. However, there are a number of new research avenues that may greatly assist in clarifying the hydrological impacts of RWH. Importantly, these options do not necessarily have to be expensive. Open source software and freely available datasets, such as those derived from satellite images, gridded rainfall datasets and soil data, have been successfully used to model the hydrology globally. It must be noted though that modelling without further data collection would not lead to further insights.

This chapter has presented a study of RWH in India, and the potential RWH offers to support sustainable management of groundwater. As highlighted here, there are a number of important issues that need to be considered when planning and implementing RWH across a watershed. In particular, this includes enabling groundwater resource users to balance their demand for groundwater in parallel with RWH construction, which must consider water balance trade-offs across the larger watershed.

References

Agarwal A, Narain S (1997) Dying wisdom. Rise, fall and potential of India's traditional water harvesting systems. Centre for Science and Environment, New Delhi, p 404

Allen RG, Pereira LS, Raes D, Smith M (1998) Crop evapotranspiration: guidelines for computing crop water requirements. FAO irrigation and drainage paper. FAO, Rome

Boers TM, Benasher J (1982) A review of rainwater harvesting. Agric Water Manag 5:145–158

Glendenning CJ, Vervoort RW (2010) Hydrological impacts of rainwater harvesting (RWH) in a case study catchment: the Arvari River, Rajasthan, India. Part 1: field-scale impacts. Agric Water Manag 98:331–342

Glendenning CJ, Vervoort RW (2011) Hydrological impacts of rainwater harvesting (RWH) in a case study catchment: the Arvari River, Rajasthan, India. Part 2: catchment-scale impacts. Agric Water Manag 98:715–730

Glendenning C, van Ogtrop F, Misra A, Vervoort W (2012) Balancing local and watershed impacts of rainwater harvesting in India. Agric Water Manag 107(1):1–13

Hashimoto T, Loucks DP, Stedinger JR (1982) Robustness of water resources systems. Water Resour Res 18:21–26

Kalf FRP, Woolley DR (2005) Applicability and methodology of determining sustainable yield in groundwater systems. Hydrogeol J 13:295–312

Kumar P, Kandpal BM (2003) Project on reviving and constructing small water harvesting systems in Rajasthan. Sida Evaluation, Sida, Department for Asia, p 98

Kumar MD, Ghosh S, Patel A, Singh OP, Ravindranath R (2006) Rainwater harvesting in India: some critical issues for basin planning and research. Land Use and Water Resour Res 6:1–17

Kumar R, Singh RD, Sharma KD (2005) Water resources of India. Curr Sci 89:794–811

Moench M, Dixit A et al (2003) The fluid mosaic. Water governance in the context of variability, uncertainty and change. A synthesis paper. Nepal Water Conservation Foundation, Kathmandu

Mudrakartha S (2007) To adapt or not to adapt: the dilemma between long-term resource management and short-term livelihood. In: Giordano M, Villholth K (eds) The agricultural groundwater revolution. Opportunities and threats to development. CAB International Publishing, Colombo

Ostrom E, Gardner R et al (1994) Rules, games and common-pool resources. University of Michigan Press, Ann Arbor

Rodell M, Velicogna I, Famiglietti JS (2009) Satellite-based estimates of groundwater depletion in India. Nature 460:999–1002

Sanford WE (2002) Recharge and groundwater models: an overview. Hydrogeol J 10:110–120

Shah T, Burke J, Villholth K (2007) Groundwater: a global assessment of scale and significance. In: Molden D (ed) Water for food, water for life: a comprehensive assessment of water management in agriculture. Earthscan, Colombo

Shah T, Roy AD, Qureshi A, Wang J (2003) Sustaining Asia's groundwater boom: an overview of issues and evidence. Nat Resour Forum 27:130–141

Shah T, Singh OP, Mukherji A (2006) Some aspects of South Asia's groundwater irrigation economy: analyses from a survey in India, Pakistan, Nepal Terai and Bangladesh. Hydrogeol J 14:286–309

Sharda VN, Kurothe RS, Sena DR, Pande VC, Tiwari SP (2006) Estimation of groundwater recharge from water storage structures in a semi-arid climate of India. J Hydrol 329:224–243

Teuling AJ, Troch PA (2005) Improved understanding of soil moisture variability dynamics. Geophys Res Lett 32:4

Authors Biography

Claire J. Glendenning worked at the International Food Policy Research Institute, in India.

Willem Vervoort is an Associate Professor in Hydrology and Catchment Management, at the Department of Environmental Sciences at the University of Sidney, Australia.

Sustainable Management of Water Quality in Southeastern Minnesota, USA: History, Citizen Attitudes, and Future Implications

Neal Mundahl, Bruno Borsari, Caitlin Meyer, Philip Wheeler, Natalie Siderius and Sheila Harmes

Abstract The water resources of southeastern Minnesota, USA, have been exploited by humans for the past two centuries. The region's sedimentary (karst) geology holds vast underground aquifers with high-quality drinking water. Springs and seeps percolate from these aquifers in valleys to produce hundreds of kilometers of coldwater trout streams. Citizens in the region place high values on these surface and groundwater resources, protecting them from potential harm by becoming informed about threats and organizing in protest over resource contamination and perceived overuse. Agriculture, ethanol production, silica sand mining and processing, and urban development have all threatened the area's water resources and prompted citizen action. Recent regional studies have examined

N. Mundahl (✉) · B. Borsari
Department of Biology, Winona State University, 175 West Mark Street,
Winona, MN 55987, USA
e-mail: nmundahl@winona.edu

B. Borsari
e-mail: bborsari@winona.edu

C. Meyer
Olmsted County Environmental Resources, 2122 Campus Drive SE, Suite 200,
Rochester, MN 55904, USA
e-mail: meyer.caitlin@co.olmsted.mn.us

P. Wheeler
Rochester/Olmsted Planning Department, 2122 Campus Drive SE, Suite 100,
Rochester, MN 55904, USA
e-mail: wheeler.phil@co.olmsted.mn.us

N. Siderius · S. Harmes
Winona County Planning and Environmental Services, 177 Main Street,
Winona, MN 55987, USA
e-mail: nsiderius@co.winona.mn.us

S. Harmes
e-mail: sharmes@co.winona.mn.us

© Springer International Publishing Switzerland 2015
W. Leal Filho and V. Sümer (eds.), *Sustainable Water Use and Management*,
Green Energy and Technology, DOI 10.1007/978-3-319-12394-3_18

long-term trends in water quality, surveyed citizen attitudes and values, and made recommendations for monitoring and protecting both surface and groundwaters in southeastern Minnesota. A culture of water stewardship will continue to grow in this region, serving as a good model to follow wherever sustainable water management practices are being developed.

Keywords Sustainable water management · Karst geology · Driftless area · Citizen engagement

1 Introduction

The multi-use water resources of the Driftless Area of the United States' Upper Midwest (Fig. 1), and specifically those in southeastern Minnesota, have been under siege since the first European settlers arrived in the early 1800s (Thorn et al. 1997), and the region's citizens have been actively engaged in protecting these waters. This riverine landscape, missed by the most recent continental glaciers, is underlain by karst geology. Surface waters can quickly enter both shallow and deep, underground aquifers via cracks, fissures, and sinkholes in the intervening layers of limestone, sandstone, and shale (Schwartz and Thiel 1963).

The groundwater aquifers of the region currently provide water for >400,000 people and the industrial and agricultural activities that support the area's economy (Fig. 2). Shallow aquifers were impacted by contaminants from surface activities decades ago, forcing reliance on deeper, more protected aquifers (Lindgren 2001; Lee 2008). These deeper aquifers eventually emerge via springs from wooded valleys to form hundreds of kilometers of coldwater trout streams, which have been restored and rehabilitated (Thorn et al. 1997) with tax-generated public funding (via amendment to the state constitution) to support multi-million dollar trout fisheries and their associated tourism (Gartner et al. 2002; Hart 2008).

Intensive agriculture and livestock grazing through the early 1900s produced heavy soil erosion, filling waterways, and extirpating native fishes, but farmers successfully adapted numerous soil conservation practices to keep the soil in place (Thorn et al. 1997; Trimble 2013). Later, chemicals and fertilizers associated with industrialized agriculture drained into aquifers, contaminating drinking waters with herbicides, pesticides, and nitrates (Fig. 2). Applications have become more efficient and better timed to reduce the likelihood of these chemicals migrating into groundwater (Randall 2003).

Urban development and growth have increased the demand for drinking water, while negatively affecting shallow groundwater supplies (via poor septic systems) and surface waters (Lee 2008; Fillmore County SWCD 2010; Fig. 2). Mandated water-conserving fixtures and appliances, rain gardens, sanitary sewer extensions, and drought-resistant landscaping have counteracted many of these problems.

Ethanol production from corn and mining activities for silica sand (needed by the oil and gas industry for hydraulic fracturing) are expanding and threatening

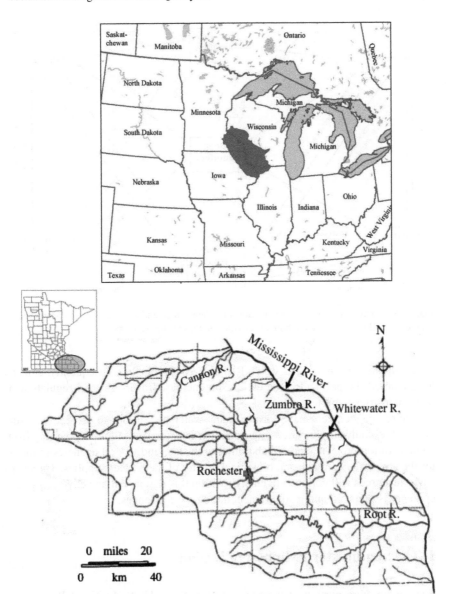

Fig. 1 Maps depicting the Driftless area (*dark shaded area*) covering portions of Minnesota, Wisconsin, Iowa, and Illinois, USA (*top map*) and the rivers and streams in southeastern Minnesota that are tributary to the Mississippi River (*lower map*). *Dashed lines* represent county borders, major rivers are labeled, and the city of Rochester is highlighted. Most of the region's designated coldwater trout streams lie within the watersheds of the Whitewater River and the Root River

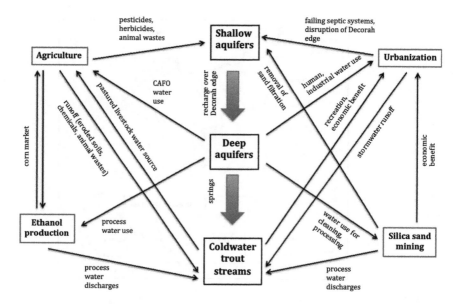

Fig. 2 Schematic depicting the interaction of groundwater and surface water in southeastern Minnesota and the influences of human activities on these water resources

water supplies (Schnoor et al. 2007; Richards 2012; Fig. 2). Consequently, citizens have entered the political arena calling for increased oversight and regulation of these industries to protect aquifers.

The region's citizens have formed many watershed groups to protect their surface and groundwater resources, participating in watershed summits, learning from demonstration projects, and collaborating with state and federal agencies to monitor the physical, chemical, and biological quality of their water supplies. Threats to water resources will continue to emerge within this region, but an actively engaged citizenry is prepared and ready to meet these new challenges.

2 The Southeastern Minnesota Region

The southeastern portion of the State of Minnesota, USA (44°N, 92°W) encompasses an area of 14,777 km² bounded on the east by the upper Mississippi River. It is part of a larger region (southeastern Minnesota, southwestern Wisconsin, northeastern Iowa, and northwestern Illinois) called the Driftless Area (Fig. 1) that was missed by the last continental glacier (Wisconsin glaciation) 15,000 years ago, but carved by its meltwaters (Fremling 2004). For 400 km within the Driftless Area, the Mississippi River flows through a gorge up to 200 m deep, carved downward through the ancient Paleozoic Plateau of sedimentary rock (limestone, sandstone, shale). Tributary streams further dissect the plateau, creating a complex of ridges, valleys, and precipitous blufflands.

The Minnesota portion of the Driftless Area spans portions of nine counties (Rice, Steele, Goodhue, Wabasha, Dodge, Olmsted, Winona, Fillmore, and Houston) encompassing four major watersheds: Cannon River, Zumbro River, Whitewater River, and Root River (Fig. 1). These systems all flow directly to the Mississippi River, carrying the runoff from urban areas, agricultural lands, and forests. Streams and rivers in this region often exhibit flashy hydrographs, rising rapidly in response to periodic heavy rainfall events and spring snowmelt to flood cities, villages, and farmlands within their valleys (Waters 1977).

Southeastern Minnesota is underlain by several hundred meters of sedimentary rocks formed under ancient seas during the Paleozoic Era 440–570 million years ago (Schwartz and Thiel 1963). Among these layers are various limestones and dolomites (Maquoketa, Dubuque, Galena, Platteville, Prairie Du Chien, St. Lawrence) with cracks and fissures that allow for rapid, vertical water movements, porous sandstones (St. Peter, Jordan, Mt. Simon) that serve as vast underground aquifers, and impervious shales (Decorah, Glenwood, Eau Claire) that restrict water movement. The region is categorized as a karst landscape, characterized by sinkholes, caves, subterranean rivers, springs, and disappearing streams (Fremling 2004). Groundwater flowing from the hundreds of springs and cave mouths gives rise to the coldwater trout streams that typify this region (Fig. 2).

The climate of southeastern Minnesota is the warmest and wettest of the entire state. Annual average temperature is 9.4 °C, with annual precipitation averaging 86.9 cm. The majority (75 %) of precipitation falls as rain from April to September. This climate results in an annual growing season of 155 days, with 890 growing degree-days (above baseline of 12.8 °C).

Agriculture is the dominant land use in southeastern Minnesota. Numbers of farms are declining, yet farms are becoming larger, with increasing field size, more soybean acreage, and decreasing acreages of small grains, forage crops, and pasture (Randall 2003). Potential impacts of these trends to water quality may include increased runoff, reduced base flows, thermal pulses, and increased nutrient, chemical, and fine sediment inputs (MN DNR 2003).

In 2010, the total human population for the nine-county region of southeastern Minnesota was 413,852. With over 106,000 people, Rochester is the area's largest city (and the third largest city in Minnesota; Fig. 1), increasing 38 % in 20 years. Four other cities (Faribault, Northfield, Owatonna, Winona) have populations between 20,000 and 30,000 people. Populations in three of the nine counties are expected to increase by >30 % by 2040 (Robertson 2012).

3 Trout Streams

Southeastern Minnesota currently has 181 designated trout streams, encompassing >1,265 km of stream length (Fig. 1). These streams support naturally reproducing populations of native brook trout and introduced brown trout, as well as put-and-take fisheries for introduced rainbow trout. The cold, clear water needed to support these trout is the result of the region's karst geology and abundant aquifers.

Trout management officially began in southeastern Minnesota in 1874 because promiscuous and largely unregulated fishing had reduced brook trout populations to low levels, requiring restricted harvest to sustain recreational fishing (Thorn et al. 1997). Brook trout were first stocked in 1878, and within the following 10 years, brown trout and rainbow trout both had been widely stocked. Despite these efforts, trout abundance plummeted in response to degrading habitat, warming waters, and greatly diminished spring flows (Thorn et al. 1997). These poor conditions and poor fisheries persisted through 1930.

A slow, gradual recovery of trout streams and their fisheries began in southeastern Minnesota in the 1930s and 1940s. After soil conservation practices were implemented to reduce erosion and flooding, stream habitat at first stabilized and then began to improve (Trimble 2013). The Minnesota Department of Conservation began to actively rehabilitate in-stream habitats in the late 1940s, completing 151 such projects involving >200 km of stream over 50 years (MN DNR 2003). Although widespread trout stocking was still necessary during the 1970s to maintain most fisheries, the need for stocking diminished during the 1980s and continues to decline as natural reproduction expands (Thorn et al. 1997). Currently, most trout populations in the region are continuing a >30-year trend of expanding abundances, with management efforts focusing on rehabilitating habitat and increasing stream access by acquiring public easements on private lands (MN DNR 2003).

Trout stream management in southeastern Minnesota is supported by three funding sources. State fishing license and trout and salmon stamp sales are a direct source of funding from within the state. The Federal Aid in Sport Fish Restoration Program provides federal money collected by taxing fishing equipment and motorboat fuel. Finally, a state constitutional amendment that dedicates a small portion of state sales taxes to within-state conservation and arts projects (Clean Water, Land, and Legacy Amendment) has provided millions of US dollars to protect drinking water sources and to protect, enhance, and restore wetlands, lakes, rivers, streams, and groundwater. The amendment required statewide voter approval in 2008, with 33 % of funds generated going specifically to the Clean Water Fund. Approximately, $7 million US from this fund has been used specifically to rehabilitate trout stream habitats in southeastern Minnesota.

These trout streams and the fishing opportunities they provide are major contributors to the economy of southeastern Minnesota. Trout anglers were estimated to use >520,000 angler-days fishing in these streams during a single year (Vlaming and Fulton 2002), spending at least $48 million US within the region to support their fishing activities (Gartner et al. 2002). Trout angler surveys suggest that region residents spend >$200 US/fishing trip and >$4,800 US/year, whereas non-region residents spend nearly $400 US/trip and >$3,700 US/year in their trout fishing pursuits (Hart 2008).

When examined in a broader context, trout fishing has an even greater economic impact on the region. For the entire Driftless Area of Minnesota, Wisconsin, Iowa, and Illinois, spending by trout anglers was estimated at $1.1 billion US/year (Hart 2008). In addition, state natural resources agencies have spent approximately $45 million US for stream restoration (725 km of stream improved, $62,000 US/km improvement cost) during the past 25 years to provide the fisheries that attract these trout anglers. Every $1 US spent on stream restoration returns $24.50 US to the regional economy and that return on investment occurs every year for the lifetime of the restoration projects (Hart 2008). Consequently, trout fishing is a significant economic driver within the Driftless Area, and protecting and enhancing these cold-water resources benefits all area residents.

4 Challenges to Maintaining Water Resources in Southeastern Minnesota

Surface waters and groundwater in southeastern Minnesota are abundant and accessible, but past human activities in the region have impaired water quality (Lee 2008; Minnesota Department of Agriculture 2012). In addition, a growing human population and greater water demands from expanding agricultural and industrial ventures threaten to consume water in volumes that may be unsustainable (Randall 2003; O'Dell 2007). The following sections summarize four ongoing challenges to sustainable water use in southeastern Minnesota (agriculture, ethanol production, silica sand mining and processing, and urban development) and describe how the region's citizens are addressing each challenge.

4.1 Agriculture

Southeastern Minnesota was the first region of the state settled by European immigrants during the mid-1800s. These early settlers removed the native vegetation (mixed hardwood forests, savannahs, and tallgrass prairies) for agriculture, lumber, and fuel (Waters 1977), creating small, subsistence-level, diversified farms dependent on oxen, horses, and humans for power (Granger and Kelly 2005). Croplands were located mostly on uplands and valley bottoms, with forested steep side slopes and livestock pastures in rolling terrain. By 1870, 80 % of Minnesota's population lived on the small farms in the southeast, and a shift to wheat monoculture had exhausted soils and forced a return to diversified farming (mixed livestock, poultry, corn, small grains, and hay). However, expanding farms and intensive livestock grazing on marginal lands through the 1920s led to severe soil erosion. Trout streams, already warmed after removal of riparian trees, became filled with mud and contaminated with livestock wastes, eliminating native brook trout and sculpin from many streams (Waters 1977).

Beginning in the 1930s and continuing to today, conservation practices such as contour farming, strip cropping, reduced tillage, improved forest management, rotational grazing, and terracing have greatly reduced soil erosion and agricultural runoff (Thorn et al. 1997). Streams have recovered and now support self-sustaining populations of brook and brown trout (Waters 1977). Despite these improvements, changing agricultural practices continue to affect the water resources of this region.

For the past half-century, agriculture in southeastern Minnesota has shifted steadily from small, diversified farms to large, row-crop (corn, soybeans) cash farming operations. Concurrently, concentrated animal feeding operations (CAFOs) expanded dramatically, especially for dairy cattle, beef cattle, and hogs. Watersheds in the region today vary from 40–70 % agriculture and 20–40 % forest, with dairy (59 %) and beef (24 %) cattle dominating livestock production and hogs and poultry (17 % combined) less common (Randall 2003). These changes in agriculture have had significant impacts on both surface and groundwaters within the karst region of southeastern Minnesota. Nutrients, pesticides, and herbicides have infiltrated streams and aquifers throughout the region, posing health hazards and economic hardships to citizens of both rural and urban areas (Fig. 2).

Nitrate concentrations have increased dramatically in groundwater aquifers in southeastern Minnesota during the past several decades (O'Dell 2007). This region is highly susceptible to groundwater contamination due to its geology (Minnesota Department of Agriculture 2012). These nitrates have been linked to inorganic fertilizers applied to corn, leakage from waste storage lagoons associated with CAFOs, and failure of residential septic systems. Since the 1980s, nitrates have been detected in nearly 100 % of well water samples tested within southeastern Minnesota, with typically 30–35 % of these samples having concentrations exceeding the Environmental Protection Agency's health risk limit for drinking water of 10 mg/L (Minnesota Department of Agriculture 2012). The widespread nature of nitrate contamination of well water has prompted health departments to issue consumption advisories, especially for infants.

The geology of the Driftless Area also makes its groundwater susceptible to contamination from fecal coliform bacteria and diseases associated with animal wastes. Manure storage and application and leaking septic systems all are potential sources of this contamination (Fig. 2). The risk of groundwater contamination is amplified by applications of manure from CAFOs to farm fields, especially those in sensitive areas near wells and sinkholes. Minnesota state laws are in place to regulate manure storage and application, although additional voluntary restrictions are required to adequately protect most aquifers from becoming contaminated (Minnesota Pollution Control Agency 2005). In addition, unexpected situations have occurred, such as sudden drainage of manure lagoons into previously unknown sinkholes, highlighting the sensitive and unpredictable nature of the region.

CAFOs within southeastern Minnesota place high demands on groundwater resources. For example, each dairy cow can consume 140–200 L/day of water (Thomas 2011). In Winona County alone there are 29,000 dairy cattle (US Department of Agriculture, 2012 Minnesota Agriculture Statistics), consuming >5 million liters per day of water, mostly from underground aquifers (Fig. 2). Additional large volumes of water are needed daily to support the necessary animal and barn cleaning operations of an operating dairy operation. A large (>1,000 animals) dairy CAFO can easily use as much water as a small community.

Surface waters in southeastern Minnesota continue to be challenged by polluted runoff (e.g., eroded soils, fertilizers and other chemicals, and animal wastes), despite long-term attempts to manage it via numerous conservation practices

(Fig. 2). State and federal agencies, working in conjunction with landowners and concerned citizens, have designed and installed water control structures, grassed waterways, buffer strips, and a myriad of other structures and practices intended to slow runoff and increase water infiltration. While individually effective, they collectively have produced only fair results. Recent modeling in a single watershed indicates that efforts to date have been effective in eliminating only 61 % of polluted runoff (Emmons and Olivier Resources, Inc., unpublished soil and water assessment tool (SWAT) model for Whitewater River watershed). Consequently, drainage from southeastern Minnesota and other farmlands throughout the United States' Upper Midwest often are implicated as the cause of the Gulf of Mexico's dead zone, the largest hypoxic zone in the United States. High-nutrient runoff from this region of intensive agriculture may be responsible for up to 70 % of the nutrient loading that reaches the Gulf of Mexico via the Mississippi River (NOAA 2009).

During the past 10–15 years, state agencies and local watershed groups have worked together to study problems in various drainages within southeastern Minnesota, with a goal of more accurately defining the problem and devising solutions that are realistic and achievable over the short term. Ultimately, this study and planning will culminate in surface waters that meet the water quality standards as set forth in the United States Clean Water Act, as enforced by the US Environmental Protection Agency. To date, both watershed-specific and region-wide total maximum daily load (TMDL) plans have been developed and approved for such pollutants as turbidity (and/or total suspended solids), fecal coliform bacteria, nutrients (nitrates, phosphates), and others. The Minnesota Pollution Control Agency is the lead agency in charge of developing such plans for surface waters in Minnesota.

In June 2013, Minnesota and US government officials announced that four Minnesota watersheds dominated by intensive agriculture had been chosen to participate in a new program, the Minnesota Agriculture Water Quality Certification Program. Using $9.5 million US from federal and state sources, the 3-year program will seek out farmers in each watershed willing to voluntarily adopt and implement precise, site-specific methods to protect surface and groundwaters from agricultural pollutants. Funds will help pay farmers to implement strategies to help mitigate their pollutant-causing activities, with participating farmers then being exempt for 10 years from new water quality regulations. The Whitewater River watershed (Fig. 1), extending across portions of three counties in southeastern Minnesota, was one of the watersheds selected for this pilot program.

4.2 Ethanol Production

There are 21 ethanol production facilities located in Minnesota, mostly using a dry mill process to produce 4.2 billion liters of ethanol/year from corn. Most of this production is used as an additive to gasoline to meet a state mandate that all gasoline sold in the state must contain at least 10 % ethanol. Minnesota was the first state in the USA to require ethanol in gasoline and will raise the minimum requirement for ethanol in gasoline to 20 % beginning in 2015. Minnesota also has the most E85 (85 % ethanol content) gasoline stations in the country, providing fuel for vehicles designed to burn this high ethanol-content fuel.

Through 2011, farmers in the United States received $6 billion US/year in federal subsidies to grow corn for ethanol production. Ethanol producers also received $0.12 US/L in tax credits to encourage ethanol production. Subsidies and tax credits both ended beginning in 2012, but ethanol production has continued to increase, largely because market demand for ethanol remains high to meet the government mandates for blended fuels.

Southeastern Minnesota has two ethanol production plants, located in Claremont and Preston. To maximize cost effectiveness, plants are sited to obtain corn from within an 80 km radius, to be near inexpensive railroad transportation and to gain access to an abundant water source. Typically, three liters of water are needed to produce each liter of ethanol. Some of this water is used once and discharged, although most water can be (but is not always) reused within the plant. A typical ethanol plant in Minnesota that can produce 150–350 million liters of ethanol/year could use up to 1.25 million liters of water/day (Fig. 2), the equivalent volume used by a city of 5,000 people (Schnoor et al. 2007).

During 2008 to 2011, a regional corporation planning to build a new ethanol production plant near the city of Eyota became embroiled in a controversy with area citizens regarding the plant's projected water use. The plant was designed to process >53,000 bushels of corn/day to produce >200 million liters of ethanol/year, while requiring >4 million liters of water/day. Citizens felt that this water demand would place the municipal water supply at risk in the long term, and discharges of warmed process waters would threaten local trout streams (Fig. 2). After state agencies determined that the plant's water needs would not harm local water resources (groundwater and surface), area citizens organized (Olmsted County Concerned Citizens) and filed a lawsuit against the state agencies. Years of legal proceedings ultimately determined that local water resources would not be harmed by the plant's water use, but plant investors were unable to raise sufficient capital to proceed with construction and the project was suspended.

4.3 Silica Sand Mining and Processing

Operations for the mining and processing of silica sands are expanding in southeastern Minnesota. Although used in many applications (e.g., water filtration, glass manufacture, industrial casting, sand blasting, and producing concrete), the current boom in the silica sand market is being driven by its use as a proppant in hydraulic fracturing for oil and gas production. Although no hydraulic fracturing occurs in Wisconsin or Minnesota, these states have the largest deposits of silica sand in the United States. Wisconsin has 60 mines and 30 processing facilities (WI DNR 2012), whereas Minnesota has only eight operating mines and a similar number of processing sites (Richards 2012).

Eight of the nine counties in southeastern Minnesota have significant deposits of silica sands near the surface, but only three mines are operational (Richards 2012). In 2012 and 2013, five of these counties imposed silica sand mining moratoria (temporary bans on the development of new mines and processing facilities) after concerns arose regarding environmental issues associated with mining and processing activities. In particular, area citizens were concerned about the potential for mining and processing activities to cause groundwater depletion, water and air pollution, increased truck traffic resulting in rapid deterioration of roads, and ultimately damage to the region's scenic beauty.

Silica sand mines and processing facilities may use hundreds of thousands to millions of liters of water/day, mostly from groundwater. Mines may use water to control dust, whereas processing facilities use water to wet sort the sand into different sizes classes for various applications. This high rate of water use concerns nearby citizens, who worry about dewatering of their own water wells, or even aquifer depletion in areas with slow groundwater recharge (Fig. 2).

Area residents also are concerned about potential water pollution problems from flocculating agents added to the water used in silica sand processing. Wash water additives such as polyacrylamides help in removing unwanted minerals and fines from the sand. Wash water containing acrylamides may infiltrate into the groundwaters when washed sands are placed in surge piles to dry (WI DNR 2012). Acrylamides will biodegrade in aerated soils, but soils beneath surge piles will be waterlogged, not aerated. The US Environmental Protection Agency has a Maximum Contaminant Level Goal of zero for acrylamides in public drinking waters, since long-term consumption of acrylamide-contaminated water can lead to blood and nervous system disorders and increased chance of developing cancer (WI DNR 2012).

After a contentious debate on silica sand mining in southeastern Minnesota, the Minnesota state legislature passed laws in 2013 regulating mining and processing activities and establishing a state-level commission to help local units of government with permitting and regulatory oversight. At the request of Trout Unlimited (a private organization with a mission to keep the United States' coldwater fisheries and their watersheds safe from environmental threats), state legislators considered setback regulations for mining to protect trout streams. Consequently, proposed silica mines within 1.6 km of designated trout streams now require additional permitting and complete hydrogeological evaluations to identify potential threats to those streams (Minnesota Statutes, section 103G.217). The state governor has stated that, if recommended by the state legislature, he would support a total ban on silica mining in southeastern Minnesota to protect sensitive water resources in this region.

4.4 Urban Development

Rochester is the largest and fastest growing city in southeastern Minnesota. The city and outlying towns that surround it comprise a metropolitan area with a population of >200,000 people. A rapidly growing city of this size has had several significant impacts on the region's water resources, ranging from increasing demands on groundwater, to constraining and controlling streams and rivers within its jurisdictional boundaries, to managing storm water runoff from hundreds of kilometers of city streets and other impervious areas (Fig. 2).

Rochester obtains its drinking water from wells tapping into deep groundwater aquifers known as the St. Peter and Prairie du Chien aquifers, or collectively as the Lower Carbonate aquifer. These aquifers contain sufficient water resources to support the continuing growth of the city (and several other communities within the region) well into the current century and beyond. The city currently uses >20 billion liters of water/year from these aquifers. The deep aquifers contain high-quality water supplies because they are protected from potentially polluting surface activities by overlying impervious layers of shale. The clay-rich Decorah shale formation is the most important of these protective layers (Schwartz and Thiel 1963), with a maximum thickness of approximately 12 m.

Because Rochester is located within a river valley surrounded by rolling hills, the Decorah shale formation often is exposed on hillsides throughout the city. Areas where the edges of the Decorah shale are exposed have been found to be important recharge zones for the deep aquifers that provide Rochester and 15 other communities (across six counties) with their drinking water (Lindgren 2001; Fillmore County SWCD 2010). Rainfall percolates down through soil and porous sedimentary rock formations of the Galena or Upper Carbonate aquifer until it reaches the Decorah shale, which prevents it from further downward movement. However, at the Decorah edge, these waters can spill out over the Decorah shale through a thin soil covering before sinking into deeper, porous sedimentary rock layers that connect to the Lower Carbonate aquifer used by Rochester. Estimates credit the Decorah edge as the site of 50–60 % of the recharge waters entering the Lower Carbonate aquifer (Lee 2008; Fig. 2).

Although the shallow-lying Upper Carbonate or Galena aquifer often contain high levels of nitrate (15–20 mg/L, higher than the current drinking water standard of 10 mg/L; Rochester abandoned use of this aquifer for drinking water in 1950 because of this contamination), nitrate concentrations of waters flowing over the Decorah edge can decrease by >90 % (Lee 2008). This denitrification is produced by a diverse community of wetland plants that exists in the saturated soils at the Decorah edge, a community with some of the highest diversity of any wetland type in the state of Minnesota. Nitrate removal by these wetland communities within Rochester alone has been valued at $5 million US/year, based on current treatment costs for removing nitrate from drinking water supplies (Lee 2008). These wetlands also remove nitrate pollutants discharging from springs and seeps to form the headwaters of the Cannon, Zumbro, Whitewater, and Root rivers, a filtering and denitrification significant enough to impact water quality in the nearby Mississippi River (Fillmore County SWCD 2010).

Prior to understanding the importance of the Decorah edge in protecting, sustaining, and purifying the drinking waters of Rochester and its neighboring communities, residential and commercial development were allowed to proceed along the Decorah edge as long as they met existing zoning and wetland ordinances. Many of these developments encountered on-site water management issues, with the many seeps and springs causing a multitude of drainage issues. Construction equipment became mired literally in muddy soils that refused to dry out, the basements of homes had continual water infiltration problems, and Rochester absorbed costs of nearly $1,000 US/household/year when basement drainage from homes built on the Decorah edge was directed into the city's sanitary sewer system (Lee 2008).

Following the lead of neighboring counties and municipalities, Rochester and Olmsted County amended their zoning and wetlands ordinances to protect the Decorah edge (Lee 2008). Restrictions were placed on development on sites with specific hydric soil types, sites near springs, seeps, streams, or waterways, sites with high water tables, and sites adjoining steep slopes. These and similar restrictions are now in place throughout the region wherever the Decorah edge is present, stretching from Rice County southeasterly for >320 km into northeastern Iowa (Fillmore County SWCD 2010).

Rochester developed in a river valley, occupying the floodplain and hillsides adjacent to the South Fork of the Zumbro River. Several tributaries of the Zumbro also join the river within the city, including Salem, Cascade, Silver, Bear, and Willow creeks. Dams were constructed on the river and creeks early in the city's history, providing power, water, and recreation. Today, the city lies along 14 km of the Zumbro River and >160 km of its tributary creeks.

Because of its location and the extreme flashy nature of the region's streams and rivers in response to sudden, heavy rain events and snowmelt (Waters 1977), Rochester has been prone to flooding since it was established in 1854. The city has a history of severe flooding, causing both loss of life and severe economic hardship. The city has fought back against this flooding by constructing (and reconstructing) various flood-control dams, levees, ditches, and storm sewers to control and redirect floodwaters around and through the city. These activities escalated after especially severe flooding in 1978. However, the vast amount of impervious surface area within the city (buildings, streets, sidewalks, parking lots) prevents infiltration of rainfall (79 cm/year average) and snowmelt (112 cm/year average), forcing it into waterways and increasing the potential for flooding.

Rochester has an annual budget of $3.1 million US for storm water management. This budget covers maintenance of 145 storm water retention ponds (and coordination on 216 retention ponds owned by other entities), 675 km of storm sewers, 15,700 storm sewer catch basins, 528 km of open roadside ditches, and 1,755 outfalls to receiving waters (Rochester Public Works Department 2013). The municipal storm water permit issued to Rochester by the state mandates the city to minimize impacts of storm water on receiving waters.

Rochester has several zoning ordinances in effect to retain storm waters on the land for later infiltration, to manage storm water flows, and to prevent flows from carrying pollutants to streams and rivers. Construction permits mandate that new commercial and residential developments retain significant proportions of their storm waters on-site in retention basins and/or rain gardens, reducing flows to surface waters during rain events and allowing time for waters to infiltrate into soils that can filter and eliminate potential pollutants (Rochester Public Works 2013).

In 2009, Rochester launched a cost-share grant program, Realize Raingardens Rochester, to promote the installation of rain gardens within the city on residential property or parcels owned by nonprofit organizations. The program provides up to 50 % of costs to design and install rain gardens to help demonstrate to the public how rain gardens can retain and treat storm water while beautifying neighborhoods and creating wildlife habitat (Realize Raingardens Rochester 2013). A private nature center partnered with the city to develop and conduct educational, how-to, rain garden classes and grant-writing workshops to help citizens with their rain garden project applications. To date, this program has been successful in establishing many highly visible rain gardens within the city, encouraging many other residents and groups to establish their own rain gardens on their own properties.

5 Trends in Surface Water Quality

Water resources within the southeastern Minnesota region, and specifically the Mississippi River-Winona watershed (which includes the Whitewater River watershed and several nearby, smaller watersheds that drain directly to the Mississippi River), have been the focus of hundreds of different projects and programs during the last century (Crawford et al. 2012). Both groundwaters and surface waters have been studied, but the greatest volume of information and the majority of investigations have been directed toward streams and rivers. Projects and programs have gathered data on nutrients, metals, nonmetals, physical variables, radiochemicals, pesticides, bacteria, invertebrates, and fish (Crawford et al. 2012).

The Mississippi River-Winona watershed covers 170,000 ha of mostly agricultural (row crops and livestock grazing) and forested lands, with some urban development. Farmers own 88 % of watershed lands, but non-farmers comprise 97 % of the area's electorate. Surface waters in the watershed are on the State of Minnesota impaired waters list for bacteria (*Escherichia coli*), nitrates, turbidity, and mercury, and some aquifers contain elevated levels of bacteria and nitrates.

A grant from Minnesota's Clean Water Fund was used to compile all existing water quality data gathered within the watershed, analyze these data for trends and other significant features, identify limitations and/or data gaps, and provide recommendations for future water quality monitoring. Historic water quality data for the watershed's surface waters were compiled from 225 different programs and analyzed. Nearly 296,000 data points from 20,000 sampling events at 136 unique monitoring sites were used to examine trends during the past 40 (for water quality data) to 80 (for water discharge data) years. Only 12 unique sites on seven stream reaches (one on Garvin Brook, six in the Whitewater River drainage) had adequate data and periods of record for long-term trend analysis (Crawford et al. 2012).

Long-term trend analysis was conducted on 10 variables across the seven stream reaches within the Mississippi River-Winona watershed: annual discharge, suspended sediment, total suspended solids (TSS), total phosphorus, ammonia, biological oxygen demand (BOD), nitrate, chloride, sodium, and sulfate. All 10 variables were not monitored at all stream reaches for durations sufficient for long-term trend analyses, but 90 % of variables were examined at two to six stream reaches each.

Within the watershed, stream discharge has remained steady or increased during the period of record. This has coincided with an increase in yearly precipitation within the region since the 1950s, although yearly precipitation was only weakly correlated with discharge at any of the stream sites (Crawford et al. 2012). Because many streams within the watershed are influenced more by groundwater discharge than by surface runoff (Schwartz and Thiel 1963), long-term increases in stream discharge may be the result of changing aquifer dynamics.

Levels of TSS, total phosphorus, ammonia, BOD, sulfate, and atrazine have declined in streams within the watershed where data are sufficient to analyze long-term trends. Several of these variables are correlated with one another (TSS, total phosphorus, ammonia, BOD), suggesting related sources. High levels of these variables were associated with past time periods characterized by severe soil erosion and runoff from livestock pastures (Trimble 2013). Improved soil conservation practices apparently have been successful in reducing concentrations of these pollutants in surface waters, even in the face of increasing precipitation. Sulfate concentrations have declined since peaking in the mid-1980s, illustrating the impact of acid rain control measures in the Clean Air Act (Crawford et al. 2012). Recent downward trends in concentrations of the herbicide atrazine and its breakdown products, even as atrazine use increases, may indicate that better application procedures have been developed and put into practice within the watershed.

In contrast to declines in some pollutants, nitrate and chloride levels have more than tripled in the watershed's rivers and streams since 1970 and sodium levels have risen significantly. Nitrate concentrations have increased from 1 mg/L prior to 1970 to >6 mg/L in 2010, with highest levels occurring during summer base flow periods (Crawford et al. 2012). Based on 2009 data, base flow nitrate levels are significantly correlated ($r^2 = 0.68$) with percent row crop agriculture upstream from sampling locations. Because base flows of these coldwater streams depend largely on spring discharges from aquifers (Waters 1977), high stream nitrate values highlight the increasing problem of contamination of aquifers by agricultural fertilizers, especially in shallow aquifers in the upper reaches of the watershed (Crawford et al. 2012).

Chloride and sodium concentrations in streams and rivers in the Mississippi River-Winona watershed have been increasing since the 1970s due to expanding use of water softener salt (NaCl), road deicing salt (NaCl), and potassium chloride (KCl) fertilizer (Crawford et al. 2012). The karst geology of the region allows salts to infiltrate aquifers from residential septic fields, roadside ditches, and agricultural lands. Average chloride concentrations appear to be stabilizing at 15–20 mg/L (Crawford et al. 2012).

Long-term water quality trends within the Mississippi River-Winona watershed indicate that several water pollutants are declining, whereas others continue to worsen. Land conservation measures within the watershed appear to have controlled significant amounts of soil erosion, reducing delivery of eroded soils and associated water pollutants via surface runoff. However, the underlying karst geology continues to allow many pollutants rapid entry into shallow aquifers that discharge directly into streams and rivers. More consistent water quality monitoring of streams and rivers is needed within the watershed to track future trends in water pollutants as additional efforts are made to further reduce surface runoff and to better manage infiltration of fertilizers and salts into shallow aquifers.

6 Citizens' Attitudes and Opinions

The citizens of southeastern Minnesota have become increasingly more concerned with the quality and availability of their water resources during the past several decades. They have taken advantage of opportunities to learn more about these resources and threats to them and have become active and engaged in protecting them. A variety of watershed-based studies, projects, and initiatives have heightened the public's awareness of water resources and potential threats, and citizens are requesting greater involvement in decision-making related to surface and groundwaters. The Minnesota Pollution Control Agency encourages citizens to become more active in water quality decision-making and maintains a Web site to help with civic engagement for watershed projects (http://www.pca.state.mn.us/index.php/water/water-types-and-programs/minnesotas-impaired-waters-and-tmdls/project-resources/civic-engagement-in-watershed-projects.html).

Watershed projects in southeastern Minnesota have become commonplace during the last 25 years, combining citizen energy, attitudes, and values with agency expertise to address water resources issues. Many citizens volunteer their time to collect basic water quality information on their neighborhood stream or river, helping to build databases that allow agency personnel to target problem areas for maximum benefit (http://www.pca.state.mn.us/index.php/water/water-types-and-programs/surface-water/streams-and-rivers/citizen-stream-monitoring-program/index.html). Minnesota has 400 volunteers monitoring >500 stream and river sites across the state's 10 major river basins.

In recent years, citizen water forums or summits have been convened to allow for direct communication between southeastern Minnesota citizens and the agency personnel charged with protecting the region's water resources. These water summits provide a face-to-face approach for informing the public about agency studies and conclusions, while providing citizens with an opportunity to speak directly with agency staff about their concerns and problems related to water issues.

As a specific example, the Mississippi River-Winona Watershed held two watershed citizen summits, spaced eight months apart, during 2012 and 2013. Attendees listened to presentations about long-term water quality trends within the watershed and the results of water opinion surveys given to watershed landowners. They also participated in a series of round table discussions involving farmers, urban residents, educators, students, local politicians, and agency staff. This diverse group of stakeholders with varying perspectives on water quality issues shared dinner together and discussed their concerns and ideas about how best to achieve their common goal of clean water. Ultimately, these and future water summits will serve to develop a vision and strategy for protecting and/or restoring the water resources within the watershed.

A second approach to engage area citizens in water issues and decision-making has involved the use of surveys given to various groups of residents in southeastern Minnesota. While lacking the face-to-face nature of the citizen summit described above, surveys, if developed and administered appropriately, can produce statistically valid data that can be used to direct and focus actions on water resource protection and management. Two surveys administered to Mississippi River-Winona Watershed residents in 2011 and 2013 will be used to illustrate this approach.

During either 2011 or 2013, 3,374 residents (from a potential pool of 18,722 households) of the Mississippi River-Winona Watershed in southeastern Minnesota were sent a six-page questionnaire via US mail to evaluate the water quality knowledge, attitudes, needs, and expectations of a diverse group of watershed residents (Wheeler 2013). Only residents in a small, select sub-watershed received surveys in 2011, whereas residents in the remainder of the watershed received them in 2013. Valid responses were received from 1,042 residents, a response rate of 30.8 %. The high rate of response and the small sample population produced a maximum response confidence interval (95 %) of ±4.3 %. For analysis, respondents were categorized as city residents, non-farm rural residents, small-farm (4–50 ha) residents, or large-farm (>50 ha) residents.

The surveys revealed six key findings about water and the residents of the watershed (Wheeler 2013):

1. An overwhelming majority (>78 %) of residents from all categories of residence want clean drinking water, streams as clean as their natural condition, and fish from local streams that are safe to eat.
2. A high proportion of private well users are uninformed about the source of their water, the safety of their well, or a source of information about well water quality.
3. Relatively high proportions of residents consider themselves somewhat or very uninformed about specific water issues within the watershed.
4. Rural residents consider the county extension services and the soil and water conservation districts as their significant information sources on water issues, preferring information in the form of printed fact sheets.

5. Residents favored local government actions, neighbor interaction, grassroots actions, and education as ways to protect water quality.
6. Despite widespread consensus on water quality issues, there were several significant differences in opinion and attitude between large-farm residents and all other respondent groups. For example, large-farm residents were much more likely to rate current stream water quality as good to excellent, whereas city residents most often rated stream water quality as fair or poor. In addition, the majority of large-farm residents listed urban runoff as the chief cause of water quality problems, whereas city dwellers listed agriculture (cropland, livestock) as the largest water quality problem.

It is apparent that both citizen summits and surveys are useful methods for obtaining information on the concerns and attitudes of watershed residents regarding water quality. Surveys can provide a wealth of useful information that can be helpful when establishing goals and objectives for protecting or restoring water quality within a watershed, while at the same time highlighting disparities that may exist among various citizen groups. Bringing these various citizen groups together in citizen summits and allowing them to present and discuss their perceptions and ideas in an informal, nonthreatening environment can be beneficial and enlightening to all citizen groups. Combining the use of both tools is the logical way for natural resource agencies to communicate with the public and to develop the citizen buy-in often required for water quality issues to be addressed successfully.

7 Conclusion

Citizens of southeastern Minnesota have been prompted to action whenever threats to their region's water resources have occurred. They value clean drinking water and high-quality streams for fishing, and they are willing to educate themselves about new activities that may threaten the quality and future availability of these resources. They are not out to squelch all projects and activities that threaten their water. Rather, they are willing to seek out solutions that allow farming, development, and other activities to continue while still protecting the valuable water resources needed by all residents.

Efforts in this watershed and this region continue toward improving and strengthening a culture of water stewardship. Citizens will continue to protect their water resources against potential threats, by staying informed on the quality of the resources and staying connected and engaged with governmental agencies charged with protecting regional waters. This engagement and partnering between the public and agencies is one of the key components leading to public buy-in and ultimately to success of projects. Successful approaches in water management within our region can be used as models when developing sustainable water management practices elsewhere.

Acknowledgments We thank Terry Lee (Olmsted County Environmental Services Coordinator) for his insight and assistance with information gathering, and Kimm Crawford (Crawford Environmental Services) for his analysis and summary of complex water quality data.

References

Crawford K, Meyer C, Lee T (2012) Mississippi river-Winona watershed water quality data compilation and trend analysis report. Olmsted County Environmental Resources, Rochester

Fillmore County Soil and Water Conservation District [SWCD] (2010) Protecting a valuable ecosystem: the Decorah edge. Fillmore County SWCD, Preston

Fremling C (2004) Immortal river: the upper Mississippi in ancient and modern times. University of Wisconsin Press, Madison

Gartner W, Love L, Erkkila D, Fulton D (2002) Economic impact and social benefits study of coldwater angling in Minnesota. University of Minnesota Extension Service, St. Paul

Granger S, Kelly S (2005) Historic context study of Minnesota farmsteads, 1820–1960. Minnesota Department of Transportation, St. Paul

Hart A (2008) The economic impact of recreational trout angling in the Driftless Area. Trout Unlimited, Driftless Area Restoration Effort, Verona

Lee T (2008) Decorah edge: a critical water supply component. Olmsted County Environmental Services, Rochester

Lindgren R (2001) Ground-water recharge and flowpaths near the edge of the Decorah-Platteville-Glenwood confining unit. Geological survey, water-resources investigations report 00-4215, Rochester, Minnesota, US, Mounds View

Minnesota Department of Agriculture (2012) Minnesota department of agriculture summary of groundwater nitrate-nitrogen data. Minnesota Department of Agriculture, St. Paul

Minnesota Department of Natural Resources [MN DNR] (2003) Strategic plan for coldwater resources management in southeast Minnesota, 2004–2015. Minnesota DNR Division of Fisheries, St. Paul

Minnesota Pollution Control Agency (2005) Applying manure in sensitive areas: state requirements and recommended practices to protect water quality. Minnesota Pollution Control Agency, St. Paul

National Oceanic and Atmospheric Administration [NOAA] (2009) Dead zones: hypoxia in the Gulf of Mexico. National Oceanic and Atmospheric Administration, Washington, DC

O'Dell C (2007) Minnesota's groundwater condition: a statewide view. Minnesota Pollution Control Agency, St. Paul

Randall G (2003) Present-day agriculture in southern Minnesota—is it sustainable? University of Minnesota Southern Research and Outreach Center, Waseca

Realize Raingardens Rochester (2013) Realize Raingardens Rochester happenings. City of Rochester, MN

Richards J (2012) Industrial silica sands of Minnesota: frequently asked questions and answers. Minnesota Department of Natural Resources, Division of Lands and Minerals, St. Paul

Robertson M (2012) Minnesota population projections 2015–2040. Minnesota State Demographic Center, St. Paul

Rochester Public Works Department (2013) Managing Rochester's storm water. City of Rochester, MN

Schnoor J, Doering O, Entekhabi D, Hiler E, Hullar T, Tilman D, Logan W, Huddleston N (2007) Water implications of biofuels production in the United States. National Academies Press, Washington, DC

Schwartz G, Thiel G (1963) Minnesota's rocks and waters: a geological story. University of Minnesota Press, Minneapolis

Thomas C (2011) Drinking water for dairy cattle: parts 1 and 2. Michigan State University Extension, Lansing

Thorn W, Anderson C, Lorenzen W, Hendrickson D, Wagner J (1997) A review of trout management in southeast Minnesota streams. N Am J Fish Manage 17:860–872

Trimble S (2013) Historical agriculture and soil erosion in the upper Mississippi valley hill country. CRC Press, Boca Raton

Vlaming J, Fulton D (2002) Trout angling in southeastern Minnesota: a study of trout anglers. University of Minnesota, St. Paul

Waters T (1977) The streams and rivers of Minnesota. University of Minnesota Press, Minneapolis

Wheeler P (2013) Mississippi-Winona watershed resident survey: summary of results. Whitewater Watershed Joint Powers Board, Lewiston

Wisconsin Department of Natural Resources [WI DNR] (2012) Silica sand mining in Wisconsin. Wisconsin Department of Natural Resources, Madison

Authors Biography

Neal Mundahl is a professor at the Department of Biology at Winona State University, USA

Bruno Borsari is an Associate Professor of Biology at Winona State University, USA

Caitlin Meyer is an environment analyst at Olmsted County Environmental Resources, in Rochester, Minnesota, USA

Philip Wheeler is a planning director at the Rochester/Olmsted Planning Department, in Rochester, Minnesota, USA

Natalie Siderius is an Economic Development and Sustainability Director at Winona County Planning and Environmental Services, in Winona, Minnesota, USA

Sheila Harmes is the Whitewater River Watershed Project coordinator at Winona County Planning and Environmental Services, in Winona, Minnesota, USA

Social, Religious, and Cultural Influences on the Sustainability of Water and Its Use

Marwan Haddad

Abstract Sustainability of water and its use in quality and quantity and in time and space is closely related not only to technical, technological, and economic aspects and influences but also to social, religious, and cultural aspects and influences. A close balance of both groups of variables is important to maintaining sustainable efficient, safe, and renewable water supply, social equity, public health, and ecosystem as well as minimizing water pollution and depletion. In this chapter, emphasis will be given to (1) detailing the influences on water sustainability and use from humanistic origin such as gender equality, involvement, and participation, unwise water use and overuse, colonization, unilateralism, and conflicts, religion and faith guidance, shortsightedness in water policies and strategies, cultural traditions and customs, and human rights and the ethic of care, and (2) proposing adaptation actions to minimize and/or reverse influences under (3) such as adapting measures for maintaining balance of water availability and use, rethinking water policies and strategies, adapting measures for public involvement and participation in water service decision making including women, adapting behavioral change measures for maintaining cultural traditions and customs and respecting related faith guidance, adapting measures for maintaining equity, equality, and justice in water use and allocation, and adapting measures for minimizing and/or resolving conflicts and disagreement related to water and its use.

Keywords Water sustainability · Gender · Religion · Cultural traditions · Social change

M. Haddad (✉)
Environmental Engineering and Civil Engineering Department, Water and Environmental Studies Institute (WESI), An-Najah National University, Nablus, Palestine
e-mail: haddadm@najah.edu; haddadm50@email.com

© Springer International Publishing Switzerland 2015
W. Leal Filho and V. Sümer (eds.), *Sustainable Water Use and Management,*
Green Energy and Technology, DOI 10.1007/978-3-319-12394-3_19

1 Introduction

Most people in the world today have an immediate and intuitive sense of the urgent need to build a sustainable future (UNESCO 1997). For water status, over one billion people are without access to safe water, while twice as many lack access to improved sanitation services (Jones and Silva 2009). Water sustainability is a common interest for the overall society being private or public, urban or rural, and poor or affluent, all economic sectors, and either at present or in the future. Severe interconnected water challenges including *increasing water and food demand because of increased population growth* and enhanced socioeconomic development, expanded irrigated agriculture, land use change, declining water quality, aging water infrastructure, climate change, and competition among sectors and countries for the resource, has led to significant impacts on regional eco-security, ecosystem service, and human health (Wang and Li 2008; Larsen 2014; Gutierrez 2014). Consequently, water management and water sustainability are major challenges facing mankind and increasingly becoming limits or obstacles to development and need to be well planned and prepared for both as a process and by all involved from governments to individuals.

However, water sustainability is multifaceted and closely related to several sub-sustainabilities including sustainable water resources for fulfilling needed supply and demand, sustainable water infrastructures, sustainable water quality, sustainable rivers/basins, sustainable agriculture, sustainable cities, sustainable environment, sustainable industry, sustainable tourism, sustainable energy supply, sustainable ecosystems, sustainable social and economic conditions, and others.

The definition of sustainability for water is not clearly understood and sometime it is complex and overloaded with unrealistic goals. Some simply defined it as the transition from thinking short term to thinking long term (Scaller 2007), while others defined it as an enduring, balanced approach to economic activity, environmental responsibility and social progress (British Standards Institution 2010), and a third group emphasized the dual productive/destructive potentials of water, indicating its inherent economic, social and environmental complexity (Allan et al. 2013). Sustainability of a water source as used in this paper would mean the continued qualitative and quantitative existence of the water source for the various uses without serious interruptions or negative impacts on environment, ecology, other natural resources, or people neither now nor in the future.

It was reported that there have been extensive efforts on measuring sustainability in the past few decades. One example is the development of assessment tools based on sustainability indicators. Several individuals and organizations have suggested various indices for assessing sustainability and on various levels (Li and Yang 2011; Juwana et al. 2012; Ioris et al. 2008; Larson et al. 2013). If we take into consideration the above sub-sustainabilities and the various water spheres and activities, we find it difficult to numerically measure or accurately identify water sustainability (Wood 2003). It may be subjectively identified.

One of the key messages achieved and agreed upon in the water security and sustainability roundtable of the Sixth World Water Forum was the lack of cooperation between sector's major actors, stakeholders, and consumers that can result in costly and ineffective water security and sustainability solutions. We must keep this in mind to avoid water losses and food waste (Global Water Framework 2012).

The main emphasis of research and development in the twentieth century including that on water was on the technical, technological/industrial, knowledge development and transfer, and economic aspects of water. Sustainability of water and its use in quality and quantity and in time and space is closely related and important not only to (1) historical known materialistic or physical aspects and influences including technical, technological, and economic but also to (2) humanistic concerns including social, religious, and traditional and cultural aspects and influences. Accordingly, combining and integrating both attributes are very essential for better and higher level water sustainability.

Orlove and Caton (2010) identified five central themes in water sustainability including value: natural resources and human rights, equity: access and distribution, governance: organization and rules, politics: discourse and conflict, and knowledge/education: local/indigenous and scientific systems. These themes were related to three water sites: watersheds, water regimes, and waterscapes.

Little was conducted and published covering societal, cultural, and humanistic concerns in water management and sustainability (Pawitan and Haryani 2011). Some studies proposed a sustainable approach to secure future for fresh water by developing a plan that draws all "new" water from better use of existing supplies and to change habits and attitudes and conserve more water (Brandes and Kriwoken 2006; Brandes and Brooks 2007), others related effective water management to tackling poverty reduction and water productivity improvement (Giordano 2009), or to human health and animal welfare (Campbell 2009).

It is important to note the need for comprehensive undertaking of aspects and influences affecting water sustainability because if damage is already happened for any water resource, it is practically difficult to fix in time and space, in quality and quantity, and from technical and financial aspects, too. There are several important and interlinked aspects and limitations to maintaining and properly achieving water sustainability including governmental role, water use, and water resource development (see Fig. 1).

In this chapter, emphasis will be given and limited to influences and aspects of water sustainability related to social including gender and human rights and the ethics to care, religious, and traditional and cultural along with some related aspects such as unwise water use and colonization, and shortsightedness in water policy.

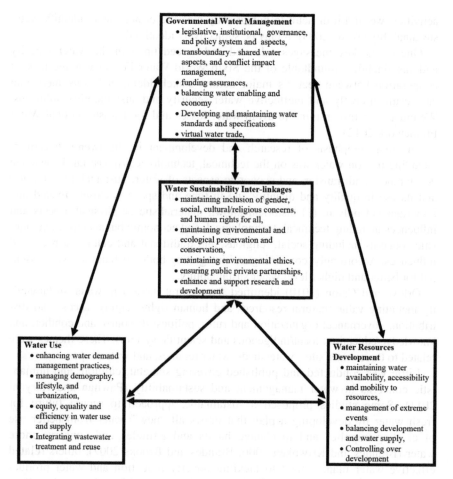

Fig. 1 Interlinked aspects and limitations to maintaining and properly achieving water sustainability

2 Humanistic Origin Influences on Water Sustainability

2.1 Gender Equality, Involvement, and Participation

Generally, women were and still are the main actors and every day labor in qualitative and quantitative water use either in domestic and/or in agricultural sectors. Some studies still claim that in specific countries and/or instances the drinking-water sector still appears insensitive to gender issues (Regmi and Fawcett 2001). However, evidence shows that the meaningful involvement of women in water resources development and management can help make water projects more sustainable, ensure that water infrastructure development yields the maximum social and economic returns, and advance progress on Millennium Development Goals set by the UN in the year 2000 and agreed upon to by all the world's countries and all the world's leading development institutions in which the third goal states that governments need to promote gender *equality* and *empowering* women by 2015 in addition to the seventh goal of ensuring environmental *sustainability* (Global Water Partnership 2006; UN-Water 2006; Millennium Developmental Goals 2000; Baguma et al. 2013).

It was noted that as water circulates through practically all domains of social life, rural as well as urban, it is managed differently by men and women (Swyngedow 2004; Swyngedouw et al. 2002; Bennett 1995; Bennett et al. 2005; Cleaver and Elson 1995; Elmendorf 1981; Harris 2005; Tortajada 2003; Orlove and Caton 2010). These differences in water management comprehend and supplement each other because men and women look at water management matters and issues from different angles and does not need to be that either one is better or worse.

Most decision and policy makers, institutions and utilities management of the water sector were, and are still dominated by men. This phenomenon is almost dominated in rural areas (Apusigah 2004). Consequently, women's role was marginalized or given minor role in resource and system development and management. In other words, women were minimally considered and allowed in the water management interactions including: acts of giving and undertaking, receiving knowledge, training, and tools, and counteracting and actively interacting, and participating. This inequality or exclusion has led to making women's attitudes toward and engagement in water and system management and development low or insignificant.

Until 1970s, no particular attention was paid to the potential contribution of women in development including water and water management (Allély et al. 2001). Starting from 1975 until present, a cycle of meetings, conferences, workshops, and research was launched and conducted focusing on this issue. Despite the large number of activities for gender inclusion in water management, research indicates that the water sector is still far from women inclusion in water management and its sustainability (Gross et al. 2000; Regmi and Fawcett 2001).

Women also are not exposed enough and/or aware of proper knowledge and extension training on water management issues. It was stated that while a lot of effort has been invested in developing gender mainstreaming materials, a major challenge facing program officers and water and gender specialists is that such information and materials are anchored in different institutions, resource centers, web sites, and organizations (UNDP 2003).

2.1.1 Unwise Water Use and Overuse

It is important to understand trends and patterns of water use and/or overuse because it helps in proper planning and achieving of water sustainability. This water use could be analyzed in quality and quantity and evaluated at domestic, country, basin/river, and at global level. Also, it is important to note that due to rapid increase in population growth, social, and economic development, water demand and resource development accelerated and consequently failure to maintain wise, controlled, balanced and as and where needed water took place (Haddad and Lindner 2001).

Examples of unwise water use and overuse include the imbalanced water distribution and/or allocation to various economic sectors or the water supply preference to urban areas compared to rural or agriculture, imbalanced population distribution in relation to available water resources, deforestation, poor water demand management practices, overexploitation of ground water resources, nonenvironmental and/or poorly planned land development, and other. In addition, on-farm agricultural water and nonwater management practices which in many areas are uncontrolled including those related to water and wastewater management are of concern for both qualitative and quantitative water sustainability.

As another example of in-equitable balance of water rights for people and the planet is optimizing business gains on the accounts of water and community sustainability. Nestle waters company practice of bottling more than Niagara Falls water capacity with little or no equitable community compensation or accountability for ecological and/or environmental impacts induced (Bean 2010).

At the domestic level, water sustainability needs much attention. It was concluded that reducing water consumption at domestic level must be tackled by changing user behavior and using multi-staged interconnected approaches. Such approaches must be applied and focused on the factors behind the various water-related activities that take place in the household including policy and decision making and integrated water resources management. Also policies, methods, and campaigns (current or possible in the near future) must be designed in view of the local cultural and social background, alongside financial and technological accessibility (Elizondo and Lofthouse 2010).

At basin/river level, water security can be jeopardized by a number of man-made factors, including river fragmentation, overgrazing, the draining of marshlands, and pollution. These problems often increase with economic development. The same factors also lead to biodiversity loss (Vörösmarty et al. 2010; Wetlands International 2010).

Globally and in discussing the land and water use conflicts along with the tradeoffs between food, fuel, and species, it was reported that by 2100, an additional 1,700 million ha of land may be required for agriculture to comply with food demands. Combined with the 800 million ha of additional land needed for medium growth bioenergy scenarios, such development threatens intact ecosystems and biodiversity rich habitats (Totten et al. 2003; Totten 2008).

2.1.2 Colonization, Unilateralism, and Conflicts

It was reported that issues of water and international conflict are linked with increasing frequency (Yoffe 2001). In many transboundary water systems, cooperation between riparians is limited and some riparians do unilateral actions altering water sustainability (system's quality and quantity) using either political, economic, and/or military overpower (Haddad 2013). It was concluded that coop-eration and unilateralism cannot coexist in the long term (Cascão 2009).

In some cases such as the Palestinian–Israeli conflict, long-term induced water use and water and land resources full control using military power has resulted in water rights loss as well as human suffering, injustice in water allocation, and long-term disputes (Haddad 2007, 2009). In addition and for the same reason, the vast majority of water supply utilities and departments in Palestine have impasse in development, in investment in water infrastructure, and in human and institu-tional capacity building (UN 1991; Haddad 1994; Haddad and Mizyed 1996).

In the same line, Fairhead and Leach (2004) argued that, colonization in Africa was a major cause in Africa's departure in their mode of natural resource man-agement. Similarly Mutembwa (1998) argued that colonization and decoloniza-tion disputes provided a "protracted conflict setting," thus raising water and river basin issues to a combustible level of conflict in southern Africa. Water was heav-ily implicated in several historical colonial efforts: Germany's colonization of South-West Africa (Namibia), particularly the push to have access to the Zambezi; Portugal's lusotropicalism, particularly the strategic imperatives behind the con-struction of the Cahora Bassa scheme in Mozambique and other river basins collab-oration works with South Africa; and apartheid's survival and sustenance in South Africa and the imperative to maintain a cordon sanitaire around the white redoubt.

In the mid 1990s, a violent environmental conflict arose between Nigeria and Cameroon in the Lake Chad river bed in response growing water and land scar-city. Even though an institution, the regional Lake Chad Basin Commission, was charged with settling the conflict, an armed clash arose (Alessa Metz 2003).

Concerning the Euphrates and Tigris Basin, Iraq had intentions in Syria, and Turkey has, at times, pressured Damascus through using the water of the Euphrates River, which runs through both countries (Lee and Ben Shitrit 2014). Iran has intensions in Iraq and use Tigris river waters and its tributaries unilater-ally and in a similar manner.

In discussing the Nile case, Wolf and Newton (2008) noted that as the Nile ripar-ians gained independence from colonial powers, riparian disputes became interna-tional and consequently more contentious, particularly between Egypt and Sudan. The Nile case was found similar to the Indus (a river basin shared by Afghanistan, China, India, Nepal, and Pakistan), over which a conflict between India and Pakistan originated during British colonial rule. The disappearance of British colonialism of turned national issues international, making agreement more difficult. Lack of water-sharing agreement lead India in April 1948 to stem flow of Indus tributaries to Pakistan. Later in 1960 and after two wars, a water agreement between Pakistan and India was negotiated and then ratified, with provisions for water conflict resolution.

It was noted that the establishment of colonies in a new country or area was determined by its water assets and access to reliable freshwater for the soon to be growing into town (Davies and Wright 2014). Therefore, colonialization either being naturally or induced by any mean cause pressures on existing water resources and its sustainability for the sake of present and future indigenous generations as well as environment and ecology.

The introduction of commercial production systems by the colonial economic and political race saw the resettlement of some Africans away from their religious and cultural systems (Paula 2004). This and in addition to reducing water availability and water use pattern it destroyed indigenous knowledge systems and undermined traditional land use and institutions.

It was demonstrated that resource struggles and conflicts are not just material challenges but emotional ones, which are mediated through bodies, spaces, and emotions. Such a focus fleshes out the complexities, entanglements, and messy relations that constitute political ecologies of resources management, where practices and processes are negotiated through constructions of gender, embodiments, and emotions (Sultana 2011).

Wagner (2008), explained how the conflict over land use and development among new and old settlers in the Okanagan valley of British Columbia has resulted in different landscape esthetics and water management. Such and similar acts would negatively impact the long-term status of water resources and affect its sustainability.

One of the key messages achieved and agreed upon in the Water and energy in Arab States roundtable of the 6th World Water Forum was Unprecedented political reform, civil unrest and ongoing conflict and military occupation within many countries in the Middle East have highlighted the need to respond to citizen demands for access to basic services such as water (Global Water Framework 2012).

During the British mandate over Palestine, in 1926, the British High Commissioner granted the Jewish owned Palestine Electricity Corporation, founded by Pinhas Rutenberg, a 70 year concession to utilize the Jordan and Yarmouk Rivers' water for generating electricity. The concession denied Palestinian farmers the right to use the Yarmouk and Jordan Rivers' water upstream of their junction for any reason, unless permission was granted from the Palestine Electricity Corporation. Permission was never granted (Isaac and Hosh 1992).

Israel decided in April 2002 to establish unilaterally a permanent barrier diverting from internationally acknowledged armistice lines between the Occupied Palestinian Territory (OPT) in the West Bank and Israel. The construction of the wall subjected Palestinians to several water vulnerabilities, including irrigation infrastructure devastation, impeded access and mobility to water and irrigation land resources, increased land aridity, and detrimental effects on community socioeconomic and migration (Haddad 2005).

2.1.3 Religion and Faith Guidance

Religion is a major influence in the world today. Religion, faith, or belief with related practices affects and shapes how people interact with natural resources including water. Over history, many different religions and belief systems have been developed of which the most followed worldwide include Animism, Bahá'í, Buddhism, Confucianism, Christianity, Hinduism, Islam, Jainism, Taoism, and Judaism. (UNESCO 2010).

Over the past two decades, the indicators of engagement on environmental issues by religions and spiritual traditions have grown markedly in addition to much publications on the subject matter were issued in the same period (Norris and Inglehart 2004). It was noted that researchers do not commonly characterize the relationship between sustainability and religion as particularly positive; however, a recent study reveals a far more complicated relationship between the two (Davis 2013; Johnston 2013). Davis (2013), emphasized the need for including the overlap subjects/issues between religious experiences and environmental sustainability such as people's values; culture and value-based institutions (religions). She explored the relationship between sustainability and the history, practices, and sacred texts of major world religions including the question of how can these theologies, concepts, and experiences be leveraged to strengthen the relationship between religion and sustainability.

It was reported that religious communities can play an important role in moving a culture toward greater sustainability—but religious ideology can also contribute to a disregard for sustainable practices. It was argued that religious traditions with an end-of-world focus can result in sloppy/careless environmental discourse (Glaser 2012).

There are controversies about the role of religion in water sustainability. World Values Survey indicated that in the data for the year 2000, 98 % of the public in *Indonesia* said that religion was very important in their lives while in China only three percent considered religion very important (Norris and Inglehart 2004). This is important from the sense of taking religion as a common influence is not the same everywhere.

White (1967) argued and conjectured that the Christian Middle Ages were the root of ecological crisis in the twentieth century because it encouraged exploitation of natural resources and spread western technology and industry around the world. Van Wensveen (2008), indicated that two insights have emerged from the debates generated by White's argument that: (1) religion particularly affects environmental sustainability by shaping human attitudes toward nonhuman nature and (2) all religions have the potential to foster both helpful and harmful attitudes. Van Wensveen recommended that that the transformation of human attitudes from ecologically harmful to ecologically fitting is a necessary, albeit insufficient, condition for environmental sustainability.

Gottlieb (2008) found that in the concept of sustainability, we find a change not only in religion's understanding of the value of the natural world including water and the need to alter its own ecological practices, but a possible awakening to the finite nature of human—including religious—existence. He indicated that most religions commit themselves to the value of their own sustainability.

Johnston (2013) presented the religious dimensions of contemporary sustainability and social movements emerging from the intersection of global environmental issues and moral traditions or religions. He investigated the differences and commonalities between secular, interfaith, and faith-based organizations that have engaged sustainability in their practices. Johnston (2013) found that (1) sustainability is now key to international and national policy, manufacture and consumption and is central to many individuals who try to lead environmentally ethical lives, (2) the inclusion of religious values in conservation and development efforts has facilitated relationships between people with different value structures, and (3) religion and sustainability present the first broad analysis of the spiritual dimensions of sustainability-oriented social movements.

It was reported that water bodies among the Akan community, a major ethnic group in Ghana, are associated with the gods or abosom and are used in accordance with structures and rules that are related to the local folks by fetish priests who are the mouthpiece of the gods. Customary laws mandate users to keep lakes and rivers pure because they are regarded as the dwelling place of the gods (Acheampong 2010).

Religious freedom is still difficult and controversial for Native Americans as many non-natives misunderstand, stereotype, and discriminate against native peoples and their spiritual beliefs and practices. For example, the people of the Winnemem Wintu nation of northern California struggle to protect their sacred sites from being again flooded by the expansion of a river dam (Nelson 2009).

An International Conference on Water, Ethics, and Religion (Stockholm 2011) was conducted to foster greater cooperation between the leaders of religious groups and the UN family, in order to improve the achievement of the Millinium Developmental Goals in relation to drinking water, sanitation and malnourishment (Ilmas et al. 2007). The basic point of this initiative was the acknowledgment of the role of religion in achieving developmental goals and water sustainability.

In author's view, all religions and in different proportions cared about the environment and natural resources when they were presented thousands of years ago. In relation to water sustainability, religions were keen toward water conservation, wise use, nonprofusion, and others. And at those times some level of sustainability was established. However, the interpretation and adaptation of those religions, teachings, and their practices to global changes since the start of the industrial revolution including resource development, consumption, land use, scientific knowledge and innovations, pollution, and other was not in place or in any parallel.

Because religion directs/guides human everyday practices and behaviors, it relates and contributes to balanced water sustainability. For example, Haddad (2000) found that in Islamic perceptions balanced water sustainability is that creates sustainable balance and inter-balance between availability of natural resources including water, their development and use for various purposes, and the consequential quality of human beings as well as the environment and ecology (see Figs. 1 and 2).

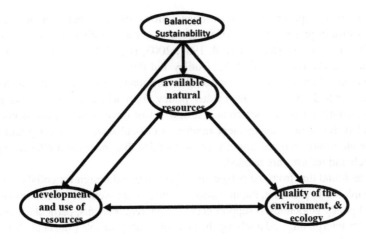

Fig. 2 Balanced and inter-balanced water sustainability elements

2.1.4 Shortsightedness in Water Policies and Strategies

In setting country's water policies and strategies, governments and policy and decision makers mostly emphasize on fulfilling human materialistic needs. Less emphasis is given to environmental and ecological concerns, integrity in equity and equality of water services among and for all public classes, habitat loss, and other social and humanistic impacts. Less involvement and participation is allowed to various stakeholders. Such disregard would in some instances lead to costly, inefficient solutions, and to impacts that is very difficult to recover or fix. For example, deforestation in Latin America has impacted landscape as well as the hydrologic cycle of the region, uncontrolled agricultural practices including the use of fertilizers, pesticides, and various hormonic and seed treatment for the sake of producing more food either in north or south America or Australia has led to polluted soils and surface groundwater that need decades to recover, the long term and ongoing desiccation of seas and lakes such as the Aral Sea in Russia and the Huleh Lake in historic Palestine, coal-mining, and metallurgical centers in many areas in the world (e.g., Poland), which have severely polluted air and water and vast areas of decimated landscape; and the draining of untreated or poorly treated wastewater in surface water bodies has resulted in qualitative degradation of those rivers for unprecedented levels such as the Volga river in Russia.

It was indicated that in the endeavor to manage water to meet increasing human food and water demands, the needs of freshwater species and ecosystems have largely been neglected, and the ecological consequences have been tragic such as the development of Amazon forests, degradation of grazing areas as a result of urbanization and the use of herbicides and pesticides, and eutrophication of water bodies by high rates of nitrogen and phosphorus release to them from agricultural fields. Human appropriation of freshwater flows must be better managed if we hope to sustain present benefits of food and water availability and freshwater biodiversity and forestry (Richter et al. 1997, 2003; IUCN 2000; Pringle et al. 2000; Stein et al. 2000; Baron et al. 2002; Dudley and Phillips 2006).

Davis and Wright (2014) reported that the management of water in Sydney, Australia, relied on a centralized water policy approach, which solved water supply and sanitation issues but created a new set of environmental problems. They concluded that water management remains a critical issue for Sydney and for its sustainable future, and accordingly, recommended that more comprehensive policy approach and reforms are needed.

It was found that from the perspective of institutional design for collaborative and sustainable water planning, the first major required improvement include provision of detailed policy guidelines to support general legal requirements, particularly practical advice for interpreting and applying the precautionary principle (Tan et al. 2012).

The inclusion of various stakeholders and communities including those in unorganized or neglected to water sustainability in national policy and strategy setting is another important aspect to be considered. Getting to lessen to communities and stakeholders concerns and engage with them in dialogues on water developmental and sustainability issues would lead to positive qualitative and quantitative changes and impacts and consequently to more sustainability.

In many instances, national water policy and strategy setting are based on short term and limited qualitative and quantitative data availability. In addition, the planning cycle of resource monitoring, data collection, analysis, and evaluation, and policy redefinition and reimprovement or re-enhancing and upgrading are not practiced. It is also important to consider decision making, planning, and policy setting under uncertainties. Such a process of data bank creation and use in policy setting and reimprovement would save a lot of energy and funds and lead to more water sustainability.

2.1.5 Cultural Traditions and Customs

As there are similarities between religion and culture and they are generally expressed and perceived as collective or group rights and characteristics, there are differences and distinctions between the two. Bonney (2004) stated that culture may be thought of as a causal agent that affects the evolutionary process by uniquely human means while religion is considered a process of revelation and contains the concept of the "faithful" who receive the message of revelation.

Human cultures are numerous and diverse—and in many cases have deep and ancient roots. They allow people to make sense of their lives and to manage their relationships with other people and the natural world (Worldwatch Institute 2010). In this regards, culture is community's habits and traditions that differentiate one from the other while religion is a way of life for one or many different cultures or communities. The World Commission on Culture and Development (1995) defined culture as "ways of living together" and argued that this made culture a core element of sustainable development and an inextricable part of the complex notion of sustainability.

The United Nations and its agencies including UNESCO have been at the forefront in highlighting the importance of respecting and protecting/conserving cultural traditions and customs in development including water resources development. Water Culture is becoming central in the new "Strategy for Water in the Mediterranean" decided by the Ministers during the Euro-Mediterranean Ministerial.

Conference on Water, held in Jordan on 22 December 2008 (Scoullos 2009). Fernald et al. (2012) found that water scarcity, land use conversion and cultural, and ecosystem changes threaten the way of life for traditional irrigation communities of the semi-arid southwestern United States. They found that there are four community coherent interlinked subsystems hold the key to future water sustainability: hydrology, ecosystem, land use/economics, and sociocultural. In this regard, managing water for agriculture, as the major water consumer, is the most important part of the solution for water scarcity/sustainability worldwide.

Among indigenous peoples of North America there is a common belief is that everything on earth and in the universe has a soul and is animated by spirit. Many of them consider land and water and everything that lives on it and in it to be sacred, a belief that often—but not always—lends itself to a sustainable lifestyle (Nelson 2009).

It was concluded that "Water Culture" is central in addressing the current and future water challenges in the Mediterranean. Initially, neither "education" nor "culture" appear in the Rio's model of Sustainable Development which was based on three pillars: Environment—Ecology, Economy and Society (UN Conference on Environment and Development, Rio de Janeiro 1992). Later, the "Delor's Commission" of UNESCO in its Report (1996) proposed "culture" as the fourth pillar of Sustainable Development, a proposal rejected by many countries out of principle but also on the ground of scientific and mostly political reasons rejecting Delor's socialist leadership and positions (Scoullos 2009). However, Delor's commission findings formed a backdrop for reflection by decision makers either at regional or national levels.

Worldwatch Institute (2010) found that important diminishing forces for sustainability include (1) the wisdom of elders who served as knowledge keepers, religious leaders, and shapers of community norms and (2) farming as one long-lived tradition. In addition, it was also found that the poor are the most vulnerable to having their traditions, relationships, and knowledge and skills ignored and denigrated ... Their culture ... can be among their most potent assets, and among the most ignored and devastated by development programs (UNESCO 2000).

2.1.6 Human Rights and the Ethic of Care

It should be noted that water sustainability ethics or the ethics to care is a new discipline that analyzes and values the water issues in regard to our continued moral obligations to future generations. In addition, Access to water is widely regarded as a basic human right and was declared so by the United Nations in 1992 (United Nations 1992). Whiteley et al. (2008) stressed that fairness in the allocation of water will be a cornerstone to a more equitable and secure future for humankind.

Gleick (1998) recommended that international organizations, national and local governments, and water providers adopt a basic water requirement standard for human needs of 50 liters per person per day (l/p/d) and guarantee access to it independently of an individual's economic, social, or political status. He argued that access to a basic water requirement is a fundamental human right implicitly and explicitly supported by international law, declarations, and State practice. By acknowledging a human right to water and expressing the willingness to meet this right for those currently deprived of it, the water community would have a useful tool for addressing one of the most fundamental failures of the twentieth century development.

It was noted that the right to water implies economic costs that must be recovered to ensure the continuity of service through the sustainability of public, private, and community-based service providers (Global Water Framework 2012). This is an important message because it is good and fair to ask for and achieve human rights to water but we should know that water sustainability has associated costs that need all to cooperate cover.

The over using of renewable and nonrenewable water resources in many areas of the world to meet the present increasing water demands has led to altered water quantity and quality and damaged environment and ecology. Consequently, human rights to water and the no care to the water needs of future generations, environment, and ecology are also altered.

Luh et al. (2013) developed an index to measure progressive realization for the human right to water and sanitation and applied it to the nondiscrimination and equality component for water. The developed index was composed of one structural, one process, and two outcome indicators and is bound between -1 and 1, where negative values indicate regression and positive values indicate progressive realization. They demonstrated that the index application can be used for all the different components of the human right.

Water sustainability should not be self-centered around the technical services provided by governments or water companies and utilities, it needs to include the compassionate-humane part and meet the ethical responsibilities and commitments to present and future generations. The compassionate-humane part combines and/or balance between technical and economic feasibility and human rights and the ethics to care and mostly need to be included in the water tariff setting, water infrastructure's and system development, and service coverage ensuring equity in allocated quantity and quality of supplied water.

3 Adaptation Actions and Measures

Overturning actions to water in-sustainability influences does mean that we need to think and rethink of all possible alternatives and continuously propose/invent/and apply some improvements, upgrades, and/or solutions. The implementation challenges facing adaptation and measures lie in cooperation/coordination, regulation, monitoring/evaluation, awareness, and continuity. Key overturning actions or adaptation measures would include but not limited to the following:

(a) *Adapting measures for maintaining balance of water availability and use*: Among the solutions to achieve water sustainability are those related to balancing water availability and allocation between sectors, and use. These solutions cannot be done abruptly in a one step-action, it should be a steady, well planned, long-term process. Meaning among others that governments, water utilities, and consumers to join hands and take adapting measures such as (1) reduce water consumption and control water demand and enhance demand management practices, (2) develop within safe yield and limits all possible conventional and nonconventional water sources, (3) supply affordable, adequate, and safe water equally to various societal groups, (4) maintain balanced planning and development of various economic sectors and urbanization, and (5) adopt breakthrough technologies and adopt water tariffs that would reflect the socioeconomic growth and classes of people allowing equitable access to water and its use.

(b) *Rethinking water policies and strategies*: These water policies and strategies rethinking should be done in a continuous-systematic cycle that would lead to continuous changes and improvements and include humanistic as well as materialistic water needs, aspects, and influences. Policies and strategies responsible teams need to include proportional professionals representing the society including gender, religious clergy, and environmentalists and ecologists. This rethinking and inclusions would lead to better public involvement and participation in water service decision making.

(c) *Adapting measures for public involvement and participation*: Public involvement and participation in water development and management decision making as a practice need to be maintained. This practice is very important for better water sustainability and need to be systematic, effective in size and level of involvement, continuous, and comprehensive including all public classes including women and youth.

(d) *Adapting behavioral change measures for maintaining water cultural traditions and customs and respecting related faith guidance*: This is a core issue to work on starting from changing public knowledge and attitude using all possible educational materials and media sources to setting rules and regulations that maintain cultural traditions and customs and respecting related faith guidance. This adaptation need to be, controlled, monitored, and accordingly improved and upgraded. Such process might be the responsibility of both governmental and nongovernmental environmental and social agencies and groups.

(e) *Adapting measures for minimizing and/or resolving conflicts and disagreement related to water and its use*: Conflicts and disagreement related to water control, allocation and its use may arise between economic sectors, societal groups, urban and rural areas, and between countries adjacent/riparian to a shared surface or groundwater basins and aquifers. For maintaining water sustainability, these conflicts and disagreements need to be given priority by governments and be tackled/managed as soon as possible and in a fair and just manner. A small disagreement now could develop with time to a conflict or armed conflict. Not resolving such conflicts and disagreements would affect water sustainability and results in energy and wealth loss as well as and public suffering of water shortages, and limits in economic and social development.

4 Discussion

Water is life for individuals, societies, ecology, and the environment and it needs to be maintained in an acceptable qualitative and quantitative availability for maintaining and accommodating good community and life. In maintaining such water qualitative and quantitative availability, the main part of water sustainability, we cannot consider only the physical/materialistic aspects and influences only but the overall picture including the cultural/religious, traditional, and environmental and ecological aspects, and influences.

Over history, humans got developed and along their development they developed needed and appropriate water management practices and water cultural/ religious/environmental and ecological traditions. This is a precious global and local richness that need to be highly conserved and preserved. The water system of Rome (*aqueducts* and pipes), water wheels in the Middle East and Persia, the gardens of Babylon's, the water canals of Oman, Roman water pipes, cisterns, and systems, the water conservation terraces in Palestine, the *Nabataeans* water conduit system in Petra, the Indus Valley, sanitation system, the Scottish, Chinese, and Japanese *water-flushing toilets*, and many others are small examples of what need to be maintained, conserved and preserved.

In this regard and in relation to water sustainability, there are endless examples of traditional/cultural agricultural practices of what and for what, where, how, and when to grow, irrigate, and manage. Since early history (Costas 2005), the preoccupation of human beings with growing and breeding wild and traditional medicinal plants, pastures, weeds, floras and faunas was in place and has cultural-traditional and materialistic values. Indians, Chinese, Egyptians, Phoenicians and Canaanites, Palestinians, Romans, Persians, Turks, red Indians, and others represent have rich documented history in water sustainability, culture, and heritage.

By emphasizing the need to include humanistic aspects and influences in water management and in water sustainability, we also emphasize the need for continuing efforts to develop materialistic aspects and influences. These two aspects represent one part and a whole and best not to be separated. Efficient and effective institutional water management system or structure of both aspects is important and a requirement for water sustainability.

It is clear that maintaining water sustainability is not simple. However, it could be considered as a chain of interrelated, interconnected, and integrated steps and processes that leads to water sustainability. The water supply chain is an example of such interrelated interconnected and integrated service provision, delivery processes, resource development, and system management (see Figs. 3 and 4).

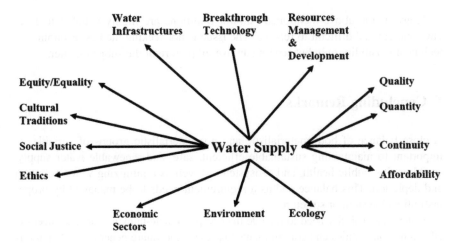

Fig. 3 Water supply chain

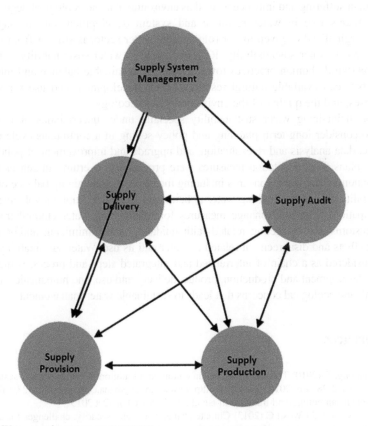

Fig. 4 Water supply system management

Many of the above-mentioned cultural traditions are highly suitable to local environment and difficult to replace. The issue is not seeking the best economic or technical feasibility only but seeking an overall picture of the subject matter.

5 Concluding Remarks

A close balance of both materialistic and nonmaterialistic groups of variables is important to maintaining sustainable efficient, safe, and renewable water supply, social equity, public health, and ecosystem, as well as minimizing water pollution and depletion. This balance and as a requirement needs to be managed by proper institutional system or structure.

Water sustainability need to include the compassionate-humane part and meet the ethical responsibilities and commitments to present and future generations. Culture is a way of life and a core element of sustainable development and water sustainability. Colonization of others land and water resources by any mean as well as practicing unilateralism in actions in joint and shared transboundary water resources is resulting in human suffering and injustice as well as environmental and ecological degradation.

Women's role in water resource and system development and management was marginalized or given minor role and the water sector is still far from women inclusion in water sustainability. Religion's role in water sustainability is needed to direct/guide human practices toward creating sustainable balance and inter-balance between available natural resources, their development and use for various purposes, and the quality of the environment, and ecology.

For maintaining water sustainability specially under uncertainties, it is important to consider long-term planning and policy setting in a continuous cycle of data update, data analysis and re-evaluation, and upgrade and improvement of polices and action plans. Five adaptation measures were proposed to overturn influences caused by nonhumanistic water concerns including measures to maintaining balance of water availability and use, rethinking water policies and strategies, public involvement and participation, behavioral change measures for maintaining water cultural traditions and customs and respecting related faith guidance, and minimizing and/or resolving conflicts and disagreement related to water and its use. Water sustainability could be considered as a chain of interrelated and integrated steps and processes including water development and production, product delivery and use, and humanistic, environmental, and ecological concerns that leads to sustainable water management.

References

Acheampong, E (2010) The role of Ghanaian culture and tradition in environmental sustainability. Posted 28 Nov 2010. Found in: http://www.modernghana.com/news/306123/1/the-role-of-ghanaian-culture-and-tradition-in-envi.html. Accessed on Oct 2013
Allan C, Xia J, Pahl-Wostl C (2013) Climate change and water security: challenges for adaptive water management. Curr Opin Environ Sustain 5(6):625–632

Allély D, Drevet-Dabbous O, Etienne, J. Francis J, Morel À L'huissier A, Chappé P. Verdelhan Cayre G (2001) Water, gender and sustainable development lessons learnt from French co-operation in sub-Saharan Africa. A report published jointly by French Ministry of Foreign Affairs, French Development Agency (AFD), and the World Bank. Found in: http://www.oecd.org/derec/france/41975983.pdf. Accessed Oct 2013

Apusigah A (2004) Indigenous knowledge, cultural values and sustainable development in Africa. Ph.D. dissertation, University for Development Studies, Wa, Ghana. Found in: http://www.academia.edu/1067839/Indigenous_Knowledge_Cultural_Values_and_Sustainble_Development_in_Africa. Accessed on Oct 2013

Baguma D, Hashim J, Aljunid S, Loiskandl W (2013) Safe-water shortages, gender perspectives, and related challenges in developing countries: the case of Uganda. Sci Total Environ 442:96–102

Baron JS, Poff NL, Angermeier PL, Dahm CN, Gleick PH, Hairston NG, Jackson RB, Johnston CA, Richter BD, Steinman AD (2002) Meeting ecological and societal needs for freshwater. Ecol Appl 12:1247–1260

Bean J (2010) Water wars: policies for sustainable water use. A power point presentation posted 1 Dec 2010. Found in: http://www.slideshare.net/JustinCBean/water-wars-policies-for-sustainable-water-use?from_search=9

Bennett V (1995) The politics of water: urban protest, gender, and power in monterrey. The University of Pittsburgh Press, Mexico

Bennett M, Peterson D, Levitt E (2005) Looking to the future of ecosystem services: introduction to the special feature on scenarios. Ecosyst 8:125–132

Bonney R (2004) Understanding the process of research. In: Chittenden D, Farmelo G, Lewenstein B (eds)Creating connections: museums and public understanding of current research. Altamira Press, California

Brandes O, Brooks D (2007) The soft path for water in a (small) nutshell. A joint publication of friends of the Earth Canada, and the POLIS project on Ecological Governance, University of Victoria, Canada, pp 16–19, (revised edition Aug 2007)

Brandes O, Kriwoken L (2006) Changing perspectives—changing paradigms: taking the "soft path" to water sustainability in the Okanagan basin. Can Water Resour J 31(2):75–90

British Standards Institution (BSI) (2010) Sustainability, the role of standards. Found in: http://www.slideshare.net/BSIStandards/sustainability-3114245?from_search=4. Accessed on Oct 2013

Campbell A (2009) Covering insecurities—water, energy, carbon, and food. Australian academy of science, November, 2009. Found in: http://www.slideshare.net/AndrewCampbell/converging-insecurities-water-energy-carbon-and-food-2424681?from_search=1. Accessed on Oct 2013

Cascão AE (2009) Changing power relations in the Nile river basin: unilateralism vs. cooperation? Water Altern 2(2):245–268

Cleaver F, Elson D (1995) Women and water resources: continued marginalisation and new policies, gatekeeper series no. 49. International Institute for Environment and Development, London

Costas T (2005) The geography of theophrastus' life and of his botanical writings. In: Karamanos AJ, Thanos CA (eds) Biodiversity and natural heritage in the Aegean, proceedings of the conference 'Theophrastus 2000', Eressos–Sigri, Lesbos, 6–8 Jul 2000. Fragoudis, Athens, pp 23–45

Davies P, Wright I (2014) A review of policy, legal, land use and social change in the management of urban water resources in Sydney, Australia: a brief reflection of challenges and lessons from the last 200 years. Land Use Policy 36:450–460

Davis N (2013) Religious perspective on sustainability. A lecture presented at the University of California, Irvine, 13 March 2013. Found in: http://www.cusa.uci.edu/events/art-perspective-on-sustainability/. Accessed on Oct 2013

Dudley N, Phillips A (2006) Forests and protected areas guidance on the use of the IUCN protected area management categories. Best practice protected area guidelines series no. 12. A report published by International union for the conservation of nature (IUCN) in cooperation with Cardiff University and World Commission on Protected Areas (WCPA). Found in: http://data.iucn.org/dbtw-wpd/edocs/PAG-012.pdf. Accessed on Oct 2013

Elmendorf W (1981) Last speaker and language change: two californian cases. Anthropol Linguist 23(1):36–49

Elizondo G, Lofthouse V (2010) Towards a sustainable use of water at home: understanding how much, where and why? J Sustain Dev 3(1):3–10

Fairhead J, Leach M (2004) False forest history, complicit social analysis: rethinking some West African environmental narratives. Environment, development and rural livelihoods. Earthscan, UK

Fernald A, Tidwell V, Rivera J, Rodríguez S, Guldan S, Steele C, Ochoa C, Hurd B, Ortiz M, Boykin K, Cibils A (2012) A model for sustainability of water, environment, livelihood, and culture in traditional irrigation communities and their linked watersheds. Sustainability 4:2998–3022

Giordano M (2009) The transboundary waters program at IWMI. a presentation made to the Strategic Foresight Group, IWMI HQ, Battaramulla. Found at: http://www.slideshare.net/IWMI_Media/2009pptmark-giordano1?from_search=16. Accessed on Oct 2013

Glaser L (2012) Religious studies important to sustainability. Lecture presented by Lane Marie Law, who at Cornell's Atkinson center for a sustainable future. 12 December 2012. Published in Cornell Chronicle, 29 Oct 2013

Gleick P (1998) The human right to water. Water Policy 1(5):487–503

Global Water Framework (2012) 6th World water forum, Marseille, 12–17 March 2012. Published June 2012. Found in: http://www.worldwaterforum6.org/en/library/detail/?tx_amswwfbd_pi2[uid]=596. Accessed on Oct 2013

Global Water Partnership (GWP), Policy Brief No. 3 (2006) Gender mainstreaming: an essential component of sustainable water management. In: Carriger S (ed) Technical Committee (TEC). Elanders 2006, Production: Svensk Information

Gottlieb R (2008) You gonna be here long? Religion and Sustainability. Worldviews 12:163–178

Gross B, van Wijk C, Mukherjee N (2000) Linking sustainability with demand, gender and poverty: a study in community-managed water supply projects in 15 countries. Published by the International Water and Sanitation Center, Dec 2000. Found in: http://www.wsp.org/sites/wsp.org/files/publications/global_plareport.pdf. Accessed on Oct 2013

Gutierrez S (2014) Toward sustainable water resource management: challenges and opportunities. Compr Water Qual Purif 1:278–287

Haddad M (1994) Principles of joint Palestinian Israeli management of shared aquifers. A paper presented at the first workshop on possible structures for joint management of shared aquifers, Jerusalem

Haddad M (2000) An Islamic approach to the environment and sustainable groundwater management. In Haddad M, Feitelson E (eds) Management of shared groundwater resources: the Israeli-Palestinian case with international perspective. Kluwer Publishing Company and Amazon.com, The Netherlands, pp 25–42

Haddad M (2005) Irrigation adaptation to changing water supply: Palestine as a case study. Paper accepted for the ASCE and world water and environmental congress and listed in conference proceeding paper, part of EWRI 2005, Anchorage, AK, 15–19 May 2005

Haddad M (2007) Politics and water management: a Palestinian perspective. In: Shuval H, Dweik H (eds) Water resources in the middle east. Springer, Berlin, pp 41–53

Haddad M (2009) Water scarcity and degradation in Palestine as challenges, vulnerabilities and risks for environmental security. In: Brauch H, Oswald U, Mesjasz C, Grin J, Kameri-Mbote P, Chourou B, Dunay P, Birkmann J (eds) Global environmental change, disaster, and security: threats, challenges, vulnerabilities and risks, Chap. 22. Springer, Berlin, pp 408–419

Haddad M (2013) The Jordan river: legal and institutional aspects. In: Kibaroglu A, Kirschner A, Mehring S, Wolfrum R (eds) Water law and cooperation in the Euphrates-Tigris region, a comparative and interdisciplinary approach Chap. 14, Martinus Nijhoff, Leiden, pp 303–333

Haddad M, Lindner K (2001) Sustainable water demand management versus developing new and additional water in the middle east: a critical review. Water Policy J 3(2):143–163

Haddad M, Mizyed N (1996) Water resources in the middle east: conflict and solutions. In: Allan T (ed) A paper published in the proceedings of the workshop on water peace and the middle east: negotiating resources of the Jordan River basin. Library of Modern Middle East, Tauris Academic Studies, New York

Harris S (2005) Belief is a content-independent process. In: Brockman J (ed) What we believe but cannot prove: today's leading thinkers on science in the age of certainty. Free Press, London

Ioris A, Hunter C, Walker S (2008) The development and application of water management sustainability indicators in Brazil and Scotland. J Environ Manag 88(4):1190–1201

Isaac J, Hosh L (1992) Roots of the water conflict in the middle east: the middle east water crisis, creative perspectives and solutions. University of Waterloo, Ontario (7–9 May 1992)

IUCN (International Union for the Conservation of Nature) (2000) Vision for water and nature: a world strategy for conservation and sustainable management of water resources in the 21st century. International Union for the Conservation of Nature, Gland, Switzerland

Johnston L (2013) Religion and sustainability: social movements and the politics of the environment. Acumen Publishing, UK, p 224

Jones S, Silva C (2009) A practical method to evaluate the sustainability of rural water and sanitation infrastructure systems in developing countries. Desalination 248(1–3):500–509

Juwana I, Muttil N, Perera B (2012) Indicator-based water sustainability assessment—a review. Sci Total Environ 438:357–371

Larsen M (2014) Global change and water availability and quality: challenges ahead. Compr Water Quality Purif 1:11–20

Larson K, Wiek A, Keeler L (2013) A comprehensive sustainability appraisal of water governance in Phoenix, AZ. J Environ Manag 116:58–71

Lee R, Ben Shitrit L (2014) Religion, society, and politics in the middle east. To be published by CQ Press, a division of SAGE in 2014, pp 210–215. Found in: http://www.cqpress.com/docs/college/Lust_Middle%20East%2013e.pdf. Accessed on Nov 2013

Li Y, Yang Z (2011) Quantifying the sustainability of water use systems: calculating the balance between network efficiency and resilience. Ecol Model 222(10):1771–1780

Llamas R, Martínez-Cortina L, Mukherji A (2007) Water ethics. Marcelino Botín water forum 2007, Santander, pp VII–VIII

Luh J, Baum R, Bartram J (2013) Equity in water and sanitation: developing an index to measure progressive realization of the human right. Int J Hyg Environ Health 216(6):662–671

Metz FA (2003) The Cameroonian-Nigerian Border conflict in the Lake Chad Region: assessment of the resource and conflict management capacities of the Lake Chad Basin commission. Found in: http://www.nai.uu.se/ecas-4/panels/21-40/panel-28/Metz-Warner-Brzoska-Full-paper.pdf. Accessed on Nov 2013

Millennium Developmental Goals (MDG) (2000) UN website front page: we can end poverty. Millennium Developmental Goals. Found in: http://www.un.org/millenniumgoals/bkgd.shtml. Accessed on Oct 2013

Mutembwa A (1998) Water and the potential for resource conflicts in Southern Africa, Feb 1998. Found in: https://www.dartmouth.edu/~gsfi/gsfiweb/htmls/papers/text3.htm. Accessed on Nov 2013

Nelson M (2009) Berkshire encyclopaedia of sustainability: the spirit of sustainability indigenous traditions—North America, pp 225–228. Posted 22 Oct 2009. Found in: http://www.nativeland.org/download/IndigenousTraditionsNA.pdf. Accessed on Oct 2013

Norris P, Inglehart R (2004) Sacred and secular: religion and politics worldwide. Cambridge University Press, New York

Orlove B, Caton S (2010) Water sustainability: anthropological approaches and prospects. Annu Rev Anthropol 39:401–415

Paula D (2004) Indigenous knowledge systems in sub-Saharan Africa: an over view. Indigenous knowledge, local pathway to global development, the World Bank, African Region

Pawitan H, Haryani G (2011) Water resources, sustainability and societal livelihoods in Indonesia. Ecohydrol Hydrobiol 11(3–4):231–243

Pringle CM, Freeman MC, Freeman BJ (2000) Regional effects of hydrologic alterations on riverine macrobiota in the new world: tropical–temperate comparisons. Bioscience 50:807–823

Regmi S, Fawcett B (2001) Men's roles, gender relations, and sustainability in water supplies: some lessons from Nepal. In: Sweetman C (ed) Men's involvement in gender and development policy and practice: beyond rhetoric. Oxford Working Papers. Oxford Found in http ://www.wateraid.org/~/media/Publications/drinking-water-sector-gender-issues-nepal.pdf. Accessed on Oct 2013

Richter BD, Braun DP, Mendelson MA, Master LL (1997) Threats to imperiled freshwater fauna. Conserv Biol 11:1081–1093

Richter B, Mathews R, Harrison D, Wigington R (2003) Ecologically sustainable water management: managing river flows for ecological integrity. Ecol Appl 13(1):206–224

Schaller D (2007) Sustainability: moving environmental protection beyond scarcity. In: Dew W (ed) U.S. EPA, Denver, Colorado, Fall 2007. Found in:http://www.readbag.com/epa-region8-ee-pdf-sustainability. Accessed on Oct 2013

Scoullos M (2009) Towards a new water culture for the Mediterranean: addressing the challenges of the future, using the lessons from the past. Paper presented at the fifth international Monaco and the Mediterranean symposium, 26–28 March 2009, Oceanographic Museum of Monaco. Found in: http://www.mioecsde.org/_uploaded_files/5th%20international%20monaco%20and%20the%20mediterranean%20symposium%20-%20scoullos%20presentation.pdf. Accessed on Oct 2013

Stein BA, Kutner LS, Adams JS (2000) Precious heritage: the status of biodiversity in the United States. Oxford University Press, New York

Stockholm Environment Institute (SEI) (2011) Understanding the nexus: background paper for the Bonn 2011 nexus conference the water, energy and food security nexus, Solutions for the green economy 16–18

Sultana F (2011) Suffering for water, suffering from water: emotional geographies of resource access, control and conflict. Geoforum 42(2):163–172

Swyngedouw E (2004) Social power and the urbanization of water. Oxford University Press, Oxford

Swyngedouw E, Kaika M, Castro E (2002) Urban water: a political-ecology perspective. Built Environ 28(2):37–124

Tan P, Bowmer K, Baldwin C (2012) Continued challenges in the policy and legal framework for collaborative water planning. J Hydrol 474:84–91

Tortajada C (2003) Professional women and water management: case study from Morocco—a water forum contribution. Water Int 28(4):39–532

Totten M (2008) Freshwater public policies and market-based actions. Paper presented at CI freshwater strategy meeting 26 Sep 2008. Found in http://www.slideshare.net/mptotten/totten-freshwater-challenges-and-opportunities-09-26-08-presentation. Slide 30. Accessed on Oct 2013

Totten M, Pandya S, Janson-Smith T (2003) Biodiversity, climate, and the kyoto protocol: risks and opportunities. Front Ecol Environ 2003 1(5):262–270, 265

The United Nations Development Program (UNDP) (2003) Resource guide on mainstreaming gender in water management: a practical journey to sustainability. Found in: http://cap-net.org/sites/cap-net.org/files/Mainstreaming%20gender%20in%20water%20management.pdf. Accessed on Oct 2013

United Nations (1991) Israeli land and water practices and policies in the occupied Palestinian and other Arab territories—a note by Secretary, General/United Nations, New York (A/46/263), p 20

United Nations Conference on Environment and Development (UNCED) (1992) Rio de Janeiro, 3–14 Jun 1992

United Nations Educational, Scientific and Cultural Organization (UNESCO) (1997) Educating for a sustainable future: a transdisciplinary vision for concerted action EPD-97/CONF.401/CLD.1, Nov 1997

United Nations Educational, Scientific and Cultural Organization (UNESCO) (2000) Culture counts, conference on financing, resources and the economics of culture in sustainable development, Organized by the government of Italy and the World Bank with the co-operation of UNESCO, Florence, Italy, 4–7 Oct 1999

United Nations Educational, Scientific and Cultural Organization (UNESCO) (2010) Module 10: culture and religion for a sustainable future. UNESCO Publishing, Paris

UN-Water (2006) Policy brief 3 on gender mainstreaming: an essential component of sustainable water management. Found in http://www.unwater.org/downloads/Policybrief3Gender.pd f. Accessed on Oct 2013

van Wensveen L (2008) Religion and ecological sustainability: beyond the technical fix. Handout religion and development policy, published for the Ministry of Foreign Affairs, Chap. 6, Utrecht, The Netherlands

Vörösmarty C, McIntyre P, Gessner M, Dudgeon D, Prusevich A, Green P, Glidden S, Bunn S, Sullivan C, Liermann C, Davies P (2010) Global threats to human water security and river biodiversity. Nature 467:555–561

Wagner J (2008) Landscape aesthetics, water, and settler colonialism in the Okanagan Valley of British Columbia. J Ecol Anthropol 12(2008):22–38

Wang R, Li F (2008) Eco-complexity and sustainability in China's water management. In: Wostl P, Moltgen K (eds) Adaptive and integrated water management, coping with complexity and uncertainty, Springer, Berlin, pp 23–39. Found in: http://bilder.buecher.de/zusatz/23/23171/2 3171103_lese_1.pdf. Accessed on Oct 2013

Wetlands International (2010) Biodiversity loss and the global water crisis—A fact book on the links between biodiversity and water security, Oct 2010. Found in: http://www.cbd.int/iyb/ doc/prints/iyb-netherlands-watercrisis.pdf. Accessed on Oct 2013

White L (1967) The historical roots of our ecologic crisis. Science 155(3767):1203–1207

Whiteley J, Ingram H, Perry R (2008) Water, place, and equity. MIT Press, Cambridge, p 336

Wolf A, Newton J (2008) Case studies of transboundary dispute resolution. In: Jerry DP, Wolf AT (eds) Managing and transforming water conflicts (Appendix C). Cambridge University Press, Cambridge

Wood W (2003) Water sustainability: science or science fiction? perspective from one scientist. Dev Water Sci 50(2003):45–51

World Commission on Culture and Development (1995) Our creative diversity. UNESCO Publishing, Paris

Worldwatch Institute (2010) A state of the world and worldwatch institute report on progress toward a sustainable society: state of the world 2010. Chapter on traditions old and new, pp 21–55. Found in: http://blogs.worldwatch.org/transformingcultures/wp-content/uploads/2013/08/SOW10-final5.pdf. Accessed on Oct 2013

Yoffe S (2001) Basins at risk: conflict and cooperation over international freshwater resources. Ph.D. dissertation, Oregon State University, 12 Oct 2001. Found in: http://www.transboundar ywaters.orst.edu/research/basins_at_risk/bar/BAR_title.pdf. Accessed on Nov 2013

Author Biography

Marwan Haddad is a full professor of environmental engineering and directing Water and Environmental Studies Institute (WESI) at An-Najah National University (ANU) in Nablus, Palestine. Haddad's main research area is in water quality and resource management. He has published over one hundred and ninety papers in his field and edited over ten international conference proceedings and refereed books. Haddad directed and acted as a team leader of ten's of major projects in his. Haddad received many national and international awards. Haddad served and is serving as an editorial board member in and a reviewer for several local and international journals in his field.

Innovative Approaches Towards Sustainable River Basin Management in the Baltic Sea Region: The WATERPRAXIS Project

Marija Klõga, Walter Leal Filho and Natalie Fischer

Abstract This paper describes the scientific background, main elements and final results of the WATERPRAXIS project, which was implemented in 2009–2012 under the Interreg IVB Baltic Sea Region Programme 2007–2013 between seven coastal countries of the Baltic Sea Region (BSR). The special focus of this project was on the reduction of excessive nutrient loads to the Baltic Sea through support in implementation of cost- and eco-efficient water protection measures in the region. The rationale behind the WATERPRAXIS project was the need to tackle the continuing eutrophication of the Baltic Sea, a phenomenon which concerns scientists and governments alike. The clear dependencies between the bad quality of river waters flowing into the sea and its ecological state are well known and are already reflected in the European Union (EU) Water Framework Directive (WFD) (Schernewski et al. in J Coast Conserv 12(2):53–66, 2008). The EU WFD requires large-scale river basin management plans (RBMP) to be developed and implemented for each river basin district, aiming to achieve at least good ecological status in all European water bodies, including coastal seas, by 2015. However, this idealistic approach is hindered in practice by several barriers, in particular the large cover of RBMP and lack of good examples of the best local practices in river basin management. The WATERPRAXIS project tried to overcome these challenges and offer examples of successful water

M. Klõga (✉)
Department of Environmental Engineering, Tallinn University of Technology,
Ehitajate Tee 5, 19086 Tallinn, Estonia
e-mail: marija.kloga@ttu.ee

W.L. Filho · N. Fischer
Faculty of Life Sciences, Research and Transfer Centre "Applications of Life Sciences",
Lohbrügger Kirschstrasse 65, 21033 Hamburg, Germany
e-mail: walter.leal@ls.haw-hamburg.de

N. Fischer
e-mail: natalie.fischer@haw-hamburg.de

© Springer International Publishing Switzerland 2015
W. Leal Filho and V. Sümer (eds.), *Sustainable Water Use and Management*,
Green Energy and Technology, DOI 10.1007/978-3-319-12394-3_20

management initiatives from several countries around the Baltic Sea (Ulvi 2011). As a concrete output of the project, four different investment plans which realise water protection measures were implemented in Poland, Lithuania, Denmark and Finland.

Keywords Water quality Baltic Sea · Eutrophication · River basin management plans · EU Water Framework Directive · Eco-efficiency

1 Introduction

1.1 Key Issues

Marine eutrophication has become a worldwide problem in many coastal areas (Ryding 1994; Smith et al. 1999). However, the Baltic Sea is especially sensitive to this process because of its very slow water exchange, while the plant nutrient loads, mainly nitrogen and phosphorus, are high delivering from a wide variety of sources within its drainage basin (Wulff et al. 1990).

Over the last 100 years, since the industrial revolution in the region, the Baltic Sea has been slowly changing from a nutrient-poor (oligotrophic), clear-water sea into a nutrient-rich (eutrophic), murky sea (Smith et al. 1999). To date, eutrophication is considered to be one of the biggest environmental problems for the Baltic Sea, leading to imbalanced functioning of the entire marine and coastal ecosystems (Lundberg et al. 2009; Ulen and Weyhenmeyer 2007). The main cause of this marginally reversible process is excessive nitrogen and phosphorus loads from various activities, such as clearing of forests, development of farms and cities and increased use of fertilisers and detergents of the approximately 85 million people living in the catchment area. According to the latest data, total input of phosphorus and nitrogen to the Baltic Sea in 2008 reached 29,000 and 859,600 tons, respectively (HELCOM 2011).

In recent decades, many sea-protective measures have been successfully implemented in the Baltic Sea Region (BSR). These include different international programmes and projects with the overall objective to prevent eutrophication of the Baltic Sea and improve the state of its nature and water quality. Furthermore, several European water legislations are now demanding concrete measures aimed at combating eutrophication in the Baltic Sea, for example, the HELCOM Baltic Sea Action Plan, the EU Strategy for the BSR, the EU Water Framework Directive (WFD) and the EU Marine Strategy Framework Directive. The WATERPRAXIS project contributed to the EU Strategy for the BSR for reducing nutrient inputs to the Baltic Sea and enhanced the implementation of the EU WFD, which aims to ensure a good water quality in all European surface waters during the next decade.

2 Sources of Nutrient Inputs to the Baltic Sea

Nutrients which cause eutrophication reach the sea mainly from various human activities in the sea's drainage basin and, in smaller extent, from natural background sources. For simplicity, the total *external* input of nutrients into the Baltic Sea can be divided into three main pathways:

1. Direct emissions into the sea from industrial and urban areas on the coast (point sources)
2. Atmospheric deposition of nutrients on the sea surface
3. River-based run-off

The river run-off originates from point sources, such as industrial or municipal wastewater plants, as well as from diffuse sources such as agriculture, scattered dwellings and atmospheric deposition within river basins. It also includes natural background sources, which mainly refers to natural erosion and leakage from unmanaged areas that would occur irrespective of human activities (HELCOM 2006). Additionally, *internal* fluxes from sediments and the fixation of atmospheric nitrogen by cyanobacteria in the sea can also be a substantial factor when calculating total nutrient supply to the Baltic Sea.

The origin of nutrients can also be described using a scheme of waterborne and airborne inputs. In this scheme, nitrogen and phosphorus sources are analysed separately to demonstrate more clearly the most significant sector of nutrients pollution.

According to HELCOM 2006, waterborne discharges are the major source of nutrient inputs to the Baltic Sea, corresponding to about 75 % of the nitrogen input and 95–99 % of total phosphorus input (Fig. 1).

Diffuse losses (mainly from agriculture, forestry and scattered dwellings) are responsible for the largest portion of waterborne nutrient inputs. Furthermore, agriculture alone contributed to about 80 % of the reported total diffuse load (HELCOM 2009).

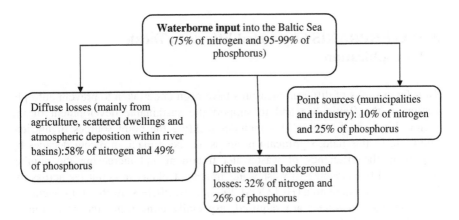

Fig. 1 Waterborne nutrient inputs into the Baltic Sea according to HELCOM 2006

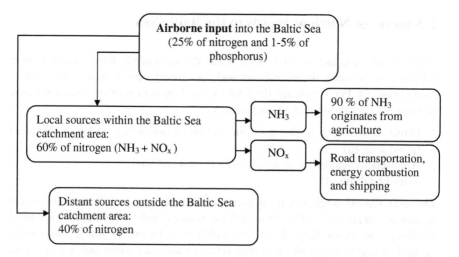

Fig. 2 Airborne nutrient inputs into the Baltic Sea according to HELCOM 2006

About 10 % of nitrogen and 25 % of phosphorus originate from point sources (municipalities and industry). The proportions of natural background losses were 32 % of nitrogen and 26 % of phosphorus.

The airborne deposition of nitrogen compounds comprises about quarter of the total anthropogenic load to the Baltic Sea (Fig. 2). The estimated airborne contribution of phosphorus is only 1–5 % of the total phosphorus load to the sea.

Nitrogen compounds are emitted into the atmosphere as nitrogen oxides and ammonia.

Road transportation, energy combustion and shipping are the main sources of nitrogen oxide emissions in the BSR; in the case of ammonia, roughly 90 % of the emissions originate from agriculture.

3 WATERPRAXIS Supported Efforts to Tackle Eutrophication

Thus far, a series of different measures have been undertaken to prevent eutrophication of the Baltic Sea and to support the sustainable development of the region. First and foremost, this includes tackling the point sources of nutrient pollution. In this field, significant progress has been made in recent decades by improving the efficiency of wastewater treatment and increasing the number of households connected to wastewater treatment plants in countries across the Baltic Sea catchment area (HELCOM 2007). Nevertheless, further improvements in wastewater treatment are required, especially concerning the reduction of phosphorus load.

However, non-point sources of pollution, which are often much more difficult to control, are the primary contributors to eutrophication in the Baltic Sea.

The list of measures for reducing the amount of nutrients from diffuse sources includes development of sustainable practices in agriculture (which, for example, include bans on the use of pesticides and fertilisers in farming, transformation of arable land into pastures, restrictions on stocking density and use of ecological farming methods), reducing pollution from transport, marine shipping and households, and pollution limitation from the energy sector. The final strategy to decrease diffuse nutrient load is fixation of nutrients after they have been discharged into the environment. This includes range of measures, such as protection of watershed forest cover and creation of buffer zones and buffer strips between streams and modern farmland (Fuerbach and Strand 2010), restoration and creation of wetlands and sedimentation pools (Rydén et al. 2003) and implementation of other measures, for example mussel cultivation in certain Baltic Sea lagoons to remove nutrients in the coastal waters (Stybel et al. 2009).

The WATERPRAXIS project also aimed to prevent the eutrophication of the Baltic Sea. The project enhanced the implementation of the EU WFD, aiming to achieve good ecological status, as a minimum, for all European waters by 2015. With the WFD, the EU specifically provides for long-term sustainable water protection management of the aquatic environment by requiring that its member states develop river basin management plans (RBMP) for each river basin district using the river basin approach instead of administrative or political boundaries (European Community 2000). The EU WFD also introduces the economic analysis of water use in order to estimate the most cost-effective combination of measures in terms of water use and requires active public participation in the development of RBMP by involvement of stakeholders, non-governmental organisations and citizens. However, applying water pollution control methods and changing land-use practices are sometimes hindered by many barriers. For example, RBMPs cover large geographical areas which are often transnational and, therefore, it is difficult to apply common public participation to the planning process and obtain joint acceptance on the local level for planned measures. Also, the cost-effective and eco-efficient calculation of measures is missing, but without proper knowledge of the environmental and economic efficiency of different water protection actions, it is virtually impossible to get sufficient political and financial support for their implementation. Furthermore, climate change has increased hydrological extremes by reducing the efficiency of water pollution control measures, and additional climate change impacts remain largely unknown (HELCOM 2010).

The WATERPRAXIS project was created to assist in overcoming these barriers and develop sustainable water management practices, as well as preparing water protection action plans and measures for selected pilot sites around the BSR. It was based on the previous Interreg BSR project Watersketch (http://www.watersketch.net/) and expanded on the results gained in other BSR projects, such as TRABANT, BERNET CATCH and ASTRA. The project partnership consisted of professionals who are specialised in river basin planning, environmental technology, environmental education and public organisations which implement water protection measures. The project was carried out in cross-national collaboration among seven coastal countries of the BSR: Finland, Denmark, Germany, Poland, Lithuania, Latvia and Sweden. Additionally, Kaliningrad, Russia, was integrated into the associated partnership status in order to secure greater Baltic coverage (Fig. 3).

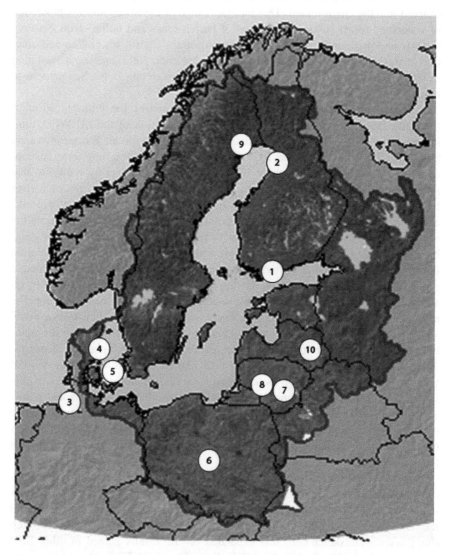

Fig. 3 Schematic view of the Baltic Sea drainage basin and location of project partners (*Source* http://www.grida.no/baltic/)

List of the partners:

1. Finnish Environment Institute, FIN
2. North Ostrobothnia Regional Environment Centre, FIN
3. Hamburg University of Applied Sciences, DE
4. National Environmental Research Institute, Aarhus University, DK
5. Municipality of Naestved, DK
6. Technical University of Łódź, PL
7. Kaunas University of Technology, LT
8. Charity and Support Fund Šešupė Euroregion, Šakiai office, LT
9. Luleå University of Technology, SW
10. Rēzekne Higher Education Institution, LV.

4 Objectives of WATERPRAXIS

The overall aim of the WATERPRAXIS project was to contribute to the efficient management of river basins to improve the ecological status of the Baltic Sea.

To move towards this strategic objective, its specific goals were identified as follows:

1. Determine and suggest improvements to current water management practices by analysing the contents and planning processes of RBMPs.
2. Establish RBMP-based action plans for pilot areas which incorporate best practices and measures for water protection and public participation.
3. Prepare investment plans (including technical and financing plans) for water protection measures at selected sites in Poland, Lithuania, Denmark and Finland.
4. Disseminate information on water management measures and best practices via publications, seminars and websites.
5. Offer training and education programmes for planners in the water management sector.

The action plans, investment plans and planning methods were prepared in close, transnational cooperation between project authorities and scientific partners. They are planned to be implemented in selected BSR countries, 2 years after project completion.

4.1 Project Pilot Areas

Four different river districts in Finland, Denmark, Lithuania and Poland were selected as the project's pilot sites for the drafting of concrete investment plans. These locations were given priority based on the urgent need for economically and environmentally feasible solutions to improve their ecological state.

All water bodies in the four pilot sites were facing water quality problems. Moreover, water quantity was considered as an important environmental issue in the pilot area in Finland. A major challenge for these sites was to create synergies which would contribute to the implementation of targets for the aquatic environment, while taking into account the social and economic needs of local communities.

When preparing concrete investments plans for these sites, cost-effective principles were also taken into consideration in order to reduce nitrogen loading at the lowest cost to society.

4.2 Project Set-up and Work Packages (WPs)

For project management, from its start to the very end, a clear project planning methodology is always needed. It describes the general structure of the project and makes every step in project implementation clear, so all project partners know exactly which objectives must be completed by what time and how this can be accomplished. WATERPRAXIS was structured as an empirical research project which tied together five different work packages (WPs).

WP1   Project management and administration. Lead partner (LP): Finnish Environment Institute (SYKE).

The main objective of this WP was general management of all project activities. WP1 was led by the main coordinator of the project with the help of management members, financial managers and members of the steering group. The coordinator monitored the overall progress and assesses the quality of work being performed. The tasks carried out included organisation of all meetings, monitoring and combining partners' activities and financial reports, preparation of periodic reports every 6 months, final reports, etc.

WP2   Communication and information. LP: Hamburg University of Applied Sciences (HAW Hamburg).

The main aim of WP2 was to facilitate effective external and internal communication among project partners and ensure that all partners and other target groups across the Baltic Sea were aware of all project activities and results. WP2 raised awareness about WATERPRAXIS by disseminating the materials and documents produced as a part of this project and promoting the project's findings using, for example, the following instruments: project website, brochures, posters and newsletters, project dissemination in external events, such as conferences, fairs and exhibitions. The overall purpose of WP2 was to achieve high recognition for decision-makers, stakeholders and end users for the long-term goal of successful implementation of project results

WP3   Reviewing RBMP and processes. LP: National Environmental Research Institute, University of Aarhus (NERI).

The main purpose of this WP was to analyse recently drafted RBMP, including their implementation processes from various BSR countries in order to identify different approaches to river basin planning. WP's work started with developing a framework and guidelines for analysing RBMP. During the analysis, the primary focus of this WP was on institutional set-up, planning approaches and procedures, institutional structure, interplay, public participation, integration with other policy goals and climate change issues. Based on the results, the best practices and solutions in river basin planning were identified, which have the potential to be widely applied throughout the BSR

WP4   From RBMP to local water protection action plans. LP: Kaunas University of Technology.

WP4   summarised the existing regional water protection action plans covering defined regional target areas and prepared consolidated recommendations for the implementation of best practices. In this transnational work, with the participation of all partners, barriers, innovative measures and funding instruments for local implementation were identified. Furthermore, improvements for existing action plans were suggested, and new plans were created for pilot areas in Finland, Denmark, Poland and Lithuania. The capacity of local stakeholders in the environmental economy was strengthened by organising a university course and cost-efficient analysis of proposed measures. The WP was consistent with the principles established by the WATERSKETCH project and was aimed at strengthening the scientific, technical and social capacity to implement sustainable water resource management

WP5   From action plans to local investments in water resource protection. LP: Technical University of Łódź.The main aim of WP5 was to create a solid bridge between action plans and local water protection investments by implementing best available water protection practices in selected BSR countries. These best applicable water protection measures at river basins, which have a significant impact on the Baltic Sea and where environmental goals are not met (pressures/impacts/mitigation measures), were provided as examples and a showcase for the general public and local politicians in charge of environment issues. The ultimate goal was to involve local politicians and, thus, secure local water protection investments for an extensive period of the project

5 Final Results from the WATERPRAXIS Project

The expected final results at the end of the WATERPRAXIS project were:

1. Examples and guidelines of the best water management practices for river basin planning at several levels (official river basin districts, single river basins and

local investments) based on previous experiences from different countries and results attained from the project's pilot studies (published as a report and online).

2. Practical examples of good investment projects (published as a report and online).
3. Training courses for regional and local planners on general river basin management focusing on environmental economy and cost-effectiveness analysis.
4. Water protection action plans for pilot areas in some partner regions.
5. Investment plans (including technical and financing plans) for water protection measures in pilot areas in Finland, Denmark, Poland and Lithuania.

Since the project's inception in January 2009, various activities have been implemented within the project framework. This includes the organisation of several workshops, training courses and symposiums as well as producing of numerous reports and publications.

The most significant training and educational events are briefly described below.

1. Workshop on land-use modelling, 11–13 November 2009, Helsinki, Finland.
   The aim of this training workshop was to provide participants with knowledge and methods of the challenges posed by climate change and land-use development on river basin planning and management. Sessions focused on hands-on exercises using the GIS-based software developed in the earlier EU Forum Skagerrak and RiverLife and Watersketch projects.
2. Symposium on climate change and sustainable water management, 9 December 2009, Lyngby, Denmark.
   The event was organised by the Hamburg University of Applied Sciences parallel to the 15th Conference of the Parties (COP 15) of the United Nations Framework Convention on climate change. Its precise aims were as follows: to discuss the links between climate change and sustainable water management, present the work of some of the organisations working in the field, introduce some of the ongoing projects and initiatives dealing with sustainable water use and sustainable river basin management, identify areas where action is needed to facilitate a better understanding of the impacts of climate change on water systems and the measures which may be adopted to promote sustainable water management.
3. Workshop on acid sulphate soils (ASS) and land use, 1–2 November 2010, Luleå, Sweden.
   The workshop was organised as a result of increasing environmental problems caused by land use in ASS for ditching and ditch cleaning. The workshop aimed to disseminate information about ASS from a scientific, administrative and practical perspective; exchange different experiences with activities on ASS; identify future research needs; and plan feasible cross-border projects for the future.
4. Training course on economical tools for WFD implementation, 13–14 January 2010, Kaunas, Lithuania.
   The training course aimed to provide river basin planners with insights and hands-on training on how economic analyses on costs and benefits can be used in water resource planning, particularly related to the implementation of the WFD.

The contents of the course were an overview of the economic requirements in the WFD, an introduction to the fundamentals of environmental economic assessments of costs and benefits, and examples of applied cost-benefit studies and cost-effectiveness studies. Additional values connected to the action plans were also analysed, for example, improved recreational possibilities and tourism alternatives.

5. Symposium on climate change challenges in river basin management, 17–19 January 2011, Oulu, Finland.

The international symposium was organised to discuss the challenges climate change poses for the use of water systems, water protection and how the EU can best tackle these challenges. During the two symposium days, presentations covered climate change challenges and adaptation from a variety of perspectives, including observed climate trends, effects on surface and ground waters, scenario studies, socioeconomic aspects and participatory tools related to water management planning.

In addition, during the project lifetime summarising final reports were produced within the WPs 3–5:

1. WP3: RBMP. Institutional framework and planning process. Cross-country analyses. The main body of the report analysed and compared the RBMPs, the planning processes and the structures and mechanisms laid down for implementation among the involved countries, namely Sweden, Finland, Latvia, Lithuania, Poland, Germany and Denmark. Based on this, the main challenges for implementation were discussed, as they could be recognised at this point in time, when RBMPs had been finalised for most countries, whereas Denmark still did not adopt the RBMPs.

The conclusions of the report are as follows:

- Institutional fit is low, a few countries opted for spatial fit
- All countries opted for coordinating bodies, and coordination seems more important than fit in all cases
- Compliance with procedures was high (except for Denmark where the focus was on implementation and financing)
- Ambitions are variable in the short distance, but high ambitions may be challenged by financial commitment
- Implementation gaps seem to be large, but learning processes may be more important for implementation successes in the long run

2. WP4:
Examples of Applied Water Management Practices in the BSR.

- WATERPRAXIS Pilots: Finland, Denmark, Poland and Lithuania.
  Water Protection Action Plans.
  Within the project local areas have been chosen in the above mentioned countries as pilot areas. For each of these areas action plans have been produced. The action plans describe existing ecological problems in the areas as well as existing management measures. Furthermore, the existing measures have been

promoted and supported and additional measures were suggested based on economic and cost-efficiency analyses.

- WATERPRAXIS Case studies. Latvian case study: Daugava River. Swedish case study: ASS. The project has also investigated important water management problems in Sweden and Latvia.

3. WP 5:

- WATERPRAXIS pilot reports of environmental, economical and social impact assessment. From Action Plans to Local Investments in Water Resources Protection. For each of the four pilot areas in Finland, Lithuania, Poland and Denmark, social, economical and environmental assessments have been conducted. The results for each pilot area are described within this report.
- Description of Investments and Investment Plans. Pilot projects from Finland, Denmark, Poland and Lithuania. In addition, the project described or established a set of investment plans for selected measures in the project's pilot areas in Poland, Lithuania, Denmark and Finland.

All these reports are available on WATERPRAXIS Web site at www.waterpraxis. net and have a wide range of information and experiences from the countries which took part in this project.

6 Conclusions

Based on the project work undertaken over the years 2009–2012, a set of conclusions can be made.

Firstly, the different challenges in water management and river basin planning in the BSR countries and also the different approaches towards meeting these challenges were identified. Therefore, it was concluded that there is still great need for further scientific cooperation between BSR countries and mutual learning is imperative in the field of water management and river basin planning.

Secondly, in the frame of the project, the current status and the needs for improvements to water management practices in the BSR countries as a whole and the pilot project areas in Poland, Lithuania, Denmark and Finland in particular were identified and some changes to these present practices were proposed.

In addition, the project has suggested improvements on water management measures and practices and prepared a set of investment plans for selected measures in the project's pilot areas in Poland, Lithuania, Denmark and Finland. Furthermore, the project has investigated important water management problems in Sweden and Latvia. WATERPRAXIS has also offered education on sustainable water management, economic analyses and land-use planning for river basin planners (Ulvi 2011).

So far, substantial progress has been made in water protection in Europe and individual European countries, as well as in tackling important issues at the European level. Nevertheless, European water bodies require continuous effort to get or keep them clean. Almost all European waters have a clear transboundary nature; thus, their sustainable use and protection can only be carried out based on hydrological boundaries and via close international cooperation between scientific communities, citizens and environmental organisations. WATERPRAXIS project succeeded to fill some existing information gaps and tried to offer examples of successful water management to BSR stakeholders at local level.

References

European Community (2000) Directive 2000/60/EC of the European Parliament and of the Council of 23 October 2000 establishing a framework for Community action in the field of water policy. OJ L 327, p 73, 22 Dec 2000

Fuerbach P, Strand J (2010) Water and biodeversity in the agricultural landscape. Environ Prot Agency, Sweden, p 50

HELCOM (2006) Eutrophication in the Baltic Sea. Draft HELCOM thematic assessment in 2006. In: HELCOM stakeholder conference on the Baltic Sea Action Plan, Helsinki, Finland

HELCOM (2007) Towards a Baltic Sea unaffected by Eutrophication. HELCOM ministerial meeting, Krakow, Poland

HELCOM (2009) Eutrophication of the Baltic Sea. Executive summary. Baltic sea environment proceedings no. 115A

HELCOM (2010) Implementation of HELCOM's Baltic Sea Action Plan (BSAP) in Finland. 17 May 2010 status report

HELCOM (2011) Activities 2011 overview. In: Baltic sea environment proceedings no. 132

Lundberg C, Jakobsson BM, Bonsdorff E (2009) The spreading of eutrophication in the eastern coast of the Gulf of Bothnia, Northern Baltic Sea—an analysis in time and space. Estuar Coast Shelf Sci 82:152–160

Rydén L, Migula P, Andersson M (2003) Environmental science: understanding, protecting, and managing the environment in the Baltic Sea Region. The Baltic University Programme, Uppsala, p 824

Ryding S-O (1994) Environmental Management Handbook. IOS Press, Netherlands, p 777

Schernewski G, Behrendt H, Neumann T (2008) An integrated river basin-coast-sea modelling scenario for nitrogen management in coastal waters. J Coast Conserv 12(2):53–66

Smith VH, Tilman GD, Nekola JC (1999) Eutrophication: impacts of excess nutrient inputs on freshwater, marine, and terrestrial ecosystems. Environ Pollut 100:179–196

Stybel N, Fenske C, Schernewski G (2009) Mussel cultivation to improve water quality in the Szczecin Lagoon. J Coast Res 5:1459–1463

Ulen BM, Weyhenmeyer GF (2007) Adapting regional eutrophication targets for surface waters—influence of the EU water framework directive, national policy and climate. Environ Sci Policy 10:734–742

Ulvi T (2011) Summary of the Waterpraxis project. www.waterpraxis.net

Wulff F, Stigebrandt A, Rahm A (1990) Nutrient balance in the Baltic—nutrient dynamics of the Baltic Sea. Ambio 19:126–133

Authors Biography

Marija Klõga is a PhD student in Department of Environmental Engineering at Tallinn University \of Technology, Estonia. The main field of her research is related to the state of the water environment and environmental protection (water quality and factors determining it, self-purification processes in rivers, methodology of water monitoring).

Walter Leal Filho (BSc, PhD, DSc, DL) heads the Research and Transfer Centre "Applications of Life Sciences" at the Hamburg University of Applied Sciences, Germany. He has over 20 years of research experience on all aspects of environmental information and education and has a particular interest on the connections between environmental management, sustainability, climate and human behaviour.

Natalie Fischer is a biologist from Research and Transfer Centre "Applications of Life Sciences" at the Hamburg University of Applied Sciences, in Hamburg, Germany. Since 2009, she has been coordinating EU projects at a national and international level.

Towards Sustainable Water Use: Experiences from the Projects AFRHINET and Baltic Flows

Walter Leal Filho, Josep de la Trincheria and Johanna Vogt

Abstract This paper presents an analysis of the subject sustainable water use and discusses its many ramifications. It also introduces two projects being undertaken at the Hamburg University of Applied Sciences, which aim to put the principles of sustainable water management into practice.

Keywords Sustainability · Water use · Baltic · Africa · Rainwater · Management

1 Introduction: Sustainable Water Use

Water is one of the essential resources for a human being. Availability of water resources has great impacts on the environmental, political and economic situations as well. According to WHO and UNICEF, it has been estimated that more than 2 billion people are affected by water shortages worldwide (WHO/UNICEF 2000 in United Nations 2003). In addition, approximately 780 million people around the globe do not have an access to clean water. This is a result of wasteful water usage that is caused amongst other reasons by improper economic incentives, underinvestment, poor management systems, obsolete equipment and failure to apply existing technologies (Pacific Institute 2014).

According to UN projections, by the year 2025, water abstraction in developed countries will increase by 18 % (United Nations 2003), whereas by 2050, at least one in four people will live in a country affected by chronic or recurring shortages of freshwater (Gardner-Outlaw and Engelman 1997 in United Nations 2003). These facts show the current and potential future risks associated with the problem of the scarcity of water resources.

W.L. Filho (✉) · J. de la Trincheria · J. Vogt
Faculty of Life Sciences, Research and Transfer Centre, Applications of Life Sciences,
Hamburg University of Applied Sciences, Lohbruegger Kirchstraße 65,
Sector S4/Room 0.38, 21033 Hamburg, Germany
e-mail: walter.leal@ls.haw-hamburg.de

© Springer International Publishing Switzerland 2015
W. Leal Filho and V. Sümer (eds.), *Sustainable Water Use and Management*,
Green Energy and Technology, DOI 10.1007/978-3-319-12394-3_21

In 1998, the issue of sustainable water use and management, as well as required measures, was addressed by the Commission on Sustainable Development (United Nations 1998). Experts define the sustainable use of water resources as the avoidance of any kind of welfare losses in the use of water resources (Bithas 2008). Sustainable water management focuses on water quality and quantity and requires society to conserve and use it more efficiently (EEA 2012b).

Until today, Europe has mostly been insulated from economic, social and environmental impacts of water shortages (EEA 2009) due to its abundance of water resources. In comparison with the global average, only around 13 % of all renewable and accessible freshwater from natural water bodies, including surface waters (rivers and lakes) and groundwater (EEA 2012b) is withdrawn to meet water demand. An average European directly uses approximately 130 L of water per day (EEA 2014c). Groundwater satisfies about 55 % of public water demand (EEA 2009 in EEA 2012b).

Despite the projections that the amount of water abstraction globally will increase, according to the EEA, in Europe, this amount is expected to decrease by about 11 % between 2000 and 2030 with pronounced decreases in western Europe. The future water demand of the "domestic sector", which includes households and small businesses, remains highly uncertain and will depend on a wide range of factors, such as income, size of households, age distribution of population and technology (EEA 2007).

In addition to households, the other main users of water are agriculture, industry and the energy sector (EEA 2014b). Discounting the disparity between regions, Europe as a whole uses around 30 % of abstracted water in agriculture, 30 % in energy production for cooling purposes, 25 % for public water supply and 15 % is used by industry (EEA 2012b).

However, prolonged periods of low rainfall or drought caused by global climate change and overabstraction due to increasing demand have significantly influenced the balance between water demand and availability, which have reached the critical level in many areas of Europe (EEA 2009). For example, one of the drivers of increasing agricultural water use across Europe over the last two decades is the Common Agricultural Policy (CAP). In some cases, the policy provides subsidies to produce water-intensive crops (EEA 2009).

The European Union requires all countries to promote sustainable water use based on long-term projection of available water resources and to ensure a balance between abstraction and recharge of groundwater (EEA 2013). These requirements are expressed in EU water legislation and policies.

The Water Framework Directive and the Sixth Environment Action Programme define the boundaries and set goals for sustainable water use and oblige the Member States to achieve "good status for surface and groundwater" of water bodies by 2015. The main goal is to prevent environmental degradation and restore or maintain sustainability via management of the combined impacts of water use and pollution pressures (Werner and Collins 2012). It covers all water categories, such as rivers, lakes, groundwater, coastal and transitional waters (EEA 2014e).

In addition, recent reforms of the CAP reduce the link between subsidies and production from agriculture and intensively promote the adoption of agri-environmental schemes, with measures related to a more sustainable use of the water resource by agriculture in future (EEA 2009).

The European Community has also signed the Watercourses and International Lakes Convention that establishes main principles and rules to develop and promote coordinated measures of sustainable use of water and related resources of trans-boundary rivers and international lakes (EEA 2013).

One of the water indicators used by UNEP, OECD and EUROSTAT (EEA 2013) that describes water status and trends and gives a general oversight of water issues in Europe is the Water Exploitation Index (WEI) (McGlade 2008). WEI is the total freshwater abstraction (i.e. water removed from any freshwater source, either permanently or temporarily, including mine water, drainage water and abstractions from precipitation) divided by the long-term average available water expressed as a percentage (Eurostat 2014). A value above 20 % indicates a stress on freshwater ecosystems from overabstraction (EEA 2012b). A value that exceeds 40 % indicates a severe scarcity (Eurostat 2014).

An example of this is in Poland between the years 2002 and 2011. In this case, the value of WEI ranged between 18 and 19.4 %, with 18.9 % in 2011. There are no available WEI values in Estonia, whereas in Germany, the only available WEI value is 18.9 % from the year 2004. Amongst the Baltic countries, Latvia and Sweden have the lowest WEI values that equal to 0.6–1.4 and 1.4 %, respectively (Eurostat 2014).

2 Some Projects Working on Sustainable Water Use

This section describes two large projects on sustainable water use being undertaken at the Hamburg University of Applied Sciences, as examples of what can be achieved.

2.1 Project 1—Baltic Flows Project—Monitoring and Management of Flowing Rain Water in Baltic Sea Catchment Areas

The Baltic Flows project is a scheme funded by the European Union Seventh Framework Programme. It concerns rainwater monitoring and management in Baltic Sea catchment areas. Rainwater, when available in large amounts, can form streams and rivers. In urban environments, heavy rain can also amount to storm water and floods. Over the years, much of this rainwater ends up in the sea. In northern Europe, the Baltic Sea conceals a history of water quality from streams, rivers and urban run-off in catchment areas. Encircled by a mix of Nordic, Central and Eastern European countries, the Baltic Sea is at the mercy of a range of national pollution and water treatment policies.

Several initiatives and projects have studied the state of the Baltic Sea and aim at improving water quality via various preservation measures. These include the Baltic Sea Action Plan by the Helsinki Commission, Finland, and the European Union's Baltic Sea Region Interreg programme.

The idea behind Baltic Flows project is that rainwater should be monitored and managed before it reaches the sea. Pollution should be detected as early as possible, preferably in high water than downstream regions. To achieve this, we should embrace three strengthening phenomena in modern society:

- increased miniaturisation of technology;
- increased citizen participation, social media; and
- understanding of urban planning.

Miniaturisation is enabling new small-size, low-cost technology. This will gradually shift the balance from individual, high-cost devices towards low-cost, small-size devices that can be installed in the masses. In future, miniature low-cost water measurement technology capable of wirelessly relaying real-time data will enable water monitoring networks of unforeseen coverage and timeliness. In addition, devices could harvest required energy from flowing water, thus eliminating the need for a power infrastructure.

In the Baltic Flows project, a total of 45 organisations will combine forces to reach a new level of world-class know-how in rainwater monitoring and management. The project consortium, comprising 14 organisations from five European regions—Estonia, Finland, Germany, Latvia and Sweden, in addition to two partner organisations collaborating via indirect representation, will form the core of the project. These will be assisted by one partner, 11 specialists, who are strongly linked to the Chinese environmental sector, and 28 supporting partners in Europe and five international regions: Russia, Belarus, China, Vietnam and Brazil. This shows that clean water is more than a European issue; it is a common concern of global magnitude.

The objectives of the Baltic Flows project have been designed to serve two top-level targets:

(a) to bring forth the technological and economic vision that will enable European regions to achieve world-class excellence and a sustainable competitive edge in the rainwater monitoring and management sector; and
(b) to fulfil the objectives of the coordination, enhancing the effectiveness of research-driven clusters in participating regions, and thus paving a smooth path towards smart specialisation, via a common trans-regional vision, strategy and realistic implementation plan.

In order to achieve real-world results, item b must follow item a; global competitiveness must be the first priority, as there is no point in interregional collaboration if the regions lack the prerequisites for potential competitiveness.

The work plan of the Baltic Flows project is designed to effectively facilitate different types of project activities carried out during the course of the project. Activities fall into four different categories: coordination and communication, network building, insight building in specific S and T area, and result consolidation. Coordination and communication work packages are essential to ensure that the fundamental goals of the project are achieved: the project is implemented as planned, and results are delivered in a manner beneficial to the European community.

The following Baltic Sea riparian states are involved in the Baltic Flows project: Finland, Sweden, Estonia, Latvia and Germany. A partner from the UK completes the partnership. The project is coordinated by the University of Turku in Finland.

1. University of Turku, Finland
2. Turku University of Applied Sciences, Finland
3. Turku Science Park Ltd., Finland
4. Regional Council of Southwest Finland V–S, Finland
5. Tallinn University of Technology, Estonia
6. Cleantech Estonia NPO, Estonia
7. City of Tallinn, Estonia
8. Hamburg University of Applied Science, Germany
9. Institute of Physical Energetics, Latvia
10. Environmental Projects Ltd., Latvia
11. Riga Planning Region, Latvia
12. University of Uppsala, Sweden
13. Upwis AB, Sweden
14. Uppsala County Administrative Board, Sweden
15. EcoTech International Ltd., UK.

Several organisations from Finland, Sweden, Estonia, Latvia, Germany, Russia, Belarus, China, Vietnam and Brazil provide input as supporting partners.

2.2 Project 2—AFRHINET

The AFRHINET project is a capacity-building project under the framework of the African, Caribbean and Pacific (ACP) Science and Technology Programme, which is funded by the European Union (EU) and implemented by ACP Secretariat.

The overall objectives of AFRHINET are twofold: to foster endogenous and self-replicable capacities in the field of RWHI management and sustainable dryland agriculture on one hand, and to boost the transfer and the adoption of research results by implementing research and technology-transfer activities and demonstration actions of innovative RWHI management on the other. This is expected to ultimately lead to improved food and water security, poverty alleviation, and socio-economic and climate resilience. The specific objectives of this project are as follows:

• To foster science and technological (S and T) capacities on RWHI, the quality of research and the capacity of the S and T communities to attract funding in this field of knowledge;

- To set-up a market-oriented research and technology-transfer framework to better capitalise and disseminate innovative research results;
- To develop the capacity of the S and T community and local communities to practically implement adequate RWHI management;
- To strengthen the link of S and T communities with the regional market, businesses/micro-enterprises, NGOs, policy-making actors and local communities;
- To establish a long-term ACP-EU network on RWHI management.

The capacity-building activities, and the transfer and demonstration of innovative RWHI management technologies envisaged under the framework of this project aim to stimulate the development and use of rainwater harvesting as a supplemental irrigation technology. This is expected to increase agricultural yields and foster the diversification of local income-generating activities for smallholder farmers through sustainable dryland agriculture, agroforestry and horticulture.

Furthermore, the AFRHINET project addresses gender equality and equal opportunities by taking into account the particular needs of women, tribes and minority groups during the design of the capacity-building courses. Efforts are being made to respect gender and ethnic balance in the project teams and in the number of participants/speakers in the capacity-building activities. AFRHINET's activities are developed combining the experience, innovations and best-practices available in eastern and southern Africa, with the concrete needs of the local context through participatory approaches, thus leading to actions that best fit the project, and results for the target countries.

In order to achieve the project objectives most efficiently, the AFRHINET project revolves around 5 groups of activity:

- Baseline study on the needs, potential and market-oriented products in the field of RWHI management
- Developing capacities on RWHI management for sustainable dryland agriculture, improved food security and poverty alleviation
- Research and technology-transfer centres on RWHI and sustainable dryland agricultural water management
- Building food, poverty and climate resilient communities: demonstration of innovative RWHI practices
- Networking, dissemination, promotion and awareness

The main outputs of the AFRHINET project are as follows:

1. Better support for the management, innovation and quality of applied research activities in the field of RWHI management for improved food security and poverty alleviation;
2. In-depth understanding of the market, as well as non-governmental, public sector and local community needs for RWHI management;
3. Reinforcement of the technical capacity to practically implement and adopt adequate and innovative RWHI management;
4. Improved market orientation of technology transfer for better capitalisation and dissemination of innovative and effective research, know-how and technologies;

5. Increased networking capacity of the S and T community with target groups and key national and international stakeholders;
6. Awareness, dissemination and promotion of RWHI and sustainable dryland agriculture management for improved food security and poverty alleviation.

The implementation methodology bears reference to achieving both long-term impacts (i.e. expertise development, transfer and adoption of research results and innovation, market-oriented science, networking, food and water security improvement and poverty alleviation) and short-term impacts (staff capacity-building, pilot and demonstration actions, a platform to network/transfer/adopt research results, etc.). Moreover, the implementation of the AFRHINET project aims at the close involvement and participation of the local stakeholders and target groups (NGOs, businesses/micro-enterprises, consultancies, public bodies and ministries, local communities, etc.) thereby important contacts to future clients and cooperation partners for research, transfer and adoption of science and technology activities will be built-up and deepened. These are vital for the successful implementation of innovative market-oriented actions in the field of RWHI and sustainable dryland agriculture.

The AFRHINET project is coordinated by the Research and Transfer Centre "Applications of Life Sciences" at Hamburg University of Applied Sciences (HAW), in Hamburg (Germany). The partner members of this project are as follows:

- Addis Ababa University, Ethiopia;
- University of Nairobi, Kenya;
- Eduardo Mondlane University, Mozambique;
- University of Zimbabwe, Zimbabwe.

The associate members of this project are as follows:

- International Crops Research Institute for Semi-Arid Tropics (ICRISAT), Zimbabwe;
- Southern and eastern Africa Rainwater Network/International Centre for Research in Agroforestry (SEARNET/ICRAF), Kenya;
- WaterAid Ethiopia, Ethiopia.

The cross-sectorial cooperation generated within this project is expected to increase the awareness and mutual understanding on the importance of qualified human resources in sub-Saharan Africa and ACP countries. In addition, these activities are also expected to strengthen the role of sub-Saharan Africa and ACP countries as producers and distributors of information, know-how and technologies. This is expected to strengthen their role as hubs of knowledge for businesses/micro-enterprises, NGOs, public institutions and policy makers as well as to enhance the environmental and technological expertise available for policy-making, integration and innovation actions. Ultimately, this may lead to a solid basis for future projects and regional development.

3 Some Areas to Look at Now and in Future

As the two projects have shown, there are some practical dimensions to sustainable water use and management. Attempts to pursue it should consider a set of economic, policy and technological instruments. These are as follows:

(a) **Water pricing and metering**

Experts see water pricing as an essential requirement for sustainable water use and management. Water prices and tariffs must internalise all external factors, including environmental and resource costs (Werner and Collins 2012). Therefore, the Water Framework Directive obliges the Member States to take account of the costs of water-related services, this would allow the environmental costs of water to be reflected in the price of water (EEA 2012b, 2014b). Thus, water services with a negative environment impact, such as pumping, weirs, dams, channels and supply systems, will be paid for by the users (e.g. agriculture, hydropower, households and navigation), based on the polluter pay principle (EEA 2014d). However, solely, full-cost pricing is not a sufficient condition for sustainable water use (Bithas 2008).

The effectiveness of the implementation of the mechanism directly depends on the volumetric pricing and metering tool, a lack of which can lead to consumers being charged a fixed amount regardless of their actual water use (Bithas 2008; Werner and Collins 2012). An example of the successful implementation of the tools is the decline in public water supply in eastern Europe since the early 1990s (EEA 2009). However, despite the legal requirements under the WFD, such practices still are not used to their full extent, including agriculture water use (European Commission 2012; EEA 2014d).

(b) **Reducing losses due to leakage**

The reduction of leakage in water systems is another possibility for increasing water-use efficiency. The problem is relevant for both Eastern and western European countries. In addition, the situation in some countries, for example in Latvia, is aggravated by not sufficiently developed centralised water supply and sanitation systems (Visockis et al. 2010).

Leakage in public water systems is a common problem that results in loss of drinking water, wasting of energy and material resources used in abstraction and treatment, and a potential risk of bacterial contamination from surrounding ground (Werner and Collins 2012). According to various studies, leakage is usually the largest component of distribution losses, which range between 5 and 8.5 m^3/km of a pipe in the supply network per day (m^3/km/day) (Werner and Collins 2012; EEA 2014c). Only in Germany, Denmark, France and Sweden, average values range from 1 to 10 m^3/km/day (EEA 2014c). Although entire elimination of leakage is an unrealistic goal, leakage reduction is a crucial part of sustainable water management (EEA 2014b).

(c) **Information and communication**

Available, reliable and up-to-date information is another tool for sustainable water use and management. It has a lot of benefits including an improved overview of the causes, location and scale of water stress, identification of trends, facilitation of the evaluation of measures implemented to address unsustainable water use and further engagement of citizens—in Europe and elsewhere—in water issues (EEA 2009).

Awareness-raising campaigns aimed at domestic and business water consumers play an important role in water conservation (Werner and Collins 2012). Over the past 10 years, the amount of information provided to consumers and agriculture regarding efficient lawn-watering and gardening practices, water conservation, water-use behaviour, water-efficiency labels for households' appliances, etc. has significantly increased (EEA 2014d).

(d) **Technical measures**

Technical measures such as installation of water saving devices, reuse of grey water and treated wastewater, and rainwater harvesting might also potentially reduce the use of publicly supplied water. For example, installation of water-efficient showerheads can save about 25 L per property per day (Waterwise 2010 in Werner and Collins 2012). Stored grey water (wastewater from baths, showers, washbasins, kitchens and washing machines) can be subsequently reused for flushing toilets and watering gardens (Werner and Collins 2012). This might have a significant impact on an amount of water used by households that typically accounts for 60–80 % of the public water supply across Europe with personal hygiene and toilet flushing that amounts to about 60 % of this share (EEA 2009).

(e) **Rainwater harvesting**

As shown in the projects Baltic Flows and AFRHINET, rainwater harvesting (RWH) can be an effective tool. It refers to the process of collecting, diverting and storing rainwater from an impervious area, such as roofs, for subsequent use (EEA 2009). It can reduce use of treated public water by households and load on urban drainage systems during heavy precipitation (Werner and Collins 2012).

The size of a rainwater harvesting system and amounts of collected water might vary significantly. There are three major types of RWH: firstly, in situ RWH: collection of the rainfall on the surface where it falls and storing in the soil; secondly, domestic RWH: water which is collected from roofs, street and courtyard run-offs; thirdly, external water harvesting: the collection of run-off originating from rainfall over a surface elsewhere and stored offside (Helmreich and Horn 2009). Water may be used for flushing toilets, watering gardens and roofs with vegetative cover, and for the replenishment of a vegetated pond (Villarreal and Dixon 2005).

Rainwater is also a means for creating green urban areas. A conventional storm system of underground pipes is substituted by surface-water drainage system designed as open channels along the street collecting water from adjacent rooftops and paved areas. One of the examples of such system is a water park in the Enköping (Uppsala, Sweden). The project was launched by local council in 1995 (Wlodarczyk 2007).

In Latvia, the use of rainwater is in the frame of the strategic aims of the country: "Careful using of nature resources and safe for next generations". The aims were established in order to follow the EU requirements regarding the decrease and optimisation of water and energy resources (Visockis et al. 2010).

In Poland, the cost of fully automatic rainwater harvesting system is relatively high in comparison with the average prices of the cubic metre of drinking water. An increase in usage of such systems in the country requires development of financial mechanisms such as subsidies and tax reliefs. Results of studies undertaken indicate that rainwater harvesting might cover between 30 and 40 %+ of the daily water consumption, depending on water consumption structure of that particular household (Mrowiec 2008).

The water use in agriculture requires special attention. One of the reasons is that on average globally, agriculture uses about 70 % of all freshwater withdrawals, out of which only 40 % contributes to crop production, whereas the remainder is lost (United Nations 2003). In some parts of southern Europe, a share of water used for the agricultural purposes reaches up to 80 % (EEA 2014a).

4 Conclusions

The sustainable use of water needs to be a top priority in the global agenda, especially in developing countries. Amongst the various technological and management measures available to increase the efficiency and sustainability of water use, the use of rainwater (project Baltic Flows) especially its use in irrigation (project AFRHINET) can play a key role. Across the European Union, potential water savings from improving conveyance efficiency are estimated at 25 % of water abstracted (WssTP 2010 in Werner and Collins 2012). In the developing counties, it can be even higher. The efficiency of irrigation depends on its type, for example, furrows, sprinklers and drip irrigation have 55, 75 and 90 % of efficiency, respectively (Werner and Collins 2012). However, rain-fed agriculture requires adequate mechanisms to reduce inherent risks (de Fraiture and Wichelns 2010).

Another measure is modification of agricultural practices, in other words, the selection of less water-intensive crop types as well as development of potential for returning irrigated land back to traditional rain-fed practices (Werner and Collins 2012).

The problem of illegal water abstraction, particularly from groundwater and often for agricultural purposes, is widespread in certain areas of Europe. This problem represents a major political and technical challenge (Werner and Collins 2012).

The next step in the European Union's efforts towards sustainable water use and management is presented in the "Blueprint to Safeguard Europe's Water" communication (European Commission 2012). The document includes reviews of the Water Framework Directive, Europe's policies on water scarcity and drought, and the water-related aspects of climate change adaptation and vulnerability. It is expected to help better integration of water objectives into EU policies and encourage water efficiency (EEA 2012a). Additional main documents in future EU water policy are the EU Biodiversity Strategy 2020 and the EU Resource Efficiency Roadmap, which aim at the efficient use of natural resources in order to support sustainable growth (EEA 2012a).

These elements all illustrate the relevance of and the need for sustainable water use and management and show how much still needs to be done, so as to make the sound use of water a reality.

References

Bithas K (2008) The sustainable residential water use: sustainability, efficiency and social equity. The European experience. Ecol Econ 68(1–2):221–229. Available at: http://linkinghub. elsevier.com/retrieve/pii/S0921800908001122. Accessed 31 May 2014

De Fraiture C, Wichelns D (2010) Satisfying future water demands for agriculture. Agri Water Manage 97(4):502–511. Available at: http://linkinghub.elsevier.com/retrieve/pii/ S037837740900239X. Accessed 24 May 2014]

EEA (2014a) Europe's water: efficient use is a must. Available at: http://www.eea.europa.eu/articles/europe2019s-water-efficient-use-is. Accessed 30 May 2014

EEA (2014b) European water resources—overview. Water resources. Available at: http://www.eea.europa.eu/themes/water/water-resources. Accessed 29 May 2014

EEA (2012a) European waters—current status and future challenges, Copenhagen. Available at: http://www.eea.europa.eu/publications/european-waters-synthesis-2012

EEA (2014c) Improving transparency in water services. Available at: http://www.eea.europa.eu/ highlights/improving-transparency-in-water-services. Accessed 30 May 2014

EEA (2012b) Part 2. Thematic indicator-based assessments. Environmental indicator report 2012—Ecosystem resilience and resource efficiency in a green economy in Europe. Available at: http://www.eea.europa.eu/publications/environmental-indicator-report-2012/ environmental-indicator-report-2012-ecosystem/part2.xhtml#chap8. Accessed 30 May 2014

EEA (2014d) Policies and measures to promote sustainable water use. Water resources. Available at: http://www.eea.europa.eu/themes/water/water-resources/policies-and-measures-to-promote-sustainable-water-use. Accessed 30 May 2014

EEA (2014e) The water framework directive structure and key principles. Waste management. Available at: http://www.eea.europa.eu/themes/water/water-management/the-water-framework-directive-structure-and-key-principles. Accessed 30 May 2014

EEA (2013) Use of freshwater resources—outlook from EEA. Available at: http://www.eea.europa. eu/data-and-maps/indicators/use-of-freshwater-resources-outlook. Accessed 30 May 2014

EEA (2007) Use of freshwater resources—outlook from EEA (Outlook 014)—Assessment published Jun 2007. Available at: http://www.eea.europa.eu/data-and-maps/indicators/use-of-freshwater-resources-outlook/use-of-freshwater-resources-outlook. Accessed 30 May 2014

EEA (2009) Water resources across Europe—confronting water scarcity and drought, Copenhagen. Available at: http://www.eea.europa.eu/publications/water-resources-across-europe

European Commission (2012) Communication from the commission to the european parliament, the council, the european economic and social committee and the committee of the regions a blueprint to safeguard Europe's water resources, Available at: http://ec.europa.eu/ environment/water/blueprint/pdf/COM-2012-673final_EN_ACT-cov.pdf

Eurostat (2014) Water exploitation index—%. Available at: http://epp.eurostat.ec.europa.eu/tgm/ web/table/description.jsp. Accessed 30 May 2014

Helmreich B, Horn H (2009) Opportunities in rainwater harvesting. Desalination 248(1–3):118–124. Available at: http://linkinghub.elsevier.com/retrieve/pii/S001191640900575X. Accessed 25 May 2014

McGlade J (2008) Towards better information for sustainable water management. Speeches. Available at: http://www.eea.europa.eu/media/speeches/towards-better-information-for-sustainable-water-management. Accessed 29 May 2014

Mrowiec M (2008) Potentials of rainwater harvesting and utilization in Polish households. In 11th international conference on urban drainage. Edinburgh, Scotland, pp 1–9. Available at: http://web.sbe.hw.ac.uk/staffprofiles/bdgsa/11th_International_Conference_on_Urban_Drainage_CD/ICUD08/pdfs/178.pdf

Pacific Institute (2014) Sustainable water management—local to global. Available at: http://pacinst.org/issues/sustainable-water-management-local-to-global/ Accessed 29 May 2014

United Nations (1998) Commission on sustainable development. Report on the Sixth Session, New York. Available at: http://www.un.org/ga/search/view_doc.asp?symbol=E/CN.17/1998/20&Lang=E

United Nations (2003) Water for people, water for life. the united nations world water development report, Paris. Available at: http://unesdoc.unesco.org/images/0012/001297/129726e.pdf

Villarreal EL, Dixon A (2005) Analysis of a rainwater collection system for domestic water supply in Ringdansen, Norrköping, Sweden. Build Environ 40(9):1174–1184. Available at: http://linkinghub.elsevier.com/retrieve/pii/S0360132304003178. Accessed 29 May 2014

Visockis E et al (2010) Research of rain water using possibilities. Eng Rural Develop 28:123–127. Available at: http://tf.llu.lv/conference/proceedings2010/Papers/22_Visockis_Edmunds.pdf

Werner B, Collins R (2012) Towards efficient use of water resources in Europe, Copenhagen. Available at: http://www.eea.europa.eu/publications/towards-efficient-use-of-water

Wlodarczyk D (2007) Sustainable rainwater management and green open space. In: Wlodarczyk D (ed) Green structure in development of the sustainable city. The Baltic University Press, Sweden, pp 1–70. Available at: http://www.balticuniv.uu.se/buuf/publications/5_buuf-greenstructures.pdf#page=44

Authors Biography

Walter Leal Fiho is a Professor at Manchester Metropolitan University (UK) and Hamburg University of Applied Sciences (Germany) where he heads the Research and Transfer Centre "Applications of Life Sciences", a centre focusing on matters related to climate change, sustainable development and renewable energy. He is a trained biologist, has supervised or cosupervised dozens of doctoral theses, had led a variety of international projects and has in excess of 300 publications to his credit.

Josep de la Trincheria is an environmental engineer and comanager of the projects AFRHINET and Baltic Flows, based at the Research and Transfer Centre "Applications of Life Sciences".

Johanna Vogt is a biologist and comanager of the projects AFRHINET and Baltic Flows, based at the Research and Transfer Centre "Applications of Life Sciences".

CPI Antony Rowe

Chippenham, UK

2016-12-12 14:51